# Restructuring the Chinese City



A sea change has occurred in China since the 1978 economic reforms. Bringing together the work of leading scholars specializing in urban China, this book examines what has happened to the Chinese city undergoing multiple transformations during the reform era, with an emphasis on new processes of urban formation and the consequent reconstituted urban spaces. With arguments against the convergence thesis that sees cities everywhere becoming more Western in form, and suggestions that the Chinese city is best seen as a multiplex city, *Restructuring the Chinese City* is an indispensable text for Chinese specialists, urban scholars and advanced students in urban geography, urban planning and China studies.

**Laurence J. C. Ma** is Professor Emeritus in the Department of Geography and Planning at the University of Akron. **Fulong Wu** is Reader in the School of Geography at the University of Southampton.

# Restructuring the Chinese City

## Changing society, economy and space

Edited by Laurence J. C. Ma and Fulong Wu

Routledge
Taylor & Francis Group
LONDON AND NEW YORK

First published 2005
by Routledge
2 Park Square, Milton Park, Abingdon, Oxon OX14 4RN

Simultaneously published in the USA and Canada
by Routledge
270 Madison Ave, New York, NY 10016

*Routledge is an imprint of the Taylor & Francis Group*



Typeset in Garamond by Wearset Ltd, Boldon, Tyne and Wear
Printed and bound in Great Britain by MPG Books Ltd, Bodmin



*British Library Cataloguing in Publication Data*
A catalogue record for this book is available from the British Library

*Library of Congress Cataloging in Publication Data*
A catalog record for this book has been requested

ISBN 0–415–33385–7

# Contents

# Illustrations

## Figures

## Tables

# Contributors

**Carolyn Cartier** is Associate Professor of Geography and Director of the Urban and Global Studies Initiative at the University of Southern California. She is the author of *Globalizing South China* (Blackwell 2001) and co-editor with Laurence J. C. Ma of *The Chinese Diaspora: Place, Space, Mobility and Identity* (Rowman & Littlefield 2003). She has also published many articles and book chapters on China and is currently working on a book about regionalism in China.

**Piper Gaubatz**, Associate Professor of Geography, University of Massachusetts, specializes in the analysis of urban form and spatial restructuring in China and Japan. She is the author of *Beyond the Great Wall: Urban Form and Transformation on the Chinese Frontier* (Stanford University Press, 1996). Her articles have appeared in *Urban Studies, Geographical Review* and *Built Environment*.

**Youqin Huang** is Assistant Professor of Geography and Planning and research associate of the Center for Social and Demographic Analysis (CSDA) at State University of New York, Albany. Her research has focused on housing, migration, gender and urbanization in China. She often uses spatial perspectives and sophisticated modeling techniques in her analysis. She has published several papers in the *Annals of American Association of Geographers*, *Environment and Planning A*, *Regional Science and Urban Economics*, *Housing Studies* and *Urban Studies*.

**Si-ming Li** is Professor in the Department of Geography and Director of the Center for China Urban and Regional Studies of Hong Kong Baptist University. He has co-edited two books on China's spatial economy and published numerous articles on China and Hong Kong in leading international journals, with recent emphasis on residential mobility in China. He is co-editor of *Asian Geographer*, on the editorial board of *China Review*, and a member of the board of directors of Urban China Research Network.

**Laurence J. C. Ma** is Professor Emeritus, Department of Geography and Planning, University of Akron, Akron, OH. He has authored, co-authored and co-edited seven books on China and published many articles on China's urban development. He is on the editorial board of six journals, including the *Annals of the Association of American Geographers*, and is a member of the board of directors of Urban China Research Network. He resides and does research in Fremont, California.

**Tianshu Pan** received his Ph.D. in anthropology from Harvard University in 2002. He is teaching anthropology at the School of Foreign Service and the Sociology Department of Georgetown University. His ethnographic research has focused on Chinese urban society, food rationing and consumption practices under state socialism, post-reform welfare systems and the relationship between nostalgia and urban development in Shanghai.

**Jianfa Shen** is Professor in the Department of Geography and Resource Management and Co-Director of Urban and Regional Program in the Hong Kong Institute of Asia-Pacific Studies, the Chinese University of Hong Kong. He has published many articles, a book and co-edited three books on China and Hong Kong-Pearl River Delta, especially migration, urbanization and regional development. He is an executive member of the editorial board of *China Review: An Interdisciplinary Journal on Greater China.*

**Xiaoping Shen** is Associate Professor in the Department of Geography, Central Connecticut State University. Her publications are centered on the regional patterns of China's rural industrial development, the analysis of suburban industrial development in Beijing using GIS, industrial productivity, and international tourism in China.

**Alan Smart** is Professor in the Department of Anthropology at the University of Calgary. He has had extensive field experience in Hong Kong and China and published widely on urban governance, transnational social networks, China's local capitalisms, property regimes, local citizenship, flexible accumulation, economic culture of the Chinese, economic relationship between Hong Kong and the Chinese mainland. In the last few years, he has been studying Hong Kong squatters after 1997.

**Wing-Shing Tang** teaches at the Department of Geography, Hong Kong Baptist University. He has published widely on China's regional and urban development. His recent research focuses on theoretical understanding of urban development and planning in Beijing, Shanghai, Guangzhou, Shenzhen and Hong Kong by interrogating Foucault's perspective of governmentality.

**Fulong Wu** is Reader in the School of Geography at the University of Southampton. He has published many papers on urban spatial structure in Chinese cities, urban housing and land development. He received in 2001 Otto Koenigsberger Prize from *Habitat International* and serves as a member of the editorial advisory board of *Environment and Planning A* and *B*. He is currently editing a book about globalization and the Chinese city.

**Weiping Wu** is Associate Professor of Urban Studies and Planning at Virginia Commonwealth University. She has served as a research follow at the Brookings Institution and a consultant to the World Bank and the Ford Foundation office in Beijing. Her research centers on China's urban development, especially migrant settlements. She has written, co-authored and co-edited four books and numerous articles on China's economic and urban reforms. She is a member of the advisory directors of Urban China Research Network.

**Anthony Gar-On Yeh** is Academician (*yuanshi*) of the Chinese Academy of Sciences. He is also Chair Professor of the Center of Urban Planning and Environmental Management, Dean of the Graduate School, and Director of the GIS Research Center and Institute of Transport Studies at the University of Hong Kong. He is also a fellow in the planning institutes of Hong Kong, the UK and Australia. He has published many books and numerous articles on China's urban development, including the use of GIS.

**Huaiting Yin** is Professor at the Department of Urban and Resource Sciences, Northwest University, Xi'an. His current research focuses on urban and regional development, urban planning and urban systems. He has co-authored four books and a number of articles on urban and regional growth in Xi'an and Shaanxi Province.

**Li Zhang** is Assistant Professor in the Department of Geography and Resource Management, the Chinese University of Hong Kong. His research areas include migration, household registration system, urbanization and urban and regional development in China. He has published a number of articles in leading international journals, including *China Quarterly, Post-Communist Economics* and *International Journal of Urban and Regional Research.*

**Tingwei Zhang** is Associate Professor in the Urban Planning and Policy Program of the University of Illinois at Chicago. He worked in Chinese universities and cities in urban planning for twenty years before coming to the United States. He has co-authored and edited several books and published numerous articles on urban development and planning in China, especially Shanghai. He writes a special column for *Chengshi guihui* (*City Planning Review*), China's leading planning journal. He is on the editorial board of *City Planning Review*, *Urban Planning Forum* and *Planners*.

**Zhe Zhao** is a graduate student in the Department of Urban and Resource Sciences, Northwest University, Xi'an.

# Acknowledgments

The authors and publisher would like to thank the following for granting permission to reproduce material in this work:

*Journal of the American Planning Association*, for Figure 7 in Yeh, A. G. O. and Wu, F., *Urban Spatial Structure in a Transitional Economy: The Case of Guangzhuo*, Vol. 65, No. 4, 1999.

Blackwell Publishing Ltd for Figure 2 in Yeh, A. G. O. and Wu, F., "The New Land Development Process and Urban Development in Chinese Cities," *International Journal of Urban and Regional Research*, 20(2), 1996.

Pion Ltd. for Figure 1 in Wu, F., "Commentary: Transitional Cities," first published in *Environment and Planning A*, 2003, 35: 1331–1338.

Tang gratefully acknowledges the financial support (research grant FRG/02–03/II–28) of Hong Kong Baptist University. Smart acknowledges the financial support of Social Sciences and Humanities Research Council Standards Research Grant, the research assistance of Shirley Lee and Karoline Guelke, and helpful comments on earlier drafts by Matt Ferchen, Zhang Li and anonymous reviewers.

# Preface

This volume is one of the early fruits of the Urban China Research Network. With support from the Andrew Mellon Foundation, a diverse group of scholars from several different countries have met regularly over the past three years with the goal of advancing our knowledge of how urban development is proceeding and how it is changing the face of China. The Network has managed a small grant program for young faculty members and for graduate students that has supported about twenty projects every year. A more ambitious initiative has been the creation of interdisciplinary working groups to assess the state of the art and mount collaborative research projects on major issues of urban change.

Fulong Wu is the leader of one of these working groups, and Laurence Ma, one of few senior statesmen in this field, has been a key participant. This edited volume springs from meetings of the group in 2002 and 2003 in Albany, Southampton and Hong Kong. The chapters focus especially on the phenomenon of spatial transformation. All of the contributors emphasize the built environment – land development, business districts, neighborhood change, housing. Chinese cities are evolving so quickly that what we know about their physical and spatial structure soon becomes urban history, and there are surprises in these chapters. In the United States, for example, there has been a considerable decentralization of city functions over the last century, though central business districts have retained a large share of business and financial service employment. In Beijing, Shanghai and Guangzhou, by contrast, new business districts were established in just the last decade, and they have already redefined the "center" of these cities.

Beyond the new portrait that these chapters draw of the Chinese city, they also provide new insights into how these changes have come about and how they affect the lives of residents. There are several themes here, every one leading to a question still to be answered. How strongly have market processes taken hold in the main coastal cities, and how do they affect the use of space? What is and what can be the role of governmental planning and regulation of land markets – or can we even distinguish between private and public spheres at this point? How do longtime residents perceive the changes in their city, and are the history and culture of city districts being swept aside or are people re-establishing connections to place? What is the extent of new social divisions – by social class and by local residence rights – and how are they expressed in where and under what conditions people live?

The contributors have not directly confronted the policy choices that are being made or that need to be made to ensure a better urban future in China. But urban policy concerns are embedded in every chapter. Rarely do scholars have the challenge of studying a system that is so evidently dynamic, where even the participants are unsure of how best to play their hands, and where our interpretive skills as social scientists are so likely to influence the future that is being made before our very eyes.

John R. Logan
University at Albany

# 1 Restructuring the Chinese city

## Diverse processes and reconstituted spaces

*Laurence J. C. Ma and Fulong Wu*

### Introduction

Cities are highly complex spatial and socioeconomic entities. Different views exist on how cities in the advanced countries of the North have changed in the last three decades. In their introductory essay to the book *Globalizing Cities: A New Spatial Order*, the editors of the volume Marcuse and van Kampen (2000: 1 and 270) raise the important question as to whether there is a new internal spatial order in today's cities which is significantly different from that of the cities of yesterday. Their answer is a decisive and surprising "No," although they do concede that there have been important and visible changes, especially greater structural and spatial divisions and increasing inequality among cities. Their conclusion carries the implication that the spatial structure of the cities of today differs from that of the recent past only in degree, not in its basic nature.

However, there are competing interpretations. Summarizing the views of a number of scholars on how the contours of the cities of Western countries have changed in the 1980s and 1990s, Hubbard and Hall (1998) conclude that there is "a new kind of Western city," post-industrial and post-modern, that are "dramatically different from its predecessor" in urban spatial structure. Similarly, MacLeod *et al.* (2003: 1655) also see the cities in the global North as having "become significantly different." Hubbard and Hall point out that an attractive new urban politics has emerged which is distinguished from the old in that entrepreneurialism has replaced managerialism and governance has supplanted government, but contradictory and repellent spaces of deprivation and despair also exist.

Relative to the cities of the advanced capitalist states, we know much less about how the internal sociospatial patterns in the cities of the former socialist countries have changed after the collapse of state socialism.[1] However, the question as to the nature of the cities in the former socialist countries has also attracted considerable attention. Referring to Hungary mainly, Szelenyi (1996: 311) observed that, without large-scale rural–urban migration, under-urbanization of the socialist era continues to be reproduced in the post-socialist period. There is, however, increasing urbanism marked by the expansion of the commercial sector, greater ethnic diversity and more urban social problems such as prostitution, homelessness, crime and social deviance. Suburbanization has also become a dominant trend as the rich find it safer and more pleasant to live in the suburbs. In post-1992 Moscow, while a considerable amount of former socialist housing space has been redeveloped for commercial use, many former socialist housing estates are in sharp physical and social decline, prompting many residents to escape from them, leaving behind the poor and ethnic minorities. In Moscow, members of the privileged elite continue to prefer to reside in the city center as central city addresses retain their appeal (Bater 2001).

A significant number of books on the Chinese city now exist (Logan 2002; Tang and Parish 2000; Dutton 1998; Lin 1997; Davis *et al.* 1995; Chan 1994), but none is devoted to its changing spatiality and related socioeconomic characteristics.[2] There is little question that the spatial order of the Chinese city today is dramatically different from that of previous eras. This volume brings together the latest scholarship on how the Chinese city has been reconfigured since the 1978 economic reforms. The studies presented focus on the impact of the diverse forces emanating from the global, national and local scales that converge in the city, emphasizing in particular the patterns of internal spatial, social and economic restructuring and the underlying processes and mechanisms. As the subject-matters covered in the studies are diverse and complex, the findings collectively may be subject to different interpretations when they are assessed against the urban changes evident in the cities of the advanced capitalist countries.[3] In what follows, we will, in the second section, provide a broad backdrop about the changing global systems of production under capitalism as well as socialism that have strongly affected the transformations of the Chinese city.[4] This will be followed in the third section by a detailed analysis of what we see as the most salient features of China's new urban realities, incorporating the findings of the chapters included in this volume. Although we recognize that each mode of accumulation generates a set of spatial and socioeconomic characteristics, we will go beyond the familiar geographic lens of seeing the spatial merely as an outcome of the political, social and economic. The fourth section will critically evaluate the validity of the convergence thesis which has relevance to the understanding of China's urban transformation. In the fifth part, we will muse on the potential problems and the appropriate ways to "script" the Chinese city. The chapter will conclude with an inquiry into the nature of the contemporary Chinese city with the suggestion that the Chinese city is best seen with "interpretative flexibility" as a "multiplex city" or a mosaic with diverse processes and heterogeneous reconstituted spaces.

## Global economic restructuring and China's multiple and simultaneous transformations

China's post-reform economic and urban transformations are a consequence of the interplay between exogenous and endogenous forces. Among the global processes of transformation and exogenous factors that have directly or indirectly affected China's post-reform socioeconomic restructuring are:

- the shift from the Fordist to post-Fordist mode of production and regime of accumulation in the capitalist countries (Amin 1994);
- the rise of more specialized systems of production and flexible accumulation based on advanced technologies to produce more variety of goods in smaller quantities to meet just-in-time the demand of the increasingly sophisticated global consumer (Piore and Sabel 1984; Harvey 1990);
- increasing speed and expanded geographic scope of the globalization of capital and the spread of transnational corporations as a consequence of "time–space compression" produced by greatly improved transportation and communication (Harvey 1990; Dicken 2003; Walters 1995); and
- the rise of neoliberalism and the adoption of neoliberal political and economic policies in a large number of the developed as well as developing countries since the early 1970s, as a consequence of the crisis of the Fordist–Keynesian accumulation regime, the demise of the Bretton Woods monetary policies and the collapse of communism. The results are multiscalar deregulation, the removal of institutional constraints, expansion of the power of the market, more privatization, greater exploitation of labor and significant liberalization of finance. Increasing transnational neoliberalism has rendered the global capitalist system more powerful at multiple spatial scales, bringing greater wealth to some people and places in "moments of creation" while intensifying sociospatial inequality, disempowerment, exclusion and injustice in "moments of destruction" (Overbeek 1993; Brenner and Theodore 2002).

These global changes have had direct or indirect impacts on China's economic and urban transformations. From the global perspective, China's economic reform is more than simply a transition from central planning to a "market economy with Chinese characteristics" as Deng Xiaoping envisioned, although this trend has been clearly evident. More accurately, China's move away from state socialism should be seen as part of a larger movement involving all former socialist nations, although China preceded the other nations by more than a decade. On the other hand, socialist countries' shift away from state socialism can be seen as part of the massive global change in the mode of regulation and the regime of accumulation that began in the advanced capitalist states in the 1970s. Viewed this way, China's urban transformation involves not just one but several global processes occurring simultaneously that have affected the production of space, urban consumption and the circulation of capital, people and technology. China's multiple transformations involve at least eight major aspects of institutional shifts:

1 Shift from a state redistributive economy under the socialist mode of regulation to an economy of market coordination (Nee 1989) whose ultimate pattern, however, remains undefined and unclear.
2 Shift from state control of economic production to state regulation of and participation in the market.
3 Shift from centralized decision-making and top-down allocation of development resources to fiscal decentralization and greater local economic autonomy.
4 Shift from extensive state industrialization focusing on heavy industries to satisfy the mandatory production quotas set by central planners to commodity production to meet the needs of the global and domestic markets.
5 Shift from industrial production for the poorly integrated and inefficient state economy to manufacturing of consumer goods for the global markets, or from the "state factory" to the "world factory."
6 Shift from resource-constrained and excessive emphasis of the manufacturing of goods deemed appropriate by the state to a demand-driven but more balanced production of consumer goods and services.
7 Shift from public (state and collective) ownership of land and its free use to paid land use right with negotiated land use price based largely on location (thus the return of rent gap).
8 Shift from virtually free provision of largely work-unit housing to commodified housing production.

Each of these shifts has strongly impacted one or more aspects of the social, economic and spatial structures of the Chinese city. The consequences of such restructuring processes are manifested in new urban realities that together constitute a new urban mosaic and spatiality for China which, despite lingering elements of the past, is visibly very different from that of the pre-reform era. The new urban realities have been created by the transfer of massive amounts of global and domestic capital to the built environment. The sunk capital is spatially fixed, and as usual, the spatial fixity of the newly reconfigured urban elements ensures their existence for at least decades to come until they become dilapidated, dated or are otherwise rendered unfit for the city. As long as it exists, the existing spatiality of the city will have reciprocal effect on the forces of economic and urban transformation that created it in the first place. As such, urban space is much more than simply a passive receptacle or a geographic container because the spatial and the economic/social is mutually constituted.

The consequences of China's multiple and simultaneous economic, social and spatial restructuring have been extremely complex. To better illustrate what we have just conceptualized, we identify below the salient features of the new urban realities that have appeared in the last quarter of a century, based in part on the empirical research presented in the chapters of this volume.

## New urban reality and spatiality

New and old are always relative. As we know relatively little about the spatial structure of the Chinese city in history (see, however, Chang 1961 and 1970; Ma 1971; Pannell 1977; Steinhardt 1990), we will focus on a comparison between the "socialist city" that existed during the first three decades of socialism in China (1949–1978) and the "post-socialist city" that has emerged since the 1978 reforms.[5] Just as the urban structure and urbanism under capitalism differ between the Fordist and post-Fordist city (Lever 2001), between the modernist and the postmodernist city (Harvey 1994), between the managerial and entrepreneurial city (Harvey 1989; Hall and Hubbard 1998), between the liberal and neoliberal city (Brenner and Theodore 2002) and between the metropolis and the postmetropolis (Soja 2000) all because the mode of regulation and the regime of accumulation have changed, so too have the cities under socialism and its aftermath, i.e., the urban patterns of state socialism are not the same as those of the post-socialist period (Pickvance 2002). But this kind of structural explanation is only partially useful as it does not take historical and local contingencies or path-dependency into consideration and it tends to emphasize selected urban elements only. With these conceptual caveats in mind and at the risk of oversimplification, we summarize below what we see as the socialist and post-socialist urban realities in China (cf. "new urban spaces" discussed in Chapter 14) without claiming that the picture sketched is a complete one. No attempt is made to repeat the extensive literature relevant to China's urban development which is available elsewhere (Ma 2002; Ma and Noble 1986; Pannell 1990; Wei 1995; Yeh 1999).

### *Urbanism and spatiality under state socialism*[6]

We begin with a brief account of the key features commonly found in Chinese cities under state socialism in the pre-reform era. Aside from the fact that the cities were seats of administration that anchored the state's territorial power and that urban form contained symbolic elements designed to glorify socialism, the city was also the center for China's extensive socialist industrialization, the core program of state socialism, where a set of socialist industrial urbanism with distinct sociospatial characteristics emerged. Shaped by centralized urban planning, the ideological/symbolic and sociospatial leitmotifs of China's socialist urban spatiality included:

1 Dual centers at different places, one political and the other commercial. The political center where the city government occupied the most imposing building was often connected to a wide avenue lined with more government buildings and with a public square used for political gatherings. The elements constituting the administrative/symbolic architectural ensemble were designed to project the greatness of socialism. By contrast, the commercial center, often some distance away, was much less impressive where state stores offered limited variety of consumer goods;

2 A large number of industries in various parts of the built-up areas and suburbs whose land was allocated free by the state. As land was free, the amount of land occupied by many urban industries was large, so much so that Fung (1981) characterized the phenomenon as "land squandering." As there was no land market, rent gap played no role in organizing land use patterns;

3 The use of state work-units *(danwei)*, such as state-owned industries, government agencies and universities, as the basic structural cell for the arrangement of urban space, with each *danwei* having a well-demarcated space, often walled, where there were different quarters for work and living. Many *danwei* were full-fledged urban communities complete with schools, stores, transportation systems, clinics and other facilities. There were some very large industrial *danwei* with tens of thousands of staff and workers. They functioned as a city within city;

4 A low level of social stratification due largely to the housing system. Fairly standardized *danwei* housing in 5–6 storey apartment buildings was provided as a welfare good almost free to *danwei* staff and workers, with only modest variations in housing quality and per capita housing space. Housing size and quality could vary by family size, employment rank and the prestige, power and wealth of a *danwei*, but income was never a factor sorting socialist urban space. Aside from *danwei* housing, there was also some municipally owned housing which tended to have greater density and lower quality.

Viewed as a whole, these spatial attributes of the Chinese city were directly and indirectly related to the city as the locale for socialist industrial development. As all industrial units were vertically integrated and controlled by their superior administrative units, typically central ministries, the city was not a well-integrated independent economic entity with vibrant economic activities of its own. Visually the city was clean but drab, orderly but dull, and standardized in urban design but monotonous in building style. Socially, the city was disciplined but humdrum, safe but lackluster, with very few social facilities and events to enliven the restrained urban aura. Personal income was strictly controlled by the state, and income levels did not vary greatly which helped curtail social and economic stratification and polarization. Except for the cities in the border regions, the overwhelming dominance of the Han people ensured the homogeneity of the urban population, although population density was extremely high.

### *New spatiality and urbanism in the post-socialist era*

As a consequence of the broad exogenous and endogenous forces of change mentioned above and the shifts in the mode of regulation and regime of accumulation in China in the reform era, new elements of urbanism and a different set of urban sociospatial attributes have surfaced in the Chinese city since 1978. Among the new urban spaces that have appeared in the Chinese city are:

1 The space of globalization typified by the creation of urban high-tech and financial districts and the construction of global image-enhancing projects such as the 2008 Olympic Games Village in Beijing (under construction) and Shanghai's 2010 World Expo (planning stage as of early 2004).
2 The space of elitist consumption represented by glittering shopping centers, malls, chain stores, supermarkets, plush hotels as well as numerous fancy restaurants. What should be noted also is the rapid development of the urban tertiary sector (Lin 2004) that has created numerous spaces of consumption for the ordinary urban residents, including regular hotels, restaurants, small shops and personal services.
3 The spaces of differentiation and marginalization exemplified by the emergence of exclusive gated communities and dilapidated migrant enclaves.

Another type of new space is what Smart and Tang (this volume) call the "space of illegality and irregularity" where illegal building has taken place, largely in previously peri-urban areas where migrants are concentrated whose demand for inexpensive housing has prompted extensive construction of irregular housing for rent by the "urban villagers." The study by Li Zhang (this volume) suggests that the emergence of such spaces is a consequence of ambiguous property rights, the proximity of such "villages" to the places of urban and suburban employment for the migrants, and local government's tolerance of their existence. Smart and Tang have called our attention to the similarities between the Chinese illegal settlements and those in other less-developed countries. What is also important is that such settlements are not pockets of urban poverty. Admittedly, crowding and poor living conditions are there, but the landlords earn a great deal of rent income and the tenants are willing to put up with the poor housing condition to minimize rent in order to save more for remittance back to their home villages. Nor should such places be mistaken for ghettoes of despair where the disadvantaged are hopelessly trapped without hope. On the contrary, migrants in such enclaves are generally positive in outlook, willing to work hard to get ahead economically, and free to return to their villages if things do not work out in the cities.

From the perspective of spatial structure, the Chinese city in the post-socialist period has also witnessed significant suburbanization (Zhou and Ma 2000) following a period of population concentration. Industrial relocation from the densely populated central city to the periphery areas, housing construction and the growing number of migrants have contributed to increasing suburbanization around the large and medium-sized cities. Also significant is the increasing trend toward polynucleation as part of urban spatial reconfiguration away from the earlier compact city to a more dispersed multi-centered metropolitan form, with new growth located in areas away from the old city center which, nonetheless, remains the dominant commercial and administrative node (Wu 1998; Wu and Yeh 1999). The new centers can be either industrial or commercial/financial in nature, as Gaubatz (this volume) and Yeh and Wu (1995) have shown in their studies of Guangzhou, Shanghai and Beijing.

Urban restructuring in the reform period can be viewed as an integral part of China's effort to rescale state power. Since the early 1980s, China has relied on the city as the engine of growth to propel national economic development. A series of administrative changes, including annexation of suburban counties and placing them under the administrative purview of the central city, have significantly enhanced the territorial power of the city (Ma 2004a). In her contribution to this volume, Cartier breaks new ground in our understanding of China's urban restructuring by looking through the analytical lens of scale relations. Building on the growing theoretical literature on scale developed in connection with the capitalist city in the West, she suggests that the "urban question" of China can be viewed as a scale question. Cartier sees scale as a set of processes of social (re)constitution, an organizational principle for territorial power, and a spatial strategy that can be adopted to realign state power among and across different geographic scales. Viewed in this light, one can argue that the urban scale has been the preferred and privileged scale in China's post-reform economic development. The contribution by Shen (this volume) broadly echoes these views.

One of the most important decisions made by the central government that has fundamentally reconfigured China's post-socialist urban landscape is the policy of paid transfer of use right for urban land while leaving state ownership of urban land intact. Introduced in 1987, this policy allows the price to obtain the use right of a piece of urban land to be negotiated between a potential user (often a developer) and the current occupier (Yeh and Wu 1996; Lin and Ho 2003; Ho and Lin 2004, 2003). As Yeh (this volume) has demonstrated, a dual urban land market has appeared based on the existence of two urban land systems – free administratively allocated land and commodified leased land. The interactions of the two systems have produced several important consequences. First, city governments and current land occupiers have been able to generate a large amount of profit from the land-leasing system, and some of the city income has been used to improve urban infrastructure and for urban renewal that have made the city a better place to live and work. Second, a great deal of urban land leased from state industries and agencies has been converted to prime commercial spaces with high-rise buildings, contributing directly to the "modern" look of many cities. Third, the dual land system has also led to a black market, resulted in urban sprawl, and caused chaotic development and incompatible land uses on administratively allocated land which is beyond the control of urban planning.

Among the most visible changes in China's post-socialist urban space is the emergence and growth of new business districts. As new spaces of consumption in the post-reform city, they are constructed with both global and local capital and their high-rise buildings, especially the signature buildings, are often designed by international architects. As such, they are the epitomes of China's urban spatial restructuring. A salient feature of this type of new urban space is that it tends to be developed in more than one place in the large cities. This is in part because of the decentralization of fiscal, economic, administrative and planning powers to the level of the urban district, and competition for investment and rivalry for prestige among city districts have been intense, as Gaubatz's study in this volume has clearly shown. In a rare study on why and how urban districts are developed differently, Tingwei Zhang (this volume) provides the development trajectories of three urban districts in Shanghai, concluding that geographic location, path-dependency and leadership quality are significant determining factors contributing to the current patterns of inter-district differentiation. It is significant that globalization seems to have played a secondary role only in the three districts' local restructuring.

The importance of path-dependency is also emphasized in Tianshu Pan's ethnographic examination of the dialectic processes of community-building and place-making in neighborhood Shanghai. Under the conceptual umbrella of "Shanghai nostalgia," he explores the extent to which the historical memory of the city's dichotomized communities is reconstructed and how new social spaces are (re)produced by an emerging class of social and economic elites in Shanghai during the reform era. His study shows how the "Shanghai nostalgia" has been reified and how the urban spatial categories of "upper and lower quarters" (*shangzhjiao* and *xiazhijiao*) developed historically have persisted and resurfaced as a consequence of the conscious efforts by the local residents, municipal officials, and real-estate agents in the processes of place-making. The spatially dichotomized communities display an acute sense of place that differs significantly from the modern generic urban landscape characterized by high-rises that seem ubiquitous but "placeless."

The growth of the space of consumption is also evident in China's inland cities. In another rare study, Yin, Shen and Zhao's chapter presents a detailed empirical account with strong cartographic evidence of how sections of the ancient city of Xi'an have undergone a major land use transformation (*tudi zhihuan*), from industrial to commercial and residential, as a consequence of the policy of paid land use, urban deindustrialization and concomitant tertiarization of the urban economy. It is clear from their study that accessibility and location matter a great deal in the city's post-reform spatial reconfiguration. This study suggests that it is domestic policies rather than global forces that have directly (re)shaped the city's spatial structure.

That China's post-reform city is significantly more differentiated than before is most clearly reflected in residential space. One of the most important factors affecting the sorting of population into different areas and the growth of suburbs is increasing residential mobility, a topic examined comprehensively by Si-ming Li in his contribution to this volume and elsewhere (Li and Wu 2004). Whereas one's institutional affiliation has a great deal to do with housing quality and residential space, social status, *hukou* and income have become increasingly important in housing tenure change and residential differentiation, as Youqin Huang's study in this book has shown. Huang's contribution reveals a unique spatial feature not found in the cities in Western countries – the existence of exclusive gated communities and dilapidated migrant enclaves in the same general area of a city. Work is needed to explain this rather unusual phenomenon in different cities. It may very well be that the reasons are local rather than national or global.

The influence of rural migrants' residential patterns on urban spatial structure is likely to increase as their numbers continue to rise. In many cities, migrants account for more than a third of the total population. Weiping Wu's earlier work (2002a and 2002b) has established that migrants' *hukou* status and their exclusion from housing reform affect their housing choice and location. Her contribution to this volume on Shanghai's migrants reveals that their spatial distribution coincides with the general trend of urban decentralization. In some suburban locations, the number of migrants exceeds the local population by a large margin, as Li Zhang's contribution to this volume indicates. In the last two decades, official policies towards rural-to-urban migration have undergone gradual but significant changes, due perhaps to the realization that migrants are essential for urban construction and the development of the urban economy. Polices toward migrants have shifted from earlier exclusion, prevention, forced eviction and discrimination in employment to more positive measures that include permitting migrants to purchase commodity housing, less discrimination in employment and the establishment of private migrant schools. To what extent the current enclaves of migrants will become permanent spaces of marginal housing remains to be seen.

## The Chinese city: a critique of the convergence thesis

The emergence of such new urban spaces as malls, gated communities, migrant enclaves and suburbs and the widespread use of automobiles in urban China raise the important theoretical question as to whether the post-socialist Chinese city has become or is becoming capitalist in form. This is a valid question as it forces us to ponder over how to best understand the nature of cities in one socioeconomic system undergoing structural transformations, in our case, from socialism to post-socialism. The question of the relationship between urban form and the mode of production is not a new one, as Sawyer (1975: 52) opined three decades ago that "urban form flows out of and must remain consistent with the basic economic structure of the society of which it is a part." In critical urban geography during the 1980s, a similar structural Marxist view examined how the capitalist mode of production engendered its own distinctive spatial forms, including the capitalist city, but thereafter and throughout the 1990s critical urban geographers gradually moved away from this perspective. One problem of Marxist structuralism, which is a metanarrative focusing on capital and labor relations, is that it does not pay sufficient attention to the importance of region-specific political, cultural and historical forces in shaping the urban. Since the 1980s, most if not all critical urban geographers have shifted away from the stance that views the spatial as the outcome of the social/economic to the position that the relationship between the two is a mutually constitutive one (Leitner and Sheppard 2003: 512).

The literature on the "convergence thesis," critically examined by Cohen (1996), can be viewed as belonging to the genre of structuralism, although the proponents may not identify themselves as Marxists or structuralists. We critique this thesis here because it has relevance to the understanding of China's urban transformation. The convergence thesis argues that cities in different parts of the world are becoming alike, converging to a set of sociospatial attributes similar to those of Western cities. The latest and most forceful argument has been made by Dick and Rimmer (1998) who claim that cities in Southeast Asia do not constitute a discreet category although they do acknowledge that the cities do have distinct elements. Instead, they argue that there should be only one "single urban discourse" as cities in Southeast Asia are becoming Western, especially American, in form. They cite the emergence of such urban elements as shopping malls, gated communities and suburban new towns resulting from globalization as evidence of urban convergence.

We question the validity of this linear, causal, simplistic and essentialist view which masks the complex reality of cities more than its selected evidence purports to represent. Dick and Rimmer's perspective is distinctly Western, based on selective urban elements. Nothing is said about the existence of indigenous urbanism in Southeast Asia (O'Connor 1995) and the ways they have interacted with or mediated the forces of globalization. They criticize McGee and others (McGee 1967; Ginsburg *et al.* 1991) for being "Orientalist," charging that their studies have "shut out First World elements." If this logic is accepted, then Dick and Rimmer's paper is decidedly "Occidentalist" in orientation as it has completely shut out Third World elements such as informal settlements, street vendors and indigenous sectors. They see urban form in Southeast Asia developed during the colonial period as "divergent" that represented merely an "unusual and transitory" (read: aberrant and temporary) deviation from the superior norms of Western cities and their urbanization logic. Clearly, they see Western urban forms as the norm while those of non-Western cities are exceptions to the norm.

Dick and Rimmer must be challenged because they seem to assume that all Western or American cities are alike, constituting one static model for other cities to converge to. In addition, they continue to apply, albeit implicitly, the much discredited notion of globalization-as-homogenization. Furthermore, they confuse similar repetitive spatial patterns with intrinsic sameness (Fincher *et al.* 2002: 35–36). The convergence thesis does not allow the possibility that similar surface features of a phenomenon (in this case, urban form) may be created by different processes in different places and that universal processes can be mediated by local forces and processes embedded in local culture, history or economic and political systems. To use selective empirical evidence and to present partial reality to represent the whole picture, or as Amin and Graham (1997: 416) have put it, to overgeneralize from a few examples and to overemphasize particular spaces, senses of time and partial representations within the city, is to commit the conceptual error of synecdoche.

## Scripting the Chinese city

As cities are an embodiment of highly complex spatial and social phenomena, any attempt to generalize their totality is bound to be selective in content and subject to the mode of thinking of the scriptor.[7] In their effort to sort out the complexities of the cities in different geographic, historical, economic and cultural settings, scholars have constructed various typologies or used typological adjectives to categorize cities, giving rise to such typologies as the preindustrial city, colonial city, capitalist city, neoliberal city, postmodern city, Third World city, socialist city, transitional city and post-socialist city, etc. (relevant references to these typologies are given in Ma 2002: 1563). This mode of analysis is useful as it brings out the most salient features of a type of cities existing in a particular period of time. However, it also has limitations as it tends to emphasize the similarities of cities belonging to a broad city category and gloss over their differences that may be very significant, even within the same country (e.g., Los Angeles differs from New York in numerous ways). In trying to understand the cities in any particular nation or region, we should take note but not be constrained, conditioned and, worst of all, overwhelmed and dictated by the features that may characterize an urban typology.

A better approach to the understanding of the nature of cities is to analyze the impacts of and the relationship between global/general processes of urban development (such as decentralization and recentralization, globalization, technological and communication improvement, migration, neoliberalization, and increasing sociospatial differentiation and exclusion) and locally produced processes at the national, subnational and city levels embedded in local political, social, economic, cultural, historical and institutional systems and structures.[8] While accepting the widely held view that a mode of regulation and the attendant regime of accumulation can generate certain broadly similar elements in the built environment of different localities, we need to be flexible enough to accept as well the possibility that the same process(es) can also produce a range of spatial elements and that similar urban elements in diverse temporal, spatial and cultural settings may be produced by very different processes and locality-specific institutional forces and structures. For example, the processes of China's suburbanization differ significantly from those of the US (Zhou and Ma 2000), and Chinese migrant settlements (Ma and Xiang 1998; Taubmann and Fan 2002) are not formed by the same institutional forces as those in other parts of the world, including Asian countries. The complexity of the relationship between form and process renders any simple essentialist conclusions about the changing patterns of a city or a category of cities questionable at best. As Beauregard and Haila (1997: 328) have reminded us: "The spatial form of the city is never a clear and perfect reflection of urban processes."

In scripting the contemporary Chinese city, a critical question is what are the major political, economic and cultural/historical systems and institutions that are specific to China and that have affected the (re)configuration of urban space and given rise to greater social differentiation discussed earlier in this chapter. We would like to suggest briefly four national political and socioeconomic attributes that are not found in countries dominated by capitalist urbanism.

First, the persistence of a very strong party-state at the central and local levels, its penetration into the market (Ma 2002; Logan 2002), and its dual roles as regulators and participants in economic production, capital accumulation and urban development. Such "institutional amphibiousness" (Ding 1994) in China's state–market relations has blurred the boundaries between the public and the private sectors of society.[9] This interpenetration of state and market is not visually apparent in the built environment, but it underlies the (re)configuration of urban China and is responsible for the appearance of many urban businesses and buildings. This is clearly the case in housing supply, especially the commodification and construction of housing, and in urban land use transformation from industrial to commercial. Many ostensible businesses are in fact owned and staffed by state work-units, agencies and their officials. Walder (1997) sees the Chinese "state as an ensemble of economic actors" and points out that government officials behave as entrepreneurs, corporate management teams, silent partners and investors in the private economy. The long arms and (in)visible hands of the state, sometimes wearing gloves, play an important role in China's urban restructuring.

The second major factor affecting China's post-socialist urban development is the widely recognized close relationship between city leaders' career advancement and the performance of their cities. The more investment capital a city can attract and the more impressive the built environment becomes, the greater are the chances for the leaders of the city to get promotion, which comes with increasing power and many tangible and intangible perks. This factor has encouraged city leaders to make their cities "entrepreneurial" (Wu 2003) and to actively promote their cities (Wu 2000) through the creation of privileged spaces of production and consumption for the investors and the rich. A common strategy to enhance a city's image is the construction of signature buildings and other types of urban physical elements (such as high-rises, public squares, waterfront development, subways, etc.) to strengthen the visual effects of the city. City leaders' desire to show their achievement has also caused competition and rivalry among cities and between city districts (Leaf 1998: 147; T. W. Zhang, this volume).

Third, the relaxation of rural–urban migration has led to massive flows of rural population to the cities whose arrival has caused numerous migrant enclaves to emerge and whose participation in urban job market with minimum income has a direct bearing on sociospatial and economic polarization and labor market segmentation (Fan 2002; Gu and Liu 2002).

Fourth, the strong inter-personal ties and networks built on the basis of kinship and provenance have been instrumental not only in the formation of migrant enclaves (Ma and Xiang 1998) but also in the emergence of informal institutions and nonstate economic activities in urban China (Tsai 2002). Most recently, a large number of highly visible buyers of urban real-estate properties, all from the city of Wenzhou, who go to various cities in groups to snatch up apartment and commercial buildings have caught the attention of the news media (Xie 2003). Their ability to finance their purchases lies in their close network ties based on their place of origin. Ma and Xiang (1998: 550) have suggested that provenance should be used to construct intra-Han subethnicity; it can also be used to shed more light on how China's urban economy and society is affected.

## Conclusion

The consequences of urban transformation and the (re)making of the restless urban landscape of China are the result of general and global forces as well as national and locality-specific factors converging in a particular city, as the empirical studies presented in this volume attest. Such forces and factors have affected China's multiple and simultaneous transformations discussed earlier in this chapter. Whereas China's post-socialist urban landscape differs dramatically from that of the socialist era, elements of the old linger as in a palimpsest because once they are fixed in space they are either too costly to be replaced in a short period of time (such as many standardized apartment buildings erected during the socialist period) or they resist erasure because they have acquired symbolic or historical values (such as the city wall and other sites under historical preservation). Urban sociospatial restructuring is an ongoing process and cities are always incomplete (Beauregard and Haila 1997) because as place they are constantly changing and "becoming" (Pred 1984). As such, the Chinese city (and perhaps cities in general) is best seen as "multiplex city" with "the co-presence of multiple spaces, multiple times and multiple webs of relations" (Amin and Graham 1997: 417) and with "an amalgam of often disjointed processes" (MacLeod *et al.* 2003: 1660). It is true that the Chinese city in the post-socialist age has witnessed the appearance of such new urban elements as gated communities and malls, but this should not be understood as evidence that the city is converging to any global model of city structure. Whereas urban typologies can help us better understand the general characteristics of a category of cities for a given period of time, we should move beyond them because they tend to emphasize similarities among the cities of a typology and downplay their differences. It would not be appropriate to label the contemporary Chinese city as transitional or as belonging to any other category of cities because cities in any category can differ significantly among themselves and because all typologies for contemporary cities have shifting boundaries, i.e., the target for comparison is a moving one (e.g., the capitalist city as a category is itself also constantly changing and there are various types of capitalism. See Hodgson 1996 and Smart 2000). The perspective of the multiplex city, which by nature cannot be a typology, is all the more appropriate when the uncertain nature of socialist transition is taken into consideration (Burawoy and Verdery 1999; Ma 2002).

Under these circumstances, it would be prudent to maintain "interpretive flexibility" (Beauregard 1989: 240) in explicating the nature of cities without harboring any preconceived notions. Global forces are not always the strongest in the formation of new urban elements and the appearance and/or evolution of an urban element may be affected by multiple types of forces combined in different proportions in different geographic and socioeconomic contexts. In the last two decades, the Chinese city under transformation has developed new elements of market capitalism and neoliberalism, both spatial and socioeconomic, but it continues to be affected by path-dependent processes and to have elements left behind from state socialism and the more distant past. Meanwhile, the Chinese city also possesses certain forms commonly found in the developing world (such as crowded streets and migrant settlements). To impose a particular categorical identity on the contemporary Chinese city based on selected empirical evidence is to commit the error of synecdoche. To unpack the complex urban condition of the Chinese city in its multiple constitutive dimensions in an eclectic, prudent but critical manner is likely to shed more light than to see the city through such fashionable but erroneous perspectives as globalization-as-homogenization and convergence.

## Acknowledgment

At various stages of preparing this chapter, we have benefitted from the critical comments of Carolyn Cartier, Deborah Davis, Cindy Fan, Piper Gaubatz, Si-ming Li, George Lin, Michael Leaf, David Meyer and Alan Smart. None of them, however, is responsible for the content of the chapter.

## Notes

1 Two points of clarification are in order here. First, in this chapter, the term "state socialism" with reference to China is used to refer to the "actually existing forms of socialism" in the pre-reform era when political, social and economic structures, relations and praxes were determined by and served the needs of the central state as defined by the Chinese Communist Party. Second, urbanism is seen as different from urbanization. Urbanism refers to the form, process, structure and function of a city, the nature of the urban economy and the urban way of life and behavior. It also includes the derived perceptions, value systems, meanings and the ways they are interpreted. For aspects of Chinese urbanism, see Tang and Parish (2000), Davis *et al.* (1995) and Dutton (1998). Urbanization is limited to the growth of urban population and the expansion of urban space relative to a base point in a city or region, and it is often measured in terms of percentage growth of urban population or built-up area. For urbanization in China, see Chan (1994) and Zhou and Ma (2000, 2003).

2 Throughout this chapter, the term "the Chinese city" is used synonymously as "cities of China." It is not a national urban typology, nor does it imply that cities in China are alike in structure and function.

3 Demographically and economically, it can be argued that it would make more sense to compare China's urban development with that of such developing countries as India and Indonesia. However, as there is a large body of readily available literature on the cities of the North which tends to dominate theoretical discourses, we will use it for comparison in this and the last chapter of the volume. Separate studies are needed to compare the Chinese city with the cities of other developing nations.

4 In this chapter, the word "transformation" is preferred over "transition" as the former has fewer teleological implications. "Transition" implies a predictable, essentialist and unilinear shift to an ideal-typical free-market capitalism with liberal democracy without leaving any room for other possible outcomes of change. For this perspective, see Stark and Bruszt (1998), Nelson *et al.* (1998) and Ma (2002).

5 The adjective "post-socialist" is used in the temporal sense only. Thus the term "post-socialist city" simply means cities developed during the post-socialist era and it does not imply that such cities share a set of common urban elements.

6 Some of the ideas presented in this and the next subsection have been discussed briefly in Ma (2004b).

7 We follow Fincher *et al.* (2002: 27) in using the term "script" here, which refers to "the ways in which cities are subject to certain explanatory modes that bring different aspects of urbanism and urbanization into view at the expense of other cities, other lives, and other processes." Thus our analysis of the Chinese city here and elsewhere is inescapably subjective, and the "knowledge" presented is necessarily "situated knowledge."

8 Gottdiener (2002) has identified deconcentration, decentralization and recentralization as general processes of contemporary urbanization.

9 We thank Michael Leaf for calling our attention to Ding's idea. For further comments on the idea, see Leaf (2003).

## References

Amin, A. (ed.) (1994) *Post-Fordism: A Reader*, Oxford: Blackwell.

—— and Graham, S. (1997) "The ordinary city," *Transactions of the Institute of British Geographers*, NS, 22: 411–429.

Bater, J. H. (2001) "Adjusting to change: privilege and place in post-Soviet central Moscow," *Canadian Geographer*, 45(2): 237–252.

Beauregard, R. A. (1989) "Urban restructuring in comparative perspective," in Beauregard, R. A. (ed.), *Atop the Urban Hierarchy*, Totowa, NJ: Rowman & Littlefield Publishers, 239–274.

—— and Haila, A. (1997) "The unavoidable incompleteness of the city," *American Behavioral Scientist*, 41(3): 327–341.

Brenner, N. and Theodore, N. (2002) *Spaces of Neoliberalism: Urban Restructuring in North America and Western Europe*, Malden, MA: Blackwell.

Burawoy, M. and Verdery, K. (eds) (1999) *Uncertain Transition: Ethnographies of Change in the Postsocialist World*, Lanham: Rowman & Littlefield Publishers.

Chan, K. W. (1994) *Cities with Invisible Walls: Reinterpreting Urbanization in Post-1949 China*, Hong Kong: Oxford University Press.

Chang, S. D. (1961) "Some aspects of the urban geography of the Chinese *hsien* capital," *Annals of the Association of American Geographers*, 51(1): 23–45.

—— (1970) "Some observations on the morphology of Chinese walled cities," *Annals of the Association of American Geographers*, 60(1): 63–91.

Cohen, M. A. (1996) "The hypothesis of urban convergence: are cities in the North and South becoming more alike in the age of globalization?," in Cohan, M. A., Ruble, B. A., Tulchin, J. S. and Garland, A. M. (eds), *Preparing for the Urban Future: Global Pressures and Local Forces*, Washington, DC: Woodrow Wilson Center Press, 25–38.

Davis, D. S., Kraus, R., Naughton, B. and Perry, E. J. (eds) (1995) *Urban Spaces in Contemporary China: The Potential for Autonomy and Community in Post-Mao China*, Washington, DC and NY: Woodrow Wilson Center Press and Cambridge University Press.

Dick, H. W. and Rimmer, P. J. (1998) "Beyond the third world city: the new urban geography of South-east Asia," *Urban Studies*, 35(12): 2303–2321.

Dicken, P. (2003) *Global Shift: Reshaping the Global Economic Map in the 21st Century*, 4th edn, New York: Guilford Press.

Ding, X. L. (1994) "Institutional amphibiousness and the transition from communism – the case of China," *British Journal of Political Science*, 24(3): 293–318.

Dutton, M. (1998) *Streetlife in China*, Cambridge: Cambridge University Press.

Fan, C. C. (2002) "The elite, the natives, and the outsiders: migration and labor market segmentation in urban China," *Annals of the Association of American Geographers*, 92(1): 103–124.

Fincher, R., Jacobs, J. M. and Anderson, K. (2002) "Rescripting cities with difference," in Eade, J. and Mele, C. (eds), *Understanding the City: Contemporary and Future Perspectives*, Oxford: Blackwell, 27–48.

Fung, K. I. (1981) "Urban sprawl in China: some causative factors," in Ma, L. J. C. and Hanten, E. W. (eds), *Urban Development in Modern China*, Boulder, CO: Westview Press, 194–221.

Ginsburg, N., Koppel, B. and McGee, T. G. (eds) (1991) *The Extended Metropolis: Settlement in Transition in Asia*, Honolulu: University of Hawaii Press.

Gottdiener, M. (2002) "Urban analysis as merchandising: the 'LA School' and the understanding of metropolitan development," in Eade, J. and Mele, C. (eds), *Understanding the City: Contemporary and Future Perspectives*, Oxford: Blackwell, 159–180.

Gu, C. L. and Liu, H. Y. (2002) "Social polarization and segregation in Beijing," in Logan, J. R. (ed.), *The New Chinese City*, Oxford: Blackwell, 198–211.

Hall, T. and Hubbard, P. (ed.) (1998) *The Entrepreneurial City: Geographies of Politics, Regime and Representation*, Chichester, NY: John Wiley & Sons.

Harvey, D. (1989) "From managerialism to entrepreneurialism: the transformation in urban governance in late capitalism," *Geografiska Annaler*, 71B: 3–18.

—— (1990) *The Condition of Postmodernity: An Enquiry into the Origin of Cultural Change*, Cambridge, MA and Oxford: Blackwell.

—— (1994) "Flexible accumulation through urbanization: reflections on 'post-modernism' in the American city," in Amin, A. (ed.), *Post-Fordism: A Reader*, Oxford: Blackwell, 361–386.

Ho, S. P. S. and Lin, G. C. S. (2003) "Emerging land markets in rural and urban China: policies and practices," *China Quarterly*, 175: 681–707.

—— (2004) "Converting land to nonagricultural use in China's coastal provinces," *Modern China*, 30(1): 81–112.

Hodgson, G. M. (1996) "Varieties of capitalism and varieties of economic theory," *Review of International Political Economy*, 3: 380–433.

Hubbard, P. and Hall, T. (1998) "The entrepreneurial city and the 'new urban politics,'" in Hall, T. and Hubbard, P. (eds), *The Entrepreneurial City: Geographies of Politics, Regime and Representation*, Chichester, NY: John Wiley & Sons, 1–26.

Leaf, M. (1998) "Urban planning and urban reality under Chinese economic reforms," *Journal of Planning Education and Research*, 18: 145–153.

—— (2003) "The culture of planning and the culture of governance: modernism and traditionalism in the rebuilding of China's cities," paper presented at the International Conference on Globalization, the State and Urban Transformation in China, Hong Kong Baptist University, December 15–17.

Leitner, H. and Sheppard, E. (2003) "Unbounding critical geographic research on cities: the 1990s and beyond," *Urban Geography*, 24(6): 510–528.

Lever, W. F. (2001) "The post-fordist city," in Paddison, R. (ed.), *Handbook of Urban Studies*, Thousand Oaks: Sage, 273–283.

Li, S. M. and Wu, F. L. (2004) "Contextualizing residential mobility and housing choice: evidence from urban China," *Environment and Planning A*, 36: 1–6.

Lin, G. C. S. (1997) *Red Capitalism in South China: Growth and Development of the Pearl River Delta*, Vancouver: University of British Columbia Press.

—— (2004) "Toward a post-socialist city? Economic tertiarization and urban reformation in the Guangzhou metropolis, China," *Eurasian Geography and Economics*, 45(1): 18–44.

—— and Ho, S. P. S. (2003) "China's land resources and land-use change: insights from the 1996 land survey," *Land Use Policy*, 20(3): 87–107.

Logan, J. R. (2002) "Three challenges for the Chinese city: globalization, migration, and market reform," in Logan, J. R. (ed.), *The New Chinese City: Globalization and Market Reform*, Oxford: Blackwell, 3–21.

Ma, L. J. C. (1971) *Commercial Development and Urban Change in Sung China (960–1279)*, Ann Arbor, MI: Department of Geography, University of Michigan, Michigan Geographical Publication No. 6.

—— (2002) "Urban transformation in China, 1949–2000: a review and research agenda," *Environment and Planning A*, 34(9): 1545–1570.

—— (2004a) "China's changing urban administrative system: spatial restructuring and local economic development," forthcoming in *Political Geography*.

—— (2004b) "Economic reforms, urban spatial restructuring and planning in China," *Progress in Planning*, 61(3): 237–260.

_____ and Noble, A. G. (1986) "Chinese cities: a research agenda," *Urban Geography*, 7(4): 279–290.

—— and Xiang, B. (1998) "Native place, migration and the emergence of peasant enclaves in Beijing," *China Quarterly*, 155: 546–581.

McGee, T. G. (1967) *The Southeast Asian City: A Social Geography of the Primate Cities of Southeast Asia*, New York: Praeger.

MacLeod, G., Raco, M. and Ward, K. (2003) "Negotiating the contemporary city: Introduction," *Urban Studies*, 40(9): 1655–1671.

Marcuse, P. and van Kempen, R. (2000) *Globalizing Cities: A New Spatial Order?* Oxford: Blackwell.

Nee, V. (1989) "A theory of market transition: from redistribution to markets in state socialism," *American Sociological Review*, 54: 663–681.

Nelson, J. M., Tilly, C. and Walker, L. (1998) *Transforming Post-Communist Political Economies*, Washington, DC: National Academy Press.

O'Connor, R. (1995) "Indigenous urbanism: class, city and society in Southeast Asia," *Journal of Southeast Asian Studies*, 26: 30–45.

Overbeek, H. (ed.) (1993) *Restructuring Hegemony in the Global Political Economy: The Rise of Transnational Neoliberalism in the 1980s*, New York: Routledge.

Pannell, C. W. (1990) "China's urban geography," *Progress in Human Geography*, 14: 214–236.

—— (1977) "Past and present city structure in China," *Town Planning Review*, 48(2): 157–172.

Pickvance, C. (2002) "State socialism, postsocialism and their urban patterns: theorizing the Central and Eastern European Experience," in Eade, J. and Mele, C. (eds), *Understanding the City: Contemporary and Future Perspectives*, Oxford: Blackwell, 183–203.

Piore, M. J. and Sabel, C. F. (1984) *The Second Industrial Divide: Possibilities for Prosperity*, New York: Basic Books.

Pred, A. (1984) "Place as historically contingent process – structuration and the time–geography of becoming places," *Annals of the Association of American Geographers*, 74(2): 279–297.

Sawyers, L. (1975) "Urban form and the mode of production," *Review of Radical Political Economics*, 7: 52–68.

Smart, A. (2000) "The emergence of local capitalisms in China: overseas Chinese investment and patterns of development," in Li, S. M. and Tang, W. S. (eds), *China's Regional Development*, Hong Kong: Chinese University Press, 65–95.

Soja, E. (2000) *Postmetropolis*, Oxford: Blackwell.

Stark, D. and Bruszt, L. (1998) *Postsocialist Pathways: Transforming Politics and Property in East Central Europe*, Cambridge: Cambridge University Press.

Steinhardt, N. S. (1990) *Chinese Imperial City Planning*, Honolulu: University of Hawaii Press.

Szelenyi, I. (1996) "Cities under socialism – and after," in Andrusz, G., Harloe, M.

and Szelenyi, I. (eds), *Cities after Socialism: Urban and Regional Change and Conflict in Postsocialist Societies*, Oxford: Blackwell, 286–317.

Tang, W. F. and Parish, W. L. (2000) *Chinese Urban Life under Reform: The Changing Social Contract*, Cambridge: Cambridge University Press.

Taubmann, W. and Fan, J. (2002) "Migrant enclaves in large Chinese cities," in Logan, J. R. (ed.), *The New Chinese City*, Oxford: Blackwell, 183–197.

Tsai, K. S. (2002) *Back-Alley Banking: Private Entrepreneurs in China*, Ithaca: Cornell University Press.

Walder, A. G. (1997) "The state as an ensemble of economic actors: some inference from China's trajectory of change," in Nelson, J. M., Tilly, C. and Walker, L. (eds), *Transforming Post-Communist Political Economy*, Washington, DC: National Academy Press, 432–452.

Walters, M. (1995) *Globalization*, 2nd edn, London and New York: Routledge.

Wei, Y. H. (1995) "Large Chinese cities: a review and research agenda," *Asian Geographer*, 14(1): 1–13.

Wu, F. L. (1998) "Polycentric urban development and land-use change in a transitional economy: the case of Guangzhou," *Environment and Planning A*, 30: 1077–1100.

—— (2000) "Place promotion in Shanghai, PRC," *Cities*, 17(5): 349–361.

—— (2003) "The (post-) socialist entrepreneurial city as a state project: Shanghai's reglobalisation in question," *Urban Studies*, 40(9): 1673–1698.

—— and Yeh, A. G. O. (1999) "Urban spatial structure in a transitional economy: the case of Guangzhou, China," *Journal of the American Planning Association*, 65(4): 377–394.

Wu, W. P. (2002a) "Migrant housing in urban China: choices and constraints," *Urban Affairs Review*, 38(1): 90–119.

—— (2002b) "Temporary migrants in Shanghai: housing and settlement patterns," in Logan, J. R. (ed.), *The New Chinese City*, Oxford: Blackwell, 212–226.

Xie, C. L. (2003) "Zhuyi Wenzhou goufantuan (Watch the housing buying groups from Wenzhou)," *Nanfang zhoumo (Southern Weekend)*, December 25, Internet news retrieved January 2, 2004.

Yeh, A. G. O. (ed.) (1999) *Bibliography on Socio-Economic Development and Urban Development in China*, University of Hong Kong, Centre of Urban Planning and Environmental Management.

—— and Wu, F. L. (1995) "Internal structure of Chinese cities in the midst of economic reform," *Urban Geography*, 16(6): 521–554.

—— (1996) "The new land development process and urban development in Chinese cities," *International Journal of Urban and Regional Research*, 20(2): 330–353.

Zhou, Y. X. and Ma, L. J. C. (2000) "Economic restructuring and suburbanization in China," *Urban Geography*, 21(3): 205–236.

—— (2003) "China's urbanization level: reconstructing a baseline from the fifth population census," *China Quarterly*, 173: 176–196.

# 2 City–space

## Scale relations and China's spatial administrative hierarchy

*Carolyn Cartier*

### Introduction

Since the onset of reform and opening in the People's Republic of China (PRC) after 1978, decentralization of state power has arguably been the most consequential transformation of the Chinese political economy, underpinning the dynamics of economic growth and state–society relations. The outstanding geographical manifestations of these processes are growth in the number and size of cities and the urban population – urbanization. How should we analyze the relationships between them? This chapter introduces scale relations as a basis for assessing the decentralization of state power and urbanization; and analyzes the "rescaling" of the Chinese state in an era of globalization.

Scale theory has emerged from a substantial literature in political economic geography to explain how specific scales of social activity, from local to global – village, city, province, region, nation-state, supra-state region and world scale – are products of social, political and economic activities and constituted through spatial processes. By contrast to the notion that the nation-state boundary and "level" of territory is neutral, even "natural," a scale relations perspective recognizes that political territories are multiple and produced through historical conditions, practices and events. In the case of China, the territorial administrative hierarchy is also the state's transhistorical administrative institution; it has maintained over the *longue durée*, and served as an important basis of political stability and territorial coherence. A scale analysis focuses attention on how actual processes work out through China's territorial administrative hierarchy, from the national capital to provinces, cities, counties and towns, and, in turn, how such political territories are constructed, mutable and dynamic.

Beginning in the early 1990s, a fast-moving project to theorize scale set forth a range of ideas and questions about dynamic political geographies (e.g. cf. Agnew 1993, 1994, 1999; Brenner 1997, 1998, 1999, 2000, 2001; Herod and Wright 2002; Howitt 1993, 1998, 2002; Marston 2000; Smith 1984, 1992, 1993, 1995, 2000; Swyngedouw 1997a, 1997b). This literature analyzes different levels of scale, how to account for their instantiation and interrelations, the changing priority of the national scale, power and resources associated with scalar processes, and, especially, the dynamics of political economic conditions that propel reconfiguration of scale relations in territorial state systems. The historical moment has a logic: by the 1990s the capitalist state had restructured in the face of globalizing economic change, losing some of its national-scale "spatial fix" (Harvey 1982, 1985). But this literature assumes an unproblematized capitalist state (e.g. Brenner *et al.* 2003), which is limited for analyzing the PRC where the state continues to articulate the spatial administrative hierarchy and its associated powers. Thus the case of the PRC prompts us to assess the continuing role of the state in China under reform, and its significance for understanding scale relations. First, let us assess common explanations about China's reform economy for its embedded assumptions about scale.

## "Central–local" relations

Between 1949–1978 the socialist state of the PRC relied on a centralized, top-down bureaucracy, in which the central state or the national level was the primary scale of decision-making. The significance of the national scale also maps on to history: the capital was the pivot of imperial China, just as its bureaucracies produced myriad texts which are essential sources for Chinese historiography. Normative scholarship, whether of the imperial era or the PRC, reflects worldviews from the center and the perspective of the empire-at-large or nation-state. The territorial administrative system of the historic and modern state has also shared a general two-level structure. A two-level hierarchy dominated the imperial territorial administrative system, in relations between the imperial capital and *xian* or county-level units (Fitzgerald 2002; Whitney 1970). Under the PRC after 1949, the "local" reemerged with the establishment of the "base units" (*jiceng danwei*) or grass-roots working units, the smallest, lowest-level units used by the Party for mass mobilization. Continued reference to the "center" and the "local" reflects this political structure: the state's contemporary terminology for relations of the territorial hierarchy is between the *zhongyang* (center) and *difang* (regional and local). So whether historic or contemporary, Chinese territorial administration has largely revolved around a two-level system, in which the center is privileged and the meaning of *difang* encompasses multiple (from village to province) scale positions. Indeed "between these two spheres of real power . . . there was much administration but little authority" (Shue 1994: 70). By contrast, under decentralization, significant state power and authority has emerged at intermediate spheres of administration; these realities are discussed below.

Significant contributions to theorizing the contemporary Chinese political economy have largely maintained the center–local dyad and the national-scale perspective. For example, prominent applications of transition theory in the area literature by Nee (1989) and Naughton (1995) explain reform from the national scale; they do not articulate with the global economy nor do they assess the uneven landscape of reform. As the normative framework for analyzing "post-socialist" states, transition theory assumes a range of expectations about the relationship between the state, national markets and emergent democratic activity and civil society. But as Walder (1996: 1068) notes about Nee's explication of "transition," "the question is not to what degree markets have emerged, but what kind of market economy is emerging in different regions." This evaluation recognizes variations in regional economy and the uneven nature of China's development.

With the onset of reform, the state focused on a highly uneven strategy of rapid development in particular zones, cities and regions, first on the south coast, and then in the coastal region generally. Transition theory, an aspatial approach, neglects these geographical contexts as well as the continuing strong role of the central state to reorder territory and territorial powers. While transition theory predicts declining state power as the power of "the market" increases, a scale analysis views decentralization in relation to the rescaling of power at different levels and how greater "porosity" may exist between them. By not examining diverse scales of political economic activity and the dynamic powers constituted "within" them, this normative scholarship on China's political economy treats the national as "natural" and the geographical as relatively fixed. In its continued reliance on the center–local configuration, it also does not reliably take into account the significant emergence of meso-level scales of activity – especially the urban scale. In lacking examination of spatial processes, this work has not been able to explain the causal, internal dynamics of decentralization. By contrast, we propose that scale relations, as a theoretical project and methodology, focus on interrelations between multiple-scale positions and resources that circulate through them, and thus spatialize the state-territorial dynamic, ultimately revealing the complex spatial processes propelling transformative change in contemporary China.

## Theorizing scale relations

Among scholars who are examining spatial questions beyond the "two-level" system, most are writing from explicitly geographical perspectives (e.g. Cartier 2001; Lin 1997; Ma 2005; Tang *et al.* 2000; Zhao and Zhang 1999; Zhou and Wu 2000). In work on regional economy in the Pearl River delta, for example, Lin (1997: 127) notes how Nanhai county, now city, achieved rapid development by "driving forward on five wheels" (*wuge lunzi yiqi zhuan*), or simultaneously at the scales of the county, township, district, village and individual. In earlier work on cities of the south China coast, I argued that scale relations is one of the most important ways of conceptualizing subnational and transboundary regional activity in relation to the global arena (Cartier 2001: xi). Geographers have also produced considerable scholarship dealing with city size and distribution and changing definitions of cities (e.g. Ma and Cui 1987; Fan 1999; Lin 2002; Zhou and Ma 2003); while this work does not explicitly deal with scale, it concerns issues and policies at various levels of scale.

Defining scale, theorizing scale relations and analyzing political economic geographies of scale are different and related subjects. In the normative terms of state administration, we refer to scales in terms of territories defined by political boundaries, i.e. towns, counties, cities, provinces or states, nation-states and world regions. This is scale as sized area. Mapped territories, then, are *representations* of these politically bounded areas. Geographical scale recognizes these political territorial scales as well as non-territorial scaled arenas of activity and at more discrete "levels," from the person as body and mobile agent, to the household, neighborhood, workplace, institution and region (Marston 2000). People, institutions and the state also act as agents at all these scales, enacting spatial practices with scale effects, in which scale is also the spatial arena where social relations are constituted. Such a view prioritizes thinking about scale through spatial processes that transcend and rework scale, rather than via a skeletal framework of ascending levels of state territory. As Swyngedouw (1997a: 169) explains, "The theoretical and political priority, therefore, never resides in a particular geographical scale, but rather in the process through which particular scales become (re)constituted." Understanding dynamic interscalar relations asks us to conceptualize how social processes work out between, transcend, and recreate scales. A scale perspective does not assume that processes necessarily move up and down a territorial hierarchy, but recognizes that social processes may move unevenly across a space and sometimes "jump" scales. This is a dialectical relationship of dynamic and irregular interrelations among and between scaled processes (Harvey 1996; Howitt 1993; Merrifield 1993).

We can see this dialectical relationship as outcomes of political economic competition and cooperation in scale reconfiguration (Smith 1992, 1993, 1995). In the recent period of global economic activity since the 1970s, known as late or advanced capitalism or disorganized capitalism, the historic national-scale "spatial fix" (Harvey 1982, 1985), which characterized the first three-quarters of the twentieth century, has shifted in part to subnational scales of activity. Internationalized neoliberal "free market" policies have led to diminished state regulation, propelling competition and economic restructuring through major growth poles, industrial districts and city-regions (Scott 1998; Storper 1997). At the global scale, these processes manifest in the expanding activities of multinational corporations and multilateral institutions. Scale restructuring is especially institutionalized in the supranational European Union. Thus political economic rescaling has coalesced "above" and "below" the more historically predictable national level of economic organization at subnational and supranational arenas. National-based modes of production have been disarticulating into multiscalar arenas and toward complex reconstitution of state activity at various scales (MacLeod and Goodwin 1999). Thus the state, while clearly national, is not specific to one scale. However, the theoretical scale literature has so closely hewn to an unexamined acceptance of the neoliberal capitalist state, in which state practices are relatively minimized, that we cannot simply apply it to the Chinese case. If we want analytical process-oriented treatment of state–scale–urban relations in China, we must consider scale relations in the context of the Chinese territorial state and state practices in contemporary China.

## The Chinese state

The Chinese territorial system forms the basis of administration and governance. It reflects persistent historical and geographical configurations of political economy, from the imperial era and the socialist planned economy to the contemporary reform system. In the past half millennium, the Chinese state has transformed from imperial to republican in the early twentieth century to revolutionary communist and in the present era to modernizing socialist with a marketizing economy. Its political ethos and system has transformed in each phase, from monarchical to republican to Leninist to liberal, and its ideological logic of societal transformation has similarly shifted from neo-Confucianst during the imperial period, to neofacist in the early twentieth century to radical Maoism, and in the late twentieth century, to pragmatic Dengism. Throughout, the administrative hierarchy has existed as the framework of state organization. As Fitzgerald (2002) has noted, this system has existed transhistorically with few major changes since the Yuan dynasty (1260–1368), when the province was introduced, and its most basic territorial component, the *xian* or county, has its origins in the Qin dynasty (265–420). Indeed the territorial system is "the most stable element in the historical annals of the Chinese state" (Fitzgerald 2002: 11). At the same time, the state has made periodic and significant alterations within the territorial system. For example, during the twentieth century, the state has changed criteria for designating settlements "urban" (whether *shi*, cities or *zhen*, towns) numerous times (Chan 1994; Ma and Cui 1987; Gu *et al.* 1997; Kirkby 1985; Zhao and Zhang 1999; Zhou and Ma 2003), which reflect state political economic planning goals. Thus in historical perspective, the Chinese state does not represent a single paradigm but its territorial structure has remained largely the same and the state has modified it to propel policy goals.

Here we will consider structural aspects of the administrative system in the second half of the twentieth century. During the Maoist era, the Chinese Communist Party (CCP) insinuated its governing regime into the territorial administrative hierarchy. The Maoist state had characteristics in common with the Soviet Union, especially features of "vertical rule" and "dual subordination," where the latter are the "multiple rather than single channels of command and information" (Shurmann 1971: 188–189). This "dual subordination" is a combination of vertical and horizontal control, in which a given bureau or agency may have to respond both to a higher-level agency and one at the same administrative level or geographical area. But the Soviet Union was more hierarchical in its organization, whereas the Chinese state structure has emphasized both "horizontal" (*shuiping*) and "vertical" (*chuizhi*) features of structural organization. This structure works through the relationship of the vertical system line (*tiaotiao*) of a government bureau or ministry in relation to the "horizontal" (*kuaikuai* lit. piece) or area territorial authority, i.e. provincial, municipal or county government (Lieberthal and Oksenberg 1988; Gong and Chen 1994; Ma 2005). So governing institutions in China, like administrative units, are also typically understood in terms of rank at the province-level, prefecture-level, county-level and so forth (Lieberthal and Oksenberg 1988: 143; Ma 2005).

The changing relationship between hierarchical and area authorities has varied significantly over time, not just during the recent era of decentralization in China under reform. In recognition of this greater spatial complexity along two axes of control, Zhao (2002) has proposed the concept of the "multi-layered centralized regime" to explain the combination of China's distinctively scaled territorial condition and the establishment of diverse bureaucracies at these different levels in the hierarchy. The vertical and horizontal conditions of the Chinese regime have arguably provided greater political and territorial control; they have also presented organizational problems, discussed below. In a comparison to Western federalism, Montinola, Qian and Weingast (1995) find that decentralization in China "depends on the political relationships among levels of government, with no reference to an explicit or constitutional basis or its promotion of individual rights and political freedom." Unlike constitutional federalism, China has a unitary political system in which the national government has ultimate authority over subnational territorial units. So whether by comparison to Soviet or Western models, analyses of the structure of the Chinese state identify the important role of the territorial administrative hierarchy and the role of the central state in continuing to articulate its powers.

We can assess the territorial administrative system more systematically by viewing the structure of the hierarchy, its spatial conditions, and the state's rearrangements of them as dynamic scale strategies. We can consider origins of the present system by comparing contemporary analysis of scale and the state to historic state structures and institutional contexts, locating the opportunity for a transhistorical state–scale analysis. However, we are not looking for linear historical models, but rather aspects of understanding contemporary state form as the product of the interaction between past conditions and present strategies adopted for their transformation. This means looking simultaneously at state territorial organization, its historical conditions, social production and political contestation.

## The territorial administrative hierarchy

Political boundary inscription, formation of administrative territories and establishment of government offices at each level of administrative territory have been fundamental elements of state practice in imperial and contemporary China. Together, these state practices make territorial administrative rank (*xingzheng jibie*) or level (*dengji*) especially important (Ma 2005). The state also periodically changes the criteria for defining administrative units, especially cities, as mentioned above, in order to promote particular political and economic goals.

A typical sketch of the administrative hierarchy belies its complexity. In the contemporary era, four main administrative levels form the hierarchy of the governmental system; these are territorial areas that are represented by their own People's Congress and government. At the highest level is the central state in Beijing. At the subnational level are provinces (*sheng*) (including the four centrally administered municipalities (*zhixia shi*, Beijing, Shanghai, Tianjin and Chongqing) and autonomous regions (*zizhi qu*, autonomously governed areas of ethnic minority groups). At the next level are cities (*shi*) of different types, followed by counties (*xian*) and county-level cities. Below counties are towns (*zhen*) and townships (*xiang*) or villages, which form the top echelon of the rural administrative hierarchy. These are not noted as part of the four-level scheme because they generally come under the jurisdiction of the county-level People's government.

What is especially interesting is the complexity of what is the city-level in the administrative hierarchy. While city-level appears to be one administrative rank in the four-level hierarchy, four different administrative ranks exist *within* the city-level. These are the centrally administered municipalities or province-level cities, subprovince-level cities, prefecture-level cities and county-level cities. The state also organizes cities by three different types of administrative characteristics or legal status, which are the province-level cities, cities with districts (subprovince- or prefecture-level) and cities without districts. Moreover, there are currently six different categories of special administrative status for select cities; these include the four province-level cities, the special economic zones, coastal open cities, and cities designated to experiment with new economic programs, among others (see Ma 2005). Clearly, some of these designations overlap. What is more interesting for our purposes is that these categories and rank orders are a dynamic territorial system, periodically altered by the state to meet diverse political and economic goals. This means that the Chinese space economy is an actively scaled territorial mosaic whose dialectical interrelations the state seeks to manage in order to spur economic development while simultaneously maintaining political control.

The urban administrative hierarchy has also enveloped incrementally smaller levels of scale, right down to the household. Under the system inherited from the Maoist era, most cities are administratively divided into urban districts (*qu*), which are further divided into street or neighborhood offices (*jiedao banshichu*) and residents' committees (*jumin weiyuanhui*). The neighborhood office is the lowest effective level of government administration in the city – even though "neighborhoods" may circumscribe several hundred thousand people. This office implements a range of state policies and programs through residents' committees, which can encompass up to 600–800 households. Even smaller or lower level arms of the residents' committees form around dozens of households, which then break out into groups with fewer than twenty members (Pannell and Ma 1983: 84–86). During the Cultural Revolution (1966–1976) the functions of residents' committees included political actions such as rallies against "class enemies" and assessing the size of residences for their potential to be divided up and shared in the name of "spatial equity." Contemporary state programs focus on social services, social order and aspects of everyday life, and have included public security, schools and childcare, health and family planning, maintenance of registered permanent residence (*hukou*) records, conflict mediation, environmental quality, and social activities as well as implementation of diverse government campaigns (Whyte and Parish 1984). They serve to integrate non-waged members of society, namely housewives, children, students and the retired, into the state system, transmitting state policy to the local level. In the Foucauldian sense, they also serve as a means of state surveillance, structuring the social space of daily life. The state has been actively building residents' committees through the 1990s as a basis of the new urban society, in the process transforming them into quasi-state–civil society organizations (Read 2000). Under reform, marketization of the economy, with its widespread effects on society, has especially transformed these lower-level arenas of state administration.

## Scale and urban hierarchy

The problematic geographical scale and its social production have become increasingly central to critical urban theory in the current period of the global economy. Contemporary economic growth has increasingly become a set of processes shaped by the relative ability of cities and regions within states to organize access to distant resources and networks (Scott 1998), which makes studies conducted at multiple scales and sites more compelling than traditional social science perspectives from the single scale of the nation-state. In this context, the urban or city–region scale is a particular focus of theorization concerning economic restructuring and the accumulation process in the world economy. In other words, theorizing scale is a step toward a more precise theoretical and empirical refinement of the spatialities of globalization.

Urban transformations in contemporary China are significant barometers of the country's economic restructuring and have far-reaching consequences, not only domestically, but in their transnational dimensions in the Asia

Pacific and globally. We are interested in processes of urban and regional restructuring in an era of globalization for their emphasis on understanding causal processes driving urban transformations and their consequent new spatial patterns in cities and metropolitan regions. Such an approach treats cities not just as spaces of transformation and places of new urban patterns, but as the *basis* of economic organization and social formation for larger urbanizing regions. Consequently, the over-arching question to be examined is how has reorganization of the state rescaled the administrative hierarchy, and how have these reorganizations and rescaling processes impacted urban growth and development? How has the "central" state reconfigured the administrative hierarchy in order to propel power and resources, and how has the "local" state adjusted its spatial practices over revenue and resource movements? In turn, how have cities interpreted and used their new powers and resources, both domestic and international, to plan and stimulate urban and regional growth and transformation?

Decentralization of power under reform is a complex and incremental process resulting in complex patterns and conditions of rescaling. In terms of vertical and horizontal state structure, decentralization of power encompasses rescaling of both decision-making and financial responsibility to lower levels of the state administration, away from central ministries in Beijing, and horizontally out to intermediate and less central bureaucracies of the state hierarchy. This means that decentralized powers are not simply "fixed" at lower levels of state administration, in cities and counties, but that they exist in vertical and horizontal *relations* among cities, that is, in constant dialectical formation. At the same time, rural and urban market reforms, including release of surplus labor from local production units, the rise of an entrepreneurial class, and consequent expansion of markets in commodities, labor and services have resituated the actual places of production and consumption. In the process, daily life at the lowest level units of social organization, in the village and *danwei*, has broken open and embraced new mobilities, in significant labor migration to towns and cities, and in social and economic connections to diverse places (Cartier 2003). Thus from what were the lowest levels of the state hierarchy, the cellular base units, social and economic life has significantly expanded in scope and forged connections with higher levels of settlement in the territorial hierarchy – to cities, regions and beyond. The resulting overall trend, from the top down and the bottom up, is convergence of state power and social and economic life at meso-levels of scale.

In addition to these general trends of rescaling, the central state has explicitly highlighted the urban scale as a means of driving economic growth under reform. In 1982, then Premier Zhao Ziyang explicitly promoted city-led development (Tang and Chung 2000: 284). Subsequently, central policies made the metropolitan state a more important arena of government and linked cities to surrounding areas to spur rural development. The central state has enhanced the role of cities in several ways: it has given them more decision-making powers to contract foreign investment independently, it has allowed them to enlarge, facilitating expanded control over land and land rents, and it has made more of them. Two particular policies have especially affected urban administration in the hinterlands of the major coastal cities. "Cities leading counties" (*shiguan xian* or *shidai xian*) has made counties come under the economic administration of cities. This territorial transformation has economic effects, in attempting to link production systems between urban and rural areas, locally collapsing vertical scale and extending horizontal reach.

Deng Xiaoping's southern excursion in 1992, which promoted intensification of reform and economic expansion, signaled further opportunity for rapid urban expansion. In 1993 and 1994, more than 700 counties came under the jurisdiction of cities each year, more than any other previous year under reform. Another new policy, "abolishing counties and establishing cities" (*chexian gaishi*) has directly facilitated economic growth by redesignating counties as cities, giving urban powers to areas formerly classified as rural. Thus counties widely vied to be reclassified "urban"; in 1993 and 1994, 53 counties were redesignated cities each year, which is over twice the average rate of previous years (Zhao and Zhang 1999: 338–339). These are both policies of "scaling up." Yet no matter the "direction" of change, these policies are also economic development and accumulation strategies, new articulations of scale and administrative geography finely tuned by the state to attract foreign direct investment and promote economic growth under reform. Thus the city has become the site of different and more complex forms of spatial knowledge necessary for successful state practices. The meso-scale has also become the site of contested power relations between cities and former counties or districts, as many of the new territorial formations experience conflicts between the city and (former) county authorities (Tang and Chung 2000).

## Historic and contemporary spatial practices

The contemporary state, the socialist state and the imperial state have all maintained territorial practices to ensure the stability of the administrative hierarchy. Stability and territorial coherence, over the *longue durée*, have arguably been the state's overriding goals. The territorial administrative system of imperial China worked to stabilize the regions and knit the empire into a coherent whole; it also periodically broke down on the margins of the Han ecumene where alternative forms of territorial order could take hold. Stability, though, was not a static geography of bounded territories but an active and highly managed basis for the imperium's massive accumulation strategies, especially taxation, by funneling grain tribute from the provinces through administrative centers to the capital (Ho 1962). It was the obligation of government officials, installed at each administrative level in the hierarchy, to propel these resources toward the center.

This spatial process engendered the commitment of officials in part because it mirrored their own desires: the imperial examination system was also a system of scaled opportunities, a three-tiered set of examinations whose sequential passage earned appointment at correspondingly higher levels in the administrative hierarchy, from the district or county level to the province, and finally, the capital. Symbolic representations of imperial power also distinguished the built environment of provincial and *xian* capitals: city walls, government administrative buildings, and temples of official cults, were reproduced in cities across the empire and worked to confirm imperial power at different scales and in diverse regional locations (Wright 1977; Feuchtwang 1992).

The Maoist state maintained or adapted several institutions from previous eras, especially the territorial administrative hierarchy, and established the modern socialist infrastructure of the contemporary state. It also shattered many traditional institutions and forms of association, especially at lower levels of society. The collectivization of agricultural production highly circumscribed mobility in rural areas, and after 1958 the state implemented the *hukou* (registered permanent residence) system, which highly circumscribed mobility by forbidding travel without permission. In this context, the territorial system became the institutional scaffolding that facilitated the "reach of the state" (Shue 1988). In cities, as discussed above, the territorial hierarchy telescoped down to districts, neighborhoods and street committees, where the state obtained its closest range of surveillance. In these ways, scale also works as the state's scaffolding of persuasion: how the state structures "mobile" aspects of state and society and the "space of flows," whether controlling the movement of fiscal resources or the human capital of migrants "up" and "down" the urban hierarchy.

The *hukou* system, which continues to exist in China under reform, inherited its spatial logic from the historic *baojia* (household administration) system. The *baojia* came into common use in the Song (960–1279) as a system of population registration. It grouped households into a scaled pyramidal network, organized to maintain local control and mutual surveillance. During the Ming dynasty the structure was ten households at the lowest level, 100 households at the intermediate level, and 1,000 households at the tertiary level. Surveillance of society was achieved by a system of periodic reports on good and bad household behavior; awards were given for praiseworthy deeds, while crimes resulted in hanging a placard in front of the offender's house. The population enumeration function of the system worked by requiring families to post a "door plate" with data on household members by name, age and particular characteristics, such as disabilities, or "outstanding contributions to the nation" (Dutton 1998). Thus the *baojia* system was a set of state spatial practices used to organize society and local territory.

After 1949 the state set up the *hukou* system in "scientific" management terms of the socialist planned state. The *hukou* system inherited its logic of control from the *baojia* system, but transferred the focus of society from the *jia* or household and the neighborhood to the *danwei* in urban areas, and to the lowest unit of the collective in rural areas (Cheng and Selden 1994). While its stated purpose was to maintain social order, it has operated as a significant form of social control, preventing rural to urban migration. The logic of human–spatial relations embedded in the *hukou* system is especially hierarchical and scaled: it created a spatial hierarchy of settlement opportunities and status in a pyramidal order, from grain-growing peasants at the bottom, to the major cities at the top. Each higher level of settlement, from small towns to prefectural cities, represents higher status and greater opportunities (Mallee 1996). Thus the system enhanced symbolic values about especially Beijing, Shanghai and Guangzhou as the most desirable cities, and cities in general. Urban *hukou* privileged access to subsidized grain, fuel, and especially housing, while farmers have had to grow their own food as well as for the urban classes at state procurement prices, which have not reflected increased farming costs. The result is an increasing gap in income distribution, through the second decade of reform, between rural and urban areas (Khan and Riskin 1999).

## Scale and state strategy

In the previous section we have focused mostly on the production of scale by the state. In order to think about scale and operating strategies of the state apparatus, it is useful to realize the difference between production of scale and politics of scale. A politics of scale involves distinct state actors and institutions, acting out goals, policies and ideologies in scale contexts. MacLeod and Goodwin (1999), for example, have urged that studies of urban rescaling should integrate a deeper concern for both the spatiality of the state and associated scalings of political economic institutions. They focus on the importance of examining state strategy as "a pattern of intervention in the economy and society which 'favors the course of [a particular] accumulation strategy and the flow of material benefits to the requisite social base'" (MacLeod and Goodwin 1999: 516). Similarly, Jessop (1997a, 1997b) advocates examining the institutional contexts of the state and through neo-Gramscian state theory "because it situates processes of accumulation in relation to specific hegemonic political or state projects" (Jessop 1997b: 52). One significant institutional arena of urban state strategy is the fiscal system.

China's fiscal system is ripe for application of a scale model. As Walder (1992: 528) has explained, "China's national budget is a nested hierarchy of independent budgets – each government unit exercises property rights over firms under their financial jurisdiction." In other words, "each government jurisdiction exercises property rights over its subordinate enterprises . . . This bureaucratic economy, far from being a monolith, is composed of thousands of government jurisdictions of varying sizes, each of which seeks to expand its revenues by capturing investments, subsidies, and grants"

(Walder 1992: 530). This bureaucratic economy is the state fiscal system, which was substantially decentralized under reform.

The fiscal system is a complex state apparatus, governed by the central state and adjusted in order to achieve economic goals at all scales. Here we outline its general contours under reform. In 1980, the state introduced the "fiscal contracting system" (*caizheng chengbao zhi*), popularly known as "eating in separate kitchens" (*fenzao chifan*), which divided revenue into central, local and shared pots. Shared revenues are negotiated, which gives provinces and cities great incentive to collect taxes. It also granted these jurisdictions authority over their own expenditures and budget management. Wealthier cities especially accumulated substantial revenue under this system while the center's share of revenue decreased. In response, the central state introduced tax reforms in 1994 in order to increase the amount of revenue flowing to the center. This recentralization strategy more than doubled the percentage of revenue received by the central state: in 1993, the central government netted only 22 percent of total revenue, whereas by 1997 its total share increased to 49 percent (Huang 1996: xix). The 1994 tax reforms worked for the central state. While cities lost shared revenue after 1994, many enhanced collection of "extra-budgetary revenue" through local taxes and fees. Such extra-budgetary funds can be as large as formal revenues, which makes for a complex system of "fiscal dualism" (Wong 1998).

In his important analysis of the Chinese political economy, Huang (1996) has evaluated central–local relations to answer why the Chinese reform economy did not experience spiraling inflation common to other post-socialist states. In this two-level model, he argues that the key to the puzzle is actually *strengthened* central control over provincial and city government fiscal responsibility, which is the continuing strong role of the state in articulating fiscal strategy. In our scale terms, the 1994 tax reforms compelled a rescaling up of fiscal responsibility. Thus contrary to the general idea of decentralization, Huang argues that China's success in avoiding runaway inflation during market reform resulted from a tight rein on political decentralization. This is the state's political strategy of economic accumulation, articulated through the territorial administrative hierarchy and its representative offices. This analysis also views the Chinese state as a federalist system, "Chinese style," in which the center has maintained control of political decision-making, while cities have controlled locally specific economic decisions. We can see scale relations centrally at stake in the complexity of state shifting on fiscal policy.

## Conclusions

The significance of scale relations in an era of globalization especially emerges around the importance of cities and regions in articulating globalizing processes in the context of nation-state territoriality. In order to access global networks, agents of urban and regional development must work through scale, in which scale is understood as the central organizational principle according to which geographical differentiation takes place at, across and transcends different spatial "levels." In the context of globalization and scale relations, the city, especially at the scale of the metropolitan region, is the spatial territory and unit of economic analysis most suited to the interaction of political, social and economic processes in an era of globalization. Thus scale is a conceptual means of geographical differentiation, a system of state administration represented by the administrative hierarchy, and a set of processes by which the local state, in its urban and regional capacities, organizes political power, society and economy.

In the case of China, application of scale relations seeks to provide a synthetic basis for research on urban restructuring and political economic change under reform. Appreciation for scale analysis also maps on to the historical state, revealing the territorial administrative hierarchy, and ideas about it, as the singular common condition of the historical empire. Scale analysis also addresses several gaps in the literature dealing with China under reform, specifically, answering how decentralization actually works through the territorial hierarchy and why the central state continues to use the urban system as a basis for political economic change. The central state under reform has enhanced administrative power of the local state by increasing decision-making and fiscal power at different levels in the administrative hierarchy, especially for cities. Scaling the state also introduces a framework for analyzing the spatial processes of urban transformation in relation to the global economy and through which researchers may pursue implications for different cities, industrial sectors, social transformations, and more, at various scales.

Further, how do conditions of the state in China inform the larger project to theorize scale relations? The conditions of the Chinese state challenge normative notions about the role of the state in so-called transitional economies, asking us instead to understand political territoriality in not only two dimensions – the hierarchical and the horizontal or areal – but in a third dimension of spatial, processural transformation between these two axes of institutional formation. State strategy remains an important force in the Chinese political economy, illustrating how we can avoid a reductionist economism in analysis of transformation under reform. The case of China reveals the project to theorize scale relations as dependent upon an under-theorized model of the capitalist state, a model that has yet to embrace the emerging recognition of different and regionally specific capitalisms.

## References

Agnew, J. (1993) "Representing space: space, scale, and culture in social science," in Duncan, J. and Ley, D. (eds), *Place/Culture/Representation*, London: Routledge, 251–271.

—— (1994) "The territorial trap: the geographical assumptions of international relations theory," *Review of International Political Economy*, 1: 53–80.

—— (1999) "The new geopolitics of power," in Massey, D., Allen, J. and Sarre, P. (eds), *Human Geography Today*, Cambridge: Polity Press, 173–193.

Brenner, N. (1997) "Global, fragmented, hierarchical: Henri Lefebvre's geographies of globalization," *Public Culture*, 10(1): 135–167.

—— (1998) "Between fixity and motion: accumulation, territorial organization, and the historical geography of spatial scales," *Environment and Planning D: Society and Space*, 16: 459–481.

—— (1999) "Beyond state-centrism? Space, territoriality, and geographical scale in globalization studies," *Theory and Society*, 28(1): 39–58.

—— (2000) "The urban question as a scale question: reflections on Henri Lefebvre, urban theory and the politics of scale," *International Journal of Urban and Regional Research*, 24(2): 361–378.

—— (2001) "The limits to scale? Methodological reflects on scalar structuration," *Progress in Human Geography*, 25(4): 591–614.

Brenner, N., Jessop, B., Jones, M. and MacLeod, M. (2003) *State/Space: A Reader*, Oxford: Blackwell.

Cartier, C. (2001) *Globalizing South China*, Oxford: Blackwell.

—— (2003) "Symbolic city-regions and gendered identity formation in south China," *Provincial China*, 8(1): 60–77.

Chan, K. (1994) *Cities with Invisible Walls: Reinterpreting Urbanization in Post-1949 China*, Hong Kong: Oxford University Press.

Cheng, T. and Selden, M. (1994) "The origins and social consequences of China's *hukou* system," *China Quarterly*, 139: 644–668.

Dutton, M. (1998) *Streetlife in China*, Cambridge: Cambridge University Press.

Fan, C. C. (1999) "The vertical and horizontal expansions of China's city system," *Urban Geography*, 20(6): 493–515.

Feuchtwang, S. (1992) *The Imperial Metaphor: Popular Religion in China*, London and New York: Routledge.

Fitzgerald, J. (2002) "The province in history," in Fitzgerald, John (ed.), *Rethinking China's Provinces*, London and New York: Routledge, 11–40.

Gong, T. and Chen, F. (1994) "Institutional reorganization and its impact on Decentralization," in Jia, H. and Lin, Z. (eds), *Changing Central–Local Relations in China: Reform and State Capacity*, Boulder, CO: Westview Press, 67–88.

Gu, C., Qiu, Y. and Ye, S. (1997) "The change of designated cities in China before and after 1949," *China City Planning Review*, 13(2): 10–23.

Harvey, D. (1982) *The Limits to Capital*, Oxford: Blackwell.

—— (1985) *The Urbanization of Capital*, Oxford: Blackwell.

—— (1996) *Justice, Nature and the Geography of Difference*, Oxford: Blackwell.

Herod, A. and Wright, M. (eds) (2002) *Geographies of Power: Placing Scale*, Malden, MA and Oxford: Blackwell.

Ho, P. (1962) *The Ladder of Success in Imperial China*, New York: Columbia University Press.

Howitt, R. (1993) "A world in a grain of sand: towards a reconceptualisation of geographical scale," *Australian Geographer*, 24(1): 33–42.

—— (1998) "Scale as relation: musical metaphors of geographical scale," *Area*, 30(1): 49–58.

—— (2002) "Scale and the other: Levinas and geography," *Geoforum*, 33: 299–313.

Huang, Y. (1996) *Inflation and Investment Controls in China: The Political Economy of Central–Local Relations during the Reform Era*, Cambridge: Cambridge University Press.

Jessop, B. (1997a) "The entrepreneurial city: re-imaging localities, re-designing economic governance, or re-structuring capital?," in Jewson, N. and MacGregor, S. (eds), *Transforming Cities: Contested Governance and New Spatial Divisions*, London: Routledge, 28–41.

—— (1997b) "A neo-Gramscian approach to the regulation of urban regimes: accumulation strategies, hegemonic projects and governance," in Lauria, M. (ed.), *Reconstructing Urban Regime Theory*, London: Sage, 51–73.

Khan, A. R. and Riskin, C. (1999) "Income and inequality in China: composition, distribution and growth of household income," *China Quarterly*, 154: 221–253.

Kirkby, R. J. R. (1985) *Urbanization in China: Town and Country in a Developing Economy 1949–2000 AD*, New York: Columbia University Press.

Lieberthal, K. and Oksenberg, M. (1988) *Policy Making in China: Leaders, Structures and Processes*, Princeton: Princeton University Press.

Lin, G. C. S. (1997) *Red Capitalism in South China: Growth and Development of the Pearl River Delta*, Vancouver: University of British Columbia Press.

—— (2002) "The growth and structural change of Chinese cities," *Cities*, 19(5): 299–316.

Ma, L. J. C. (2005) "China's changing urban administrative system: spatial restructuring and local economic development," forthcoming in *Political Geography*.

—— and Cui, G. (1987) "Administrative changes and urban population in China," *Annals of the Association of American Geographers*, 77(3): 373–395.

MacLeod, G. and Goodwin, M. (1999) "Space, scale and state strategy: rethinking urban and regional governance," *Progress in Human Geography*, 23(4): 503–527.

Mallee, H. (1996) "Reform of the *hukou* system: introduction," *Chinese Sociology and Anthropology*, 29(1): 3–14.

Marston, S. A. (2000) "The social construction of scale," *Progress in Human Geography*, 24(2): 219–242.

Merrifield, A. (1993) "Place and space: a Lefebvrian reconciliation," *Transactions of the Institute of British Geographers*, 18(4): 516–531.

Montinoa, G., Qian, Y. and Weingast, B. R. (1995) "Federalism, Chinese style: the political basis for economic success in China," *World Politics*, 48: 50–81.

Naughton, B. (1995) *Growing out of the Plan: Chinese Economic Reform, 1978–1993*, Cambridge: Cambridge University Press.

Nee, V. (1989) "A theory of market transition: from redistribution to markets in state socialism," *American Sociological Review*, 54: 663–681.

Pannell, C. W. and Ma, L. J. C. (1983) *China: The Geography of Development and Modernization*, New York: John Wiley.

Read, B. (2000) "Revitalizing the state's urban nerve tips," *China Quarterly*, 163: 806–820.

Scott, A. J. (1998) *Regions and the World Economy: The Coming Shape of Global Production, Competition, and Political Order*, Oxford and New York: Oxford University Press.

Shambaugh, D. (2000) "The Chinese state in the post-Mao era," in Shambaugh, D. (ed.), *The Modern Chinese State*, Cambridge: Cambridge University Press, 161–187.

Shue, V. (1988) *The Reach of the State*, Stanford: Stanford University Press.

—— (1994) "State power and social organization in China," in Migdal, J. S., Kohli,

A. and Shue, V. (eds), *State Power and Social Forces: Domination and Transformation in the Third World*, Cambridge: Cambridge University Press, 65–88.

Shurmann, F. (1971) *Ideology and Organization in Communist China*, 2nd edn, Berkeley, CA: University of California Press.

Smith, N. (1984) *Uneven Development: Nature, Capital and the Production of Space*, Oxford: Basil Blackwell.

—— (1992) "Geography, difference and the politics of scale," in Doherty, J., Graham, E. and Melek, M. (eds), *Postmodernism and the Social Sciences*, Houndsmills: Macmillan, 57–79.

—— (1993) "Homeless/global: scaling places," in Bird, J. *et al.* (eds), *Mapping the Futures, Local Cultures, Global Change*, London: Routledge, 87–119.

—— (1995) "Remaking scale: competition and cooperation in prenational and postnational Europe," in Eskelinen, H. and Snickers, F. (eds), *Competitive European Peripheries*, Berlin: Springer Verlag, 59–74.

—— (2000) "Scale," in Johnston, R. J., Gregory, D., Pratt, G. and Watts, M. (eds), *The Dictionary of Human Geography*, Oxford: Blackwell, 724–727.

Storper, M. (1997) *The Regional World: Territorial Development in a Global Economy*, New York: Guilford.

Swyngedouw, E. (1997a) "Excluding the other: the production of scale and scaled politics," in Lee, Roger and Wills, Jane (eds), *Geographies of Economies*, London: Arnold, 167–176.

—— (1997b) "Neither global nor local: 'globalization' and the politics of scale," in Cox, Kevin R. (ed.), *Spaces of Globalization: Reasserting the Power of the Local*, New York: Guilford Press, 137–166.

Tang, W. S. and Chung, H. (2000) "Urban–rural transition in China: beyond the *desakota* model," in Li, S. M. and Tang, W. S. (eds), *China's Regions, Polity and Economy: A Study of Spatial Transformation in the Post-Reform Era*, Hong Kong: Chinese University Press, 275–308.

Tang, W. S., Li, S. M. and Kwok, R. Y. W. (2000) "Space, place, and region and the study of contemporary China," in Li, S. M. and Tang, W. S. (eds), *China's Regions, Polity, and Economy: A Study of Spatial Transformation in the Post-Reform Era*, Hong Kong: Chinese University Press, 3–34.

Walder, A. G. (1992) "Property rights and stratification in socialist redistributive economies," *American Sociological Review*, 57: 524–539.

—— (1996) "Markets and inequality in transitional economies: toward testable theories," *American Journal of Sociology*, 4: 1060–1073.

Whitney, J. B. R. (1970) *China: Area, Administration and Nation Building*, Chicago: University of Chicago, Department of Geography Research Paper No. 123.

Whyte, M. K. and Parish, W. L. (1984) *Urban Life in Contemporary China*, Chicago and London: University of Chicago Press.

Wong, C. (1998) "Fiscal dualism in China: gradualist reform and the growth of off-budget finance," in Brean, D. J. S. (ed.), *Taxation in Modern China*, New York: Routledge, 187–208.

Wright, A. (1977) "The cosmology of the Chinese city," in Skinner, G. W. (ed.), *The City in Late Imperial China*, Stanford: Stanford University Press, 33–74.

Zhao, H. (2002) *Political Regime of Contemporary China*, Lanham, MD: University Press of America.

Zhao, X. B. and Zhang, L. (1999) "Decentralization reforms and regionalism in China: a review," *International Regional Science Review*, 22(3): 251–281.

Zhou, Q. and Wu, C. T. (2000) "Regional implications of central vs. local resource control under Chinese reforms," in Li, S. M. and Tang, W. S. (eds), *China's Regions, Polity and Economy*, Hong Kong: Chinese University Press, 187–216.

Zhou, Y. and Ma, L. J. C. (2003) "China's urbanization levels: reconstructing a baseline from the fifth population census," *China Quarterly*, 173: 176–196.

# 3 Space, scale and the state

## Reorganizing urban space in China

*Jianfa Shen*

## Introduction

With rapid economic development and urbanization since 1978, there have been significant changes in Chinese cities (Fan 1999; Ma 2002; Pannell 2002; Gu and Shen 2003). Urban change in China is clearly related to the decentralization process of the state in China. Before 1978, the power of the hierarchical administrative system in China was highly centralized. Since the early 1980s, there has been a clear tendency of decentralization in economic and financial administration and decision-making. Various scales and forms of state activity have been rearticulated through complicated processes involving the twin issues of governance and scale.

Such changes in China are parallel to the changing scalar relations in North America and Western Europe (Brenner 1999). But the processes involved are not necessarily the same. The triple processes in Western countries are denationalization, destatization and internationalization that have resulted in the rescaling of regulatory codes, norms and institutions downward or upward (MacLeod and Goodwin 1999; Swyngedouw 1997). The scale theory has been developed to describe such scalar reorganization of capitalism.

The transition process from state socialism to market economy in China involves the interaction of three processes: decentralization, marketization and globalization (Lin 1997; Lau and Shen 2000; Y. Wei 2001). It is argued that the general principle of scale theory also applies to socialist transitional economies where scalar configuration has been rearticulated by liberalization and globalization. This chapter will use the scale theory for a political–economic analysis of the urban space reorganization in post-reform China.

The reorganization of urban space, i.e., urban administrative area, has become an essential measure in promoting urbanization in China (Wang 2001). It refers to the designation of new cities and towns and the constant adjustment of city and town boundaries. The role and function of the state have been contested and reconfigured consequently at various spatial levels in this process. In scalar analysis, the global forces, the central state and the local actors in a city are taken into consideration. In Europe, it is found that little can happen subnationally without the national state's cooperation, acquiescence or benign ignorance (MacLeod and Goodwin 1999). Recognizing the role of the central state and global forces, this chapter argues that the local government becomes increasingly important in the urban dynamics after decentralization and marketization in China.

Previous studies concerning the application of "urban growth machine" and "urban regime theory" developed in the Western context are relevant in the explanation of urban transformation. However, adopting the scalar perspective, this chapter emphasizes the interaction between different levels of government and between governments, firms and residents in cities. Thus the chapter moves beyond the constraints of fixed-scale approach in the local focused studies under the notions of "urban growth machine" and "urban regime theory." The chapter attempts to integrate different spatial scales (central state, local state, firms and individuals) within a single analytical framework. Through a multiscaled analysis of urban changes in China, the chapter contributes towards theorizing the urban space reorganization as a politically constructed process so that the state strategy, the roles of local governments and local communities can be analyzed using a political economy perspective.

## Brief review of scale theory and scale processes

It is now popular to assume that the role of the central state is on the decline while that of the local state is rising. The key concepts in this debate of sociospatial transformation are scale, territorialization, de-territorialization and rescaling of scalar relations. According to Smith (2000: 725), "specific geographical scales can be conceived as platforms for specific kinds of social activity." Smith recognized "a loose hierarchy of geographical scales, from that of the BODY, the home and the COMMUNITY through the local, regional, national and global." The concern on scale is on the "production of scale" rather than the "scale" itself. The production and construction of geographical scale involve four major processes.

First, scaling refers to "the establishment of geographical differences according to a metric of scales – etches a certain order of empowerment and containment into the geographical landscape" (Smith 2000: 726). Second, rescaling refers to the shifting in the power and control over scales, such as from the national scale to the urban scale or to the global scale. Thus rescaling process involves two or more scales. Rescaling the state means re-territorializing state power onto multiple spatial scales that do not converge with the national scale (Brenner 1999). Territorialization and de-territorialization are involved simultaneously in such rescaling process. A general term for such a process may be rescaling of territoriality. In China, the introduction of the "system of city-leading-county" is an example of rescaling of territoriality. Territorialization is a concept related to the economic organization in the discussion of globalization. According to Storper (1997:

21), "an activity is fully territorialized when its economic viability is rooted in assets (including practices and relations) that are not available in many other places and that cannot easily or rapidly be created or imitated in places that lack them." Concept of territorialization is also applicable to cities and states (Brenner 1999). One important form of de- or re-territorialization is the shrinking or expansion of urban space.

The third process of the production of scale is the interaction of scales. For example, "jumping scales" refers to the case that political power established at one geographical scale is expanded to another scale (Smith 2000). "Penetrating scales" refers to the case that actors in one particular scale attempt to exert their influence in other scales. Overlapping of scales refers to the case that the lower scale has certain power and property of the upper scale. For example, a semi-provincial city (fu *shengjishi*) in China is under the administration of a province, but it also possesses certain provincial power.

The fourth process of the production of scale is the territorialization taking place at the same scale. Sociospatial transformation within the same scale is also significant in the contemporary world. In China, spatial reconfiguration of counties and cities has long-term implications for urban and regional development.

## Rescaling the state: changing central–local relations

In Western countries, local government has emerged as a major actor in urban restructuring (Brenner 1999). Local growth coalitions have been formed by local actors in American cities. Business interests are central in the local coalition. The entrepreneurial regime has become popular. There is a change in urban government from running the daily functions such as transportation (managerialism) to developing active risk-taking strategies to promote the economic growth (entrepreneurialism) (Goldsmith 1995).

Similarly, local urban governments have assumed an increasing role in local development in China. Various scales and forms of state activity have been rearticulated. In this chapter, the four scales of concern are the national scale, the urban scale, enterprises and individuals (Figure 3.1). This section will focus on the rescaling of the central and local states at national and urban scales. The rescaling and interaction of four scales will be analyzed in subsequent sections.

### *Decentralization and marketization*

The decentralization and marketization processes are perhaps the most influential in the rescaling of political economy in China. In contrast to the tight control by the central state in the pre-reform socialist command economy, the residents, firms and local states have been given much autonomy in their pursuit for development and prosperity since 1978 (Oi 1992; Walder 1992;

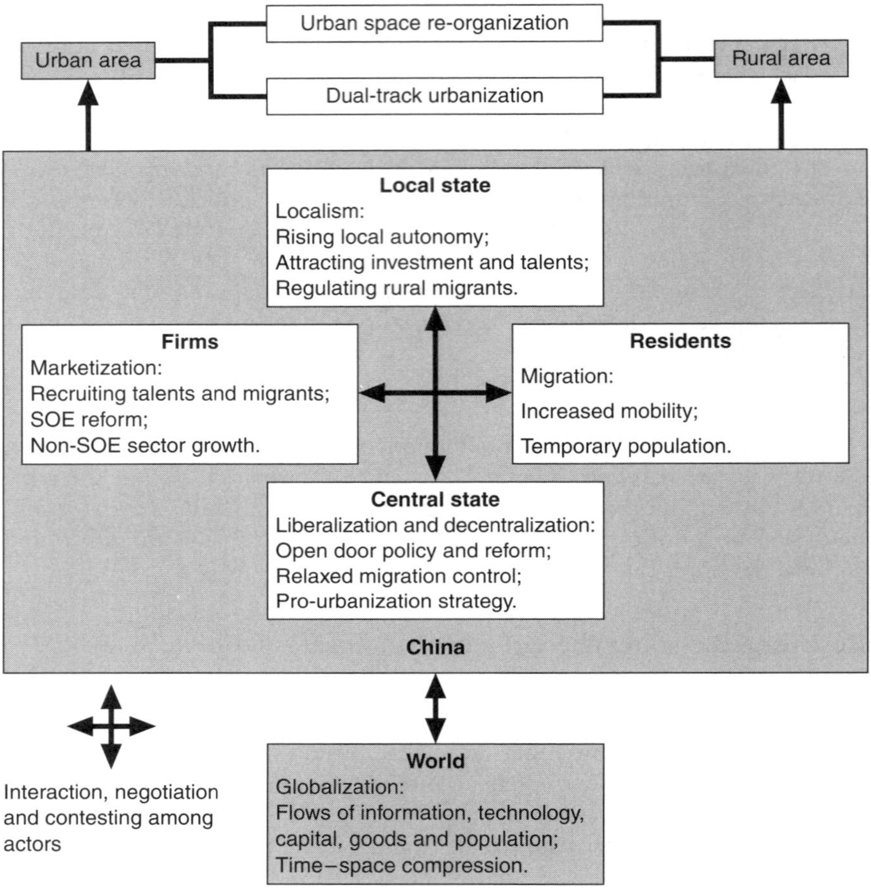


*Figure 3.1* Rescaling of political economy in post-reform China.

Gu *et al.* 2001). Y. Wei (2001: 7) argued that "China's economic reforms have restructured the relationship between the central and local states, between plan and market, and between domestic and international forces." The central planning system has gradually been replaced by market mechanisms in China. By 1998, most industrial and agricultural production was based on market conditions. Only nine agricultural products were still produced under guided plan of the state while partial output of twelve industrial products was under directive planning of the state (Wang *et al.* 1998).

The central–local relation has been rearticulated. The power of the central state has been steadily decentralized with increasing power going to local governments. It is useful to examine how the power has been rescaled between central and local states, especially on fiscal matters.

### *Reforming central and local fiscal relations*

A highly centralized fiscal system prevailed in China under state socialism before 1978 (Lu and Sit 1997). The most extreme of central control was in the period 1949–1953 when revenues and expenditures were totally controlled by the central government, with the central government taking 44 percent of the total fiscal revenue. Various forms of central and local fiscal arrangement were tried in the 1970s and the share of central fiscal revenue was reduced to only 13 percent in 1976. The central government faced a poor fiscal situation. A fiscal contracting system was introduced which improved continuously the fiscal conditions of the central state in the period 1980–1994.

There have been four major reforms in the fiscal system (Chan *et al.* 2002). The first reform was implemented in the period 1980–1984. The scopes of revenue collection for central and local governments were divided following the idea of "eating in separate kitchens." For local governments with fiscal surplus, the proportion of the local surplus to be remitted to the central government was fixed. For local governments with fiscal deficit, a fixed quota was set up for subsidy from the central government. The proportion or quota would be fixed for five years. These measures had the effect of encouraging local governments to stimulate economic development to increase their revenue and to control their expenditure. Subsequent reforms are consistent with this reform spirit.

The second reform was implemented in the period 1985–1987. Based on tax-for-profit economic reform, tax revenue was divided into three parts: items for central revenue, items for local revenue and shared items for central and local revenue. The previous contracting system based on proportion and quota was maintained using the local revenue and expenditure in 1983 as the basis.

The third reform was implemented in the period 1988–1993. Sixteen province-level units would remit a proportion of their total revenue or increased revenue to the central government. Some sixteen provinces and autonomous regions would receive fixed subsidies from the central government. Guangdong and Hunan would remit revenue to the central government based on a fixed growth rate. Shanghai, Shandong and Heilongjiang would remit a fixed amount of revenue to the central government.

Similar to the central and provincial fiscal relations, a fiscal contracting system was implemented between a province and its subordinate units (Oi 1992; Wong *et al.* 1995; Chan *et al.* 2002). Fiscal contract was signed between a provincial government and each of its prefecture-level administration or city government. The fiscal relations between the two levels of governments had tremendous varieties and were subject to negotiation. Some used an overall ratio so that a lower-level government kept a percentage of fiscal revenue (say, 80 percent) and the remaining (say, 20 percent) was sent to the higher-level government. Some local governments only needed to hand over a fixed quota of fiscal revenue to its higher-level government. For example, the fiscal contract stipulated that the revenue to be turned over to the provincial government of Guangdong should increase 7 percent a year for cities of Foshan, Jiangmen, Shaoguan and Maoming, and 6 percent for Zhanjiang beginning in 1985 (Lam 1999).

These fiscal reforms based on proportion and quota drove the local government to stimulate economic development and increase its revenue. But two problematic measures were often adopted. First, local governments offered much tax concession to enterprises as the tax revenue was shared with the central government (Walder 1992). Such concessions were often at the expense of the tax revenue of the central government (Wang *et al.* 1998). As a result, the fiscal revenue of the central government grew slowly despite rapid economic growth. Second, local governments were keen to expand economic sectors that offered high profit and tax revenue, resulting in widespread construction of similar industries all over the country. For example, due to high tax rate of alcohol and tobacco, many small factories were established to produce these products (Chan *et al.* 2002). Thus the central state made several attempts in 1989–1991 and 1993 to recentralize and rectify the rampant tax concessions offered by local states. This shows that the central–local relation was subject to adjustment and negotiation (Hsu 2000).

In the fourth reform, a tax assignment system was formally introduced in 1994 aiming to correct the problems mentioned above. Circulation tax, i.e., industrial–commercial tax including product tax, value-added tax and business tax, became a major source of tax revenue. Different kinds of taxes were designated as revenue for central or local governments. The power of local governments to offer tax concessions was reduced and the revenue of the central government increased. After the reform, the share of the central government in fiscal revenue increased from 22 percent in 1993 to 55.7 percent in 1994. A tax rebate system was also introduced for the central government to return the extra tax collected to local governments due to the new tax assignment system. However, for any 1 percent growth in such extra tax in the coming years, only 0.3 percent increase would be returned to local governments, ensuring fast growth of the tax revenue of the central government (Chan *et al.* 2002). But local governments became dependent on the revenue transfer from the central government that controls a large share of tax revenue. In 1991, the total fiscal revenue of local governments (229.6 billion yuan) within the budget (*yusuannei*) was close to the total fiscal expenditure (221.1 billion yuan) (Editorial committee 1992, 2001). In 2000, the total fiscal revenue of local governments (640.6 billion yuan) was just 61.3 percent of their total fiscal expenditure (1,036.7 billion yuan).

Many local governments were keen to bargain with the central government for more revenue support and more investment projects. Table 3.1 presents the distribution of fiscal revenue and expenditure among local governments in 1991 and 2000. It is clear that the provincial and county-level

*Table 3.1* Distribution of fiscal revenue and expenditure among local governments in China, 1991 and 2000 (%)

| *Item* | *1991* | *2000* |
|---|---|---|
| *Fiscal revenue* | | |
| Provincial governments | 22.3 | 22.4 |
| Prefecture-level units | 40.1 | 36.5 |
| County-level units | 22.9 | 25.1 |
| Towns/townships | 14.7 | 16.0 |
| *Fiscal expenditure* | | |
| Provincial governments | 30.8 | 29.3 |
| Prefecture-level units | 28.3 | 30.6 |
| County-level units | 30.1 | 29.0 |
| Towns/townships | 10.8 | 11.2 |

Sources: Editorial committee (1992: 151; 2001: 93).

Note
The majority of prefecture-level units are prefecture-level cities.

governments had a larger share of fiscal expenditure than that of revenue while other low-level governments had a larger share of revenue than that of expenditure. The provincial governments depended on the revenue transfer from either the low-level governments or the central government. The local governments were also induced to collect taxes more seriously for local fiscal balance.

## *The rising of local states*

Generally, the new tax assignment system increased the share of central government in the fiscal revenue and provided a sound institutional framework in the division of fiscal revenue for local and central governments. But local governments still possess overwhelming autonomy in economic administration. By the late 1990s, local governments, firms and the non-state sector had emerged as major agents of investment and the central government is playing a less significant role in capital accumulation and investment allocation (Y. Wei 2001; Lam 1999). Walder (1992: 309) pointed out that:

> In Chinese cities revenues come almost exclusively from taxes on enterprise profits. The proceeds are divided with the provincial or the central government, according to a negotiated formula. With its share, the city funds its social services, public works, and infrastructure development.

Local states have adopted a number of strategies to capture financial gains within their boundary. First, the local states are keen to improve their infrastructure such as highways, electricity supply and communication facilities for local development. But they pay much less attention to improve the living environment and social security provisions that are less tangible but need a large amount of financial resources. For example, less than 50 percent of wastewater is treated in most cities (Wang 2001).

Second, tax concessions and other policy incentives are offered by the local states to attract foreign investment. Cheap and even free land has been offered to foreign investors including those from Hong Kong, Macao and Taiwan. For example, a Hong Kong based firm, China Travel International (CTI) which is actually owned by the Chinese government, was given $4.4\,km^2$ of free land to develop the Overseas Chinese Town in Shenzhen (Zhu 1999). CTI's investment was regarded as investment from Hong Kong. It made substantial profit due to property boom in the mid-1990s. Although the local government gained 950 million yuan in taxes in return from this project in the period 1986–1995, it is likely that the government lost substantial land revenue that could otherwise be used to improve public facility and social welfare. The CTI has benefitted the most from the project.

Third, local states have developed close relations with local firms by direct or indirect participation and intervention conceptualized as "Local State Corporatism" (Oi 1992; Walder 1992). An informal local growth coalition is formed between local states and enterprises so that local states subsidize local enterprises at the expense of the central revenue or public assets while enterprises pay tributes to the local states in forms of various fees, payments and donations (Zhu 1999). For example, eighty-five kinds of fees were imposed on real-estate development in Guangzhou accounting for 25–30 percent of total development cost. "Relationship (*guanxi*)" with the local officials has had important impact, formal and informal, on the local business operation (Oi 1992).

Rising localism has led to widespread local market protection in China. This is partly due to inadequate regulation of the central state to ensure fair competition and trade. Some local governments have adopted local product standards or discriminative policies to protect local firms. For example, consumers buying non-local cars have to pay higher fees than those buying locally produced cars (Chan *et al.* 2002). This may be regarded as an overstretched strategy of territorialization (Storper 1997).

Under the protection of local governments, many firms have been established to produce similar goods for the local market resulting in overinvestment, market segmentation, duplicated construction and surplus production capacity. In 1998, 8,835 Chinese factories produced one-third of the world's cement. But some 75 percent of the factories were small in scale, each producing less than 60,000 tons of low-quality cement. According to the national industrial census in 1995, among 285 main industrial products, the utilization rate of production capacity was below 50 percent for ninety products (H. Wei 2001).

Fourth, the commodification and marketization of land and housing property have created substantially landed interest since 1988 (Zhu 1999;

Zhang 2002). The local states acquire significant land revenue, due to huge rent gap, when agricultural land is turned into industrial and urban land for commercial buildings, offices and housing. Thus local states have every incentive to control a large amount of land and to convert much arable land for industrial and urban development. This is a major force driving the urban space expansion in China.

## Rescaling the urbanization process

The rescaling of urbanization has significant impact on urban development and urban space reorganization in China. This section focuses on population migration and urban population growth from the perspective of rescaling urbanization over various scales (Figure 3.1). Heavy contestation is involved in this scalar structuration that is producing dual-track urbanization in China – state-sponsored urbanization and spontaneous urbanization that Ma and his collaborators (Ma and Lin 1993; Ma and Fan 1994) analyzed earlier. Spontaneous urbanization is a new track of urbanization from below in the post-reform period. However, state-sponsored urbanization that prevailed in pre-reform China has continued although it has been reconfigured under the emerging market economy.

### *The role of the central state*

With the rescaling of central and local states in post-reform China detailed in the previous section, the role and function of the central and local states in the process of urbanization have been contested and reconfigured. The central state is busy in adjusting its urban policies to accommodate the local interests without disrupting the urban order in an increasingly decentralized, liberal and urbanizing society.

In pre-reform China, the central state, via the political regime of central planning and the household registration (*hukou*) system, controlled the speed of urbanization to pursue industrialization (Chan and Zhang 1999). Under the *hukou* system, the household registration status of agricultural population is different from that of non-agricultural population. Almost all urban enterprises in the pre-reform period were state-owned or owned by urban collectives, and they had no autonomous power to recruit migrants but had to accept employees allocated by the government. Urban population or non-agricultural population were fully supported by the state in terms of education, housing, employment and other social benefits. This is called state-sponsored urbanization (Shen *et al.* 2002).

The urban policy of the central state has changed significantly since the early 1980s. Facing the condition of oversupply of manufactured goods and serious deflation in the national economy, a pro-urbanization strategy has been considered as an effective way of stimulating demand and economic development since 2000 (Figure 3.1) (State Council of China 2000; Wang

2001). Tight control on the growth of urban non-agricultural population has been relaxed. The central state has also reconfigured state-sponsored urbanization by reducing formal support to urban non-agricultural population (Smart and Smart 2001; Shen *et al.* 2002). In the meantime, migration control has been relaxed (Shen and Huang 2003).

### *The role of local states*

Local governments also have great power in regulating the urban population under their jurisdiction. First, urban governments have attempted to collect various management fees from enterprises and rural migrants, which act as a barrier to migration. In one city, an enterprise is required to pay 20 yuan a month to the local Labor Bureau for each non-local migrant it employs (Chan *et al.* 2002).

Second, the registration for the temporary population (rural migrants) is compulsory for people who move away from their place of household registration. Each migrant is issued a Temporary Residency Card by the local Office of Public Security. A rural migrant is also required to get a work permit from local Labor Bureau. Migrants' access to some attractive occupations is restricted by tight regulations (Yu and Hu 1998). For example, in 1999, the labor and social security bureau of a city listed 104 occupations that were open to temporary population but 103 occupations that were not (Chan *et al.* 2002).

Third, rural migrants without local *hukou* are not eligible for social services and welfare reserved for local residents (Shen and Huang 2003). As they have no access to subsidized housing and education, most rural migrants leave their children and spouses in their home villages. There is much tension between the local state and rural migrants in many cases (Zhang 2001).

Fourth, some cities like Shenzhen and Shanghai have attempted to attract talents from outside to support their economic development by offering formal *hukou* or quasi non-agricultural population status (Kong 2001). Shanghai and other cities have introduced "blue-stamp" *hukou* that is available, after paying an urban construction fee, to rural migrants with stable urban employment, income and housing. The fee is set by individual cities and can be as high as 10,000 yuan in some cases. The holders of blue-stamp *hukou* have the same education and welfare rights as the local urban residents. About 30,000 people acquired "blue-stamp" *hukou* in Shanghai in the period 1994–2000. About 20,000 of them obtained the *hukou* through their investment in housing property. But such "blue-stamp" *hukou* is valid only locally in a city. It is different from the usual "red-stamp" *hukou*. Similar locally valid *hukou* has been available to some towns since 1997 and to the urban proper of county-level cities and all designated towns since 2001 (State Council of China 2000). These "quasi non-agricultural population" statuses are localized under the principle of "local need, local benefit, local responsibility and local validity."

In the reform period, rural urbanization led by the township and village enterprise (TVE) growth takes place at the level of local designated towns, market towns and townships with active participation and support of local governments. Their growth and expansion has been an important part of locally driven urbanization (Ma and Fan 1994; Shen 1995). There were 28.3 million TVE employees in 1978, but the number jumped to 52.1 million in 1984 and 130.9 million in 2001 (NBS 2002). Together with a temporary population of over 120 million, they accounted for about half of the total 498.8 million rural laborers in China. Without these TVEs, much more rural population would have moved into cities.

### *The role of enterprises and individuals*

At the enterprise level, the increasing autonomy of firms in the emerging market economy is important in the urbanization process. Now firms are willing to employ skilled talents and cheap migrants from outside which may be consistent with or contradictory to the agenda of urban governments. At the individual level, the rescaling of urbanization empowers residents to participate in the urbanization process. A "non-*hukou* population," i.e., "floating population," has emerged at various destinations. According to official population surveys and censuses, the number of "non-*hukou* population" in China increased from 6.6 million in 1982 to 48.4 million in 1995 and 121.1 million in 2000 (Shen and Huang 2003).

### *Interaction among the central state, the local state, enterprises and the individuals*

It is interesting to note that the central state, the local state, enterprises and the individuals do not take actions separately. Indeed, there has been intense interaction and negotiation among these actors reflecting the rescaling process of urbanization in China (Figure 3.1). The central state's pro-urbanization strategy at the national scale has been instrumental in forming the dual-track urbanization in post-reform China. The reconfiguration of the state-sponsored urbanization by reducing the welfare support to SOE (state-owned enterprise) staff has also met controversy and resistance (Smart and Smart 2001). Instead of outright dismissal, redundant SOE workers who do not need to work anymore are paid a fraction of their wages to stay home.

For another example, facing increasing number of non-*hukou* migrants in urban China, the state had to introduce the "temporary population" status to give a legal status to rural migrants in 1985. But the local state has attempted to charge various excessive fees while registering the temporary population (Yu and Hu 1998). There has been intense conflict between rural migrants and city governments (Zhang 2001; Gu and Shen 2003). Recently, the central state has attempted to regulate the relation between rural migrants and the local government. For example, the Office of State Council

(2003) has issued a notice instructing local governments that occupation restrictions on rural migrants should be abolished.

## Reorganizing urban space in China

### *Cities as important administrative units*

According to the "Law on the Organization of Local Governments" revised in 1995 (Editorial Committee 1995), the People's Congress of a city is the organization with the state power. The city government is not only the executive arm of the People's Congress of a city but also the local administrative arm of the state. Indeed, a government department such as the Department of Civil Affairs in a city government is not only responsible to the city government but also to the Ministry of Civil Affairs under the central government. Thus the city government was responsible to both the People's Congress of the city and the state administrative organization at higher levels.

The current process of city designation and boundary changes is stipulated by relevant laws. According to the revised constitution passed by the Fifth National People's Congress in 1982, the designation of a municipality needs to be approved by the National People's Congress and the designation of a prefecture-level or a county-level city and its boundary demarcation needs to be approved by the State Council. Cities could be designated at provincial, prefecture or county level.

One major event of rescaling the city is the designation of semi-provincial cities introduced in 1994. The status of a semi-provincial city is higher than a prefecture-level city but lower than a municipality under the direct administration of the central government. The key consideration is to facilitate the autonomy of some larger cities for economic development. These cities are given a privileged position in the national plan that their economic plans are listed separately from their provinces. In 1997, there were fifteen such cities (Dai 2000).

One major form of urban territorialization refers to the expansion of the administrative area of a city to enclose both landed and other economic interests. It takes place when a city's territory expands to include more areas under its control. The city gains control over the land and economic interest in the new area. It takes place in various forms. One important form that cities expand their influence is the designation of new cities and towns. Local governments are keen to negotiate with the central state for urban status for economic, land and strategic interests. A city has several advantages over a county. First, when a county is designated as a city, it acquires greater administrative power. Second, a city is listed separately in the provincial plan. Third, a county-level city can use 7 percent of the local fiscal revenue for urban maintenance and construction while a county can only use 5 percent (Wang 2001). Fourth, a city is much easier to promote its economy and products.

## *Changing regulations for city designation*

The central government has been influential in city designation via various regulations before and after 1978. The first regulation on the designation of cities and towns was announced by the State Council in 1955 which stipulated that settlements with a population of over 100,000 and other important places could be designated as cities. During the period 1949–1978, the designation of cities was tightly controlled and the total number of cities was increased from 132 to 193 only.

Considering the importance of urban development, the criteria of city designation were revised in 1983, 1986 and 1993 respectively to speed up the process (Liu and Wang 2000). Before 1983, a settlement must normally have 100,000 non-agricultural population to be designated as a city. In 1983, a county with a population of less than 500,000 could be designated as a city if the seat of county government had over 80,000 non-agricultural population and its manufacturing output was over 200 million yuan. The 1986 criteria included a new section for upgrading a town to a city as well as the condition for the introduction of the "system of city-leading-county." A town could be designated as a city if it had a non-agricultural population over 60,000 and the gross domestic product (GDP) was over 200 million yuan. A medium-sized city with over 250,000 non-agricultural population and a GDP of over 1 billion yuan could administer other counties. A set of more sophisticated criteria for the designation of county-level cities and prefecture-level cities was approved by the State Council in 1993. It was the first time that the criteria for the designation of prefecture-level city were stipulated. To be designated as a prefecture-level city, the main urban settlement, the seat of the city government, should have a non-agricultural population over 200,000 and a strong economy.

## *Major forms of urban designation and expansion*

"Separating a built-area from a rural county" is one form of city designation used mainly before 1978. Under this system, urban and rural areas were completely separated into a city and a county. For example, part of the original Jinjiang County in Fujian Province was cut off to establish a new Shishi City in 1987. The city administered an urban subdistrict (*jiedao*) and three towns with an area of 160 km$^2$ and a population of 250,000 (Anonymous 1993). The remaining Jinjiang County still administered fifteen towns with an area of 649 km$^2$ and a population of 900,000. Jinjiang County was also designated as a county-level city in 1992. This kind of city designation can create serious governance problems due to overlapping of governments. The county will either lose its momentum of urbanization or become the rival of the city by developing a new urban center in its own territory.

The second form of city designation is called "redesignation of a whole county as a city." A whole county is redesignated as a city if it meets the criteria for city designation. Over 70 percent of the current cities in China are designated in this form. A county-level city often consists of several towns and townships. For example, Chonghua County in Guangdong was designated as a county-level city in 1994 (DCA and MPH 2000). Its largest town, Jiekou, only had 100,000 people in an area of $49\,km^2$. In the remaining fourteen towns, each only had a non-agricultural population of less than 5,000.

Urban expansion is an important strategy to enhance a city's strength. Urban territorialization takes place when a city annexes a county or a county-level city as its urban district and when two or more towns or cities are merged. Again, the process involves sophisticated negotiation and interaction between the central and local states. When a county or a county-level city was designated as an urban district, it was totally absorbed by the prefecture-level city. Such a move often met stiff resistance of the county or county-level city involved, such as in the case of Panyu (Dai 2000). After the change, an urban district has less power but less conflict with the city. Generally, a county or county-level city is an independent political unit having much political and economic power. An urban district is part of a city with little independence. Many political and economic matters are administered by the city government directly while the government of the urban district is the lowest government level in a city that often only performs the function of providing public services to its own residents.

With changing urban policies, many townships have been designated as towns. The number of towns increased rapidly, from 2,173 in 1978 to 20,312 in 2000. One major problem is that many towns are small in scale. In 1996, the average area of 16,124 designated towns (excluding the central towns of counties) in China was only $2.43\,km^2$ with a total population of 4,520 and a non-agricultural population of 2,072 (Dai 2000). Another major problem is the low efficiency of land use. Urban built-up area per resident is as large as $149\,m^2$ in designated towns, 38 percent greater than the urban average of $108\,m^2$ (Table 3.2). It is argued that, for better social and

*Table 3.2* Urban built-up area per resident by size of settlement in China

| *Size of settlement* | *Built-up area per resident ($m^2$)* |
|---|---|
| Super-large city | 75 |
| Large city | 99 |
| Medium city | 105 |
| Small city | 132 |
| Designated town | 149 |
| Village | 168 |

Source: Wang (2001: 35).

economic efficiency, resource utilization and environmental protection, the minimum population of a town should be 50,000. The policy of changing townships to designated towns was suspended by the State Council in 2002 because the criteria used for the designation of towns were considered too low. Coupled with rapid economic growth, too many small and economically weak townships were designated as towns. Some provinces have taken measures to merge small towns. In Jiangsu Province, the number of towns/townships was reduced by 508, from 1,974 in 1998 to 1,466 in 2000 (Wang 2001).

Vertical rescaling and horizontal territorialization can also take place simultaneously. Vertical rescaling refers to a change in the administrative status of a city while horizontal territorialization refers to the expansion of the administrative area of a city. For example, when the system of "city-leading-county" is introduced, a city may be scaled up from county-level to prefecture-level city with an expanded administrative area in most cases. By 1994, the "city-leading-county" system was implemented in all provincial units except Hainan Province in China. A county under the administration of a city may later qualify for city status and be designated as a county-level city. Thus a system of "city-leading-city" also emerged. The system is designed to enhance the integration of a central city and its surrounding hinterland. But in some cases, there are intense conflicts between the governments of the central city and the county-level units (Dai 2000). When a county is under the administration of a province, it has much autonomous power. But when it is governed by a prefecture-level city, the city tightens its administration on the county. A central city may give priority to its urban proper in bidding for large construction projects, foreign investment and the right for direct foreign trade. Under such a condition, the system is regarded as "city exploiting county" instead of "city helping county" (Liu and Wang 2000). The mayors of about twenty county-level cities share this view in a survey (DRE and JICA 2001).

The case of Foshan City and Nanhai City in Guangdong Province is an example. Nanhai was designated as a county-level city in 1992 under the jurisdiction of Foshan, a prefecture-level city. By 2000, Nanhai had become even stronger than Foshan's urban proper in economic terms. Nanhai's GDP was 33.9 billion yuan while the GDP of the latter was only 15.9 billion yuan in 2000. GDP per capita was close in the two cities, 31,076 and 33,434 yuan respectively (Statistical Bureau of Guangdong 2001). Thus it is difficult to coordinate the urban infrastructure that falls under the jurisdiction of two competing city governments. The solution taken by the provincial government was to convert Nanhai and other county units into urban districts under Foshan City. Through such administrative adjustment, the power of the central city was increased while the county-level local state suffered economically and politically. It may help coordinate development in a region, but if urban development is not properly planned, unhealthy core-periphery tension and conflict may result.

The urban system has changed dramatically after above changes. The total number of cities increased from 193 in 1978 to 245 in 1982 and 668 in 1997. It was reduced to 663 in 2000 due to the merging of cities mentioned before. The number of prefecture-level cities increased from 112 in 1982 to 207 in 1997 and 244 in 2000.

### *The role of central and local governments*

The central government has substantial power in the process of urban rescaling and territorialization. First, the National People's Congress and the State Council are empowered by the nation's constitution to approve the designation of cities and their boundary changes. Second, the State Council has the power to set up and revise criteria and guidelines for city designation and boundary changes. Third, the power of the central government has been implemented through the hierarchical administrative system, as local governments at provincial, prefecture and county levels are made responsible to both the People's Congress of their territory and the governments at higher level.

City designation and boundary changes are a matter of local and central governments. Local governments are actively engaged in the process. But public participation is rare in China although the State Council stipulated as early as 1961 that changes in administrative divisions must consult the public. For example, the merging of towns in Jiangsu has resulted in many problems involving the interests of residents and local places (Wang 2001). Where will the merged towns be located? What name will a merged town take? How to compensate residents who sustain an economic loss due to the merger? Clearly, community participation is important but is poor under the socialist market system in China (Zhang 2002).

## Conclusion

The scale theory has been developed to describe the scalar organization of capitalism. It is argued that the general principle of the theory also applies to socialist transitional economies where scalar configuration has also been rearticulated by liberalization and globalization. This chapter uses the scale theory for a political–economic analysis of urbanization and urban space reorganization in post-reform China.

The rescaling of central and local states is a politically constructed process. It does not just mean the declining of the central state and the rise of the local state. The central government has introduced recentralization policies from time to time to gain control over financial resources and administrative power. As argued by Zhang (2002), the urban regime theory is applicable to the analysis of urban governance in Chinese cities. But the local growth coalition is marked by problems in the rescaling of the central and local states, i.e., the transitional or muddling nature of the society and economy with immature and inadequate regulation mechanism (Smart 2002). With its increasing fiscal and administrative power and inadequate regulation of the central state in an immature market economy, a not well-regulated "local growth machine" has emerged in China.

With the rescaling of central and local states in post reform China, rescaling of urbanization takes place among various scales: the central state, the local state, the enterprises and the residents. This has major implications to urban development.

Industrial and land development in China can bring huge financial returns to local governments such as land revenue and industrial–commercial tax. Decentralization has led to the emergence of localism with increasing competition instead of cooperation among localities. Every local state is interested in attracting investment and concentrating the development in its own administrative area, a kind of urban territorialization. To meet increasing needs for the expansive industrial and land development, acquiring urban status and expanding the spatial boundary under the jurisdiction of local states have become a new strategy for local growth, resulting in massive urban space reorganization in Chinese history.

The role and function of the state have been contested and reconfigured consequently at various spatial levels in this process. At the individual level, the spatial fixing function of *hukou* has been shaken resulting in increasing mobility of peasant migrants across the "invisible wall" separating city and country. At the community and township level, local industrialization and small towns are expanding rapidly. At the city and county level, local governments are keen to acquire urban status and to expand their urban space. In the meantime, the central state is busy adjusting its urban policies to accommodate the local interests while at the same time trying to maintain the urban order in an increasingly decentralized, liberal and urbanizing society. Within China's hierarchical administrative system, the central government has substantial power in the process of urban rescaling.

## Acknowledgments

The work for this chapter was fully supported by a Direct Research Grant of the Chinese University of Hong Kong, Project Code 2020659. Thanks are due to Larry Ma, Fulong Wu and two referees for their helpful comments on an earlier draft of the chapter.

## References

Anonymous (1993) *Fujian sheng ditu ji (The Atlas of Fujian Province)*, Fuzhou: Fujian ditu chubanshe.

Brenner, N. (1999) "Globalisation as reterritorialisation: the re-scaling of urban governance in the European Union," *Urban Studies*, 36: 431–451.

Chan, D. Q., Ying, W. Q. and Cang, Y. R. (2002) *Dapo difang shichang fenge* (*Eliminating Local Market Fragmentation*), Beijing: Zhongguo jihua chubanshe.

Chan, K. W. and Zhang, L. (1999) "The *hukou* system and rural–urban migration in China: processes and changes," *China Quarterly*, 160: 18–55.

Dai, J. L. (2000) *Zhongguo shizhi (City System of China)*, Beijing: Zhongguo ditu chubanshe.

DCA and MPH (Department of Civil Affairs of Guangdong Province and Map Publishing House of Guangdong) (eds) (2000) *Guangdong sheng zhengqu tuce* (*Atlas of Administrative Divisions of Guangdong Province*), Guangzhou: Guangdong ditu chubanshe.

DRE (Department of Regional Economy of the State Development Planning Commission of China) and JICA (Japan International Cooperation Agency) (2001) *Chengshihua: Zhongguo xiandaihua de zhuxuanlü* (*Urbanization: the Main Theme of China's Modernization*), Changsha: Hunan renmin chubanshe.

Editorial Committee (1992) *Zhongguo caizheng nianjian 1992* (*Finance Yearbook of China 1992*), Beijing: Zhongguo caizheng zazhishe.

—— (1995) *Zhongguo difang zhengfu jigou gaige* (*Institutional Reform of Local Government in China*), Beijing: Xinhua chubanshe.

—— (2001) *Zhongguo caizheng nianjian 2001* (*Finance Yearbook of China 2001*), Beijing: Zhongguo caizheng zazhishe.

Fan, C. C. (1999) "The vertical and horizontal expansions of China's city system," *Urban Geography*, 20: 493–515.

Goldsmith, M. (1995) "Local government," in Pierre, J. (ed.), *Urban and Regional Policy*, Brookfield, VT: E. Elgar, 49–66.

Gu, C. L., Shen, J. F., Wong, K. Y. and Zhen, F. (2001) "Regional polarization under the socialist-market system since 1978 – a case study of Guangdong Province in South China," *Environment and Planning A*, 33: 97–119.

—— (2003) "Transformation of urban socio-spatial structure in socialist market economies: the case of Beijing," *Habitat International*, 27: 107–122.

Hsu, S. P. (2000) "Local tax effort in Shanghai, 1987–93: decentralization vs. local fiscal balance," *Issues & Studies*, 36: 116–138.

Kong, Y. (2001) *Jiedu Shanghai 1990–2000* (*Understanding Shanghai 1990–2000*), Shanghai: Shanghai renmin chubanshe.

Lam, T. (1999) "Institutional constraints, leadership and development strategies," in Chung, J. H. (ed.), *Cities in China: Recipes for Economic Development in the Reform Era*, London: Routledge, 256–295.

Lau, C. M. and Shen, J. F. (ed.) (2000) *China Review 2000*, Hong Kong: Chinese University Press.

Lin, G. C. S. (1997) *Red Capitalism in South China: Growth and Development of the Pearl River Delta*, Vancouver: UBC Press.

Liu, J. D. and Wang, Y. M. (2000) *Zhidu yu chuangxin: Zhongguo chengshi zhidu de fazhan yu gaige xinlun* (*Systems and Innovation: On the Development and Reform of the City System in China*), Nanjing: Dongnan daxue chubanshe.

Lu, D. D. and Sit, F. S. (eds) (1997) *Zhongguo quyu fazhan baogao 1997* (*China Regional Development Report 1997*), Beijing: Shangwu yinshuguan.

Ma, L. J. C. (2002) "Urban transformation in China, 1949–2000: a review and research agenda," *Environment and Planning A*, 34: 1545–1569.

—— and Fan, M. (1994) "Urbanization from below: the growth of towns in Jiangsu, China," *Urban Studies*, 31: 1625–1645.

—— and Lin, G. C. S. (1993) "Development of towns in China: a case study of Guangdong province," *Population and Development Review*, 19: 583–606.

MacLeod, G. and Goodwin, M. (1999) "Space, scale and state strategy: rethinking urban and regional governance," *Progress in Human Geography*, 23: 503–527.

NBS (National Bureau of Statistics) (2002) *China Statistical Yearbook 2002*, Beijing: China Statistics Press.

Office of State Council (2003) *Guanyu zuohao nongmin jincheng wugong jiuye guanli he fuwu gongzuo de tongzhi* (*A Notice about Employment Management and Service Provision to Rural Migrants in Cities*), Beijing.

Oi, J. (1992) "Fiscal reform and the economic foundations of local state corporatism in China," *World Politics*, 45: 99–126.

Pannell, C. W. (2002) "China's continuing urban transition," *Environment and Planning A*, 34: 1571–1589.

Shen, J. F. (1995) "Rural development and rural to urban migration in China 1978–1990," *Geoforum*, 26: 395–409.

—— and Huang, Y. (2003) "The working and living space of the 'floating population' in China," *Asia Pacific Viewpoint*, 44: 51–62.

——, Wong, K. Y. and Feng, Z. (2002) "State sponsored and spontaneous urbanization in the Pearl River Delta of South China, 1980–1998," *Urban Geography*, 23: 674–694.

Smart, A. (2002) "The Hong Kong/Pearl River Delta urban region: an emerging transnational mode of regulation or just muddling through?," in Logan, J. R. (ed.), *The New Chinese City: Globalization and Market Reform*, London: Blackwell, 92–105.

—— and Smart, J. (2001) "Local citizenship: welfare reform urban/rural status, and exclusion in China," *Environment and Planning A*, 33: 1853–1869.

Smith, N. (2000) "Scale," in Johnston, R. J., Gregory, D., Pratt, G. and Watts, M. (eds), *The Dictionary of Human Geography*, Oxford: Blackwell, 724–727.

State Council of China (2000) *Guanyu cujin xiaochengzhen jiankang fazhan de ruogan yijian* (*Some Views on Promoting the Health Development of Small Cities and Towns*), Beijing.

Statistical Bureau of Guangdong (2001) *Guangdong Statistical Yearbook 2001*, Beijing: China Statistics Press.

Storper, M. (1997) "Territories, flows and hierarchies in the global economy," in Cox, K. R. (ed.), *Spaces of Globalization*, New York: Guilford Press, 19–44.

Swyngedouw, E. (1997) "Neither global nor local: 'glocalization' and the politics of scale," in Cox, K. R. (ed.), *Spaces of Globalization*, New York: Guilford Press, 137–166.

Walder, A. (1992) "Local bargaining relationships and urban industrial finance," in Lieberthal, K. G. and Lampton, D. M. (eds), *Bureaucracy, Politics and Decision Making in Post-Mao China*, Berkeley, CA: University of California Press, 308–333.

Wang, Y. (ed.) (2001) *"Shiwu" chengzhenhua fazhan guihua yanjiu* (*Studies on the Development Planning of Urbanization in the Tenth Five-year Plan*), Beijing: Zhongguo jihua chubanshe.

Wang, Y. K., Liu, X. T. and Lang, P. J. (eds) (1998) *Zhongguo xingzhen tizhi gaige ershi nian* (*Administrative System Reform in China in past 20 Years*), Zhengzhou: Zhongzhou gujie chubanshe.

Wei, H. K. (2001) *Cong chongfu jianshe zouxiang youxu jingzheng* (*From Duplicated Construction to Orderly Competition*), Beijing: Renmin chubanshe.

Wei, Y. D. (2001) "Decentralization, marketization and globalization: the triple processes underlying regional development in China," *Asian Geographer*, 20: 7–23.

Wong, C. P. W., Heady, C. and Woo, W. T. (1995) *Fiscal Management and Economic Reform in the People's Republic of China*, Oxford: Oxford University Press, 81–135.

Yu, D. X. and Hu, X. Y. (1998) *Jiaoxia de changlu: Zhongguo mingong xianxiang touxi* (*Long March under the Foot: an Analysis of Migrant Workers in China*), Beijing: Jingji kexue chubanshe.

Zhang, L. (2001) *Strangers in the City: Reconfigurations of Space, Power, and Social Networks within China's Floating Population*, Stanford: Stanford University Press.

Zhang, T. W. (2002) "Urban development and a socialist pro-growth coalition in Shanghai," *Urban Affairs Review*, 37: 475–499.

Zhu, J. M. (1999) "Local growth coalition: the context and implications of China's gradualist urban land reforms," *International Journal of Urban and Regional Research*, 25: 534–548.

# 4 Dual land market and internal spatial structure of Chinese cities

*Anthony Gar-On Yeh*

## Introduction

The open-door policy and economic reform in 1978 has profound influences on the urban development and internal spatial structure of Chinese cities. Social and economic development are often considered to be the main factors affecting the internal structure of cities (Bourne 1982; Knox and Pinch 2000). However, it is found that the form of urban development of a city is also greatly influenced by its land development process which in turn is influenced by its socioeconomic and political structures (Healey and Barrett 1990; Healey 1991). This is especially true in China because of the transitional economy with the coexistence and co-functioning of leased land that operates according to market principle and administratively allocated land that is given to state enterprises and agencies free of charge.

Before land reform in 1988, land was owned by the government and is normally administratively allocated free to users without any charges. Although experiments of charging land use fees were carried out in Shenzhen, Guangzhou and Foshan (Walker 1991; Yeh 1985; Yeh and Wu 1995), they were mainly limited to land use involving foreign investment. It was not until the First Session of the Seventh People's Congress in 1987 that paid transfer of land use rights (*tudi youchang zhuanrang*) was made official. The clause "The right to the use of land may be transferred in accordance with law" was added to Article 10, Section 4 of the constitution which states that "No organization or individual may seize, buy, sell land or make any other unlawful transfer of land." This amendment was approved by the National People's Congress on April 12, 1988 (Tang 1989). This opens a new era of lawful transactions of urban land. In addition, in 1988, the State Council announced the "Provisional Regulations of the People's Republic of China on Land Use Tax in Cities and Towns" which enabled cities and towns to collect land use taxes. These mark the end of free land use and the establishment of urban land markets in China (Dowall 1993).

Although land reform has introduced land leasing in China, the previously administratively allocated land still exists and the administrative land allocation system still continues (Xie *et al.* 2002; Zhang 1997). Because of this, land reform has created a dual land market in China that consists of leased and administratively allocated land. It is different from the Western societies where there is only one land market with free transactions. The interactions of the dual land market and system have much impact on urban development in China. Apart from urban economic, housing and land reform and economic globalization (Wu 2001; Yeh and Wu 1995), they are one of the most influential forces that are shaping Chinese urban development and internal spatial structure of Chinese cities.

## Land development process under dual land market

The dual land market system has introduced a new land development process in China (Yeh and Wu 1996). There are three main types of land ownerships in China. The first type is *rural land* which is collectively owned by farmers. The second is urban land which is owned by the state in name but occupied by various state work-units. Land is allocated to them through the administrative allocation method. This can be referred to as *administratively allocated land.* The third is urban land that is owned by the state but the land use rights are transferred to the users through the paid transfer of land use rights. This can be referred to as *leased land.*

Urban land acquired through the paid transfer of land use rights only occupies a small portion of the existing urban land and land acquired from rural land. Yet, the transfer of land use rights caused a profound change on the land system in China. A dual land market system has been set up. As mentioned above, most of the land is allocated through the administrative allocation method, while a small portion of land is leased by the state to the users through paid transfer of use rights. The city governments monopolize the supply of leased land. Under the Land Administration Act, any transactions of land ownership directly from farmers are illegal. Rural land can only be acquired either by the state work-units or the municipality to ensure that the urban land belongs to the state. The transactions between users and farmers are prohibited by the system of land use certificates. Development has to obtain land use certificate from the municipal land administration bureau before it can proceed. Developments which acquire land directly from farmers could not obtain the certificate.

Urban land is owned by the state and managed by the municipality. The state work-units acquire land through administrative allocation method and "occupy" the state land. Yet, they do not have the land use rights that could be transferred further to other users. State enterprises can exchange land with each other on a voluntary basis. The exchange is treated as a special way of administrative allocation and the land acquired by this method is not allowed to transfer its land use rights to other users, except state enterprises. The implication of the policy is that only land acquired through paid transfer of land use rights could be transacted. The dynamics of urban land transactions in China can be generalized as the following major types (Figure 4.1).

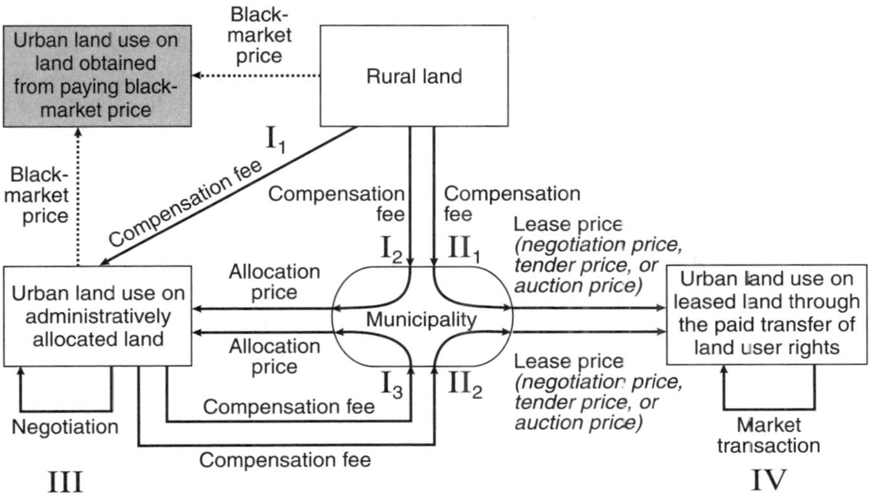


*Figure 4.1* Dual land market transactions and black land market in Chinese cities (source: Yeh and Wu 1996).

### *Type $I_1$ – Acquisition of rural land by work-units for project-specific development*

This is the dominant source of land supply for urban development in China. The typical way to acquire this type of land is through project-specific development. A work-unit which needs land for development can apply a land acquisition permit from the municipality. It can then acquire rural land by paying a standard compensation fee to the farmers. According to the Land Administration Act 1986, compensation amounts to three to six times of the average annual agricultural production of the rural land in the last three years before land acquisition. In addition, payments have to be made on buildings and agricultural products attached to the rural land at the time of the acquisition. Compensation will be paid if acquisition involves the relocation of people.

Before the 1978 economic reform, economic activities were organized mainly through various ministries in the central government. Municipalities played a relatively unimportant role. Urban land was allocated freely to land users. Thus, work-units tended to occupy more land than they actually needed (Fung 1981). There is a misconception that the municipalities acquire rural land, compensate farmers and allocate the land to work-units without any charges. In fact, the involvement of the municipality before 1978 was minimal apart from approving the land acquisition permit. Land acquisition and compensation were directly dealt with between the work-units and the farmers. After the establishment of the State Land

Administration Bureau in 1986 and various municipal land administration bureaus in the following year, the power of the municipality was strengthened. Land acquisition in the jurisdiction of the city is approved by the city government. The 1986 Land Administration Act stipulates that acquisition of more than 1,000 mu (67 ha) of cultivated land or more than 2,000 mu (133 ha) of other types of land needs to be approved by the State Council. Acquisition of land less than 3 mu (0.2 ha) of cultivated land and less than 10 mu (0.7 ha) of other types of land should be approved by local government at the county level. However, municipalities can hardly reject the investment from the sectoral departments. The control of development is not through local physical plans which are prepared by the municipalities but rather through nationwide or province-wide economic plans. The principle of this type of land supply is that only state work-units, usually large industrial enterprises, could acquire land through the administrative allocation method and that such type of land could not be leased to foreign investors.

### *Type $I_2$ – Acquisition of rural land by the municipal government for comprehensive development*

Municipalities play an important role in this type of land transaction because they directly acquire land from farmers, develop it comprehensively, and then allocate it to users. This type of transaction was mainly initiated after 1978. The purpose of comprehensive development is to deal with land acquisition more efficiently and quickly because the municipality, instead of the users, acquires land directly from farmers through a standard compensation fee. The negotiation with the farmers could proceed more easily and the community facilities could be shared by several work-units. Comprehensive development when first introduced in the late 1970s did not involve much commercial consideration (Yeh and Wu 1999). Most of the land was supplied to state work-units. But later, some land was also supplied to housing development companies for building commercial houses. These development companies usually belong to the municipal housing bureau or construction bureau. In the early stages of housing commercialization, most commercial housing was on this type of land because land could be obtained cheaply compared to land obtained from urban redevelopment which involved heavy compensation. By obtaining land from the municipality, the price of housing could be relatively low because the development companies only need to pay the farmers compensation for agricultural land rather than land price for residential land which is more expensive. However, like the first type of land (Type $I_1$), the user can use the land but could not transfer it freely to other users, especially to foreign buyers. Commercial housing built on this type of land is mainly for use by Chinese buyers in contrast to commercial housing built on leased land which can be used by both Chinese and foreign buyers. Comprehensive development, originated from the cooperation of several work-units to build shared housing apartments, is thus a kind of development of the preliminary market mechanism, i.e. a property market without a land market.

### *Type $I_3$ – Acquisition of existing administratively allocated urban land for urban infrastructure by the municipal government*

Usually this type of transaction is for the development of urban infrastructures. The municipal government negotiates with the occupiers of the previously administratively allocated land that is required for the construction of urban infrastructures. Theoretically, land is owned by the state. As this type of transaction only involves the change of land users that might both belong to the state, it seems that no compensation is needed. However, as mentioned above, existing occupiers have to pay compensation to farmers and the cost of leveling and formation of the land to the municipal government when they acquire the administratively allocated land. Thus, the municipal government needs to compensate the occupiers for their loss and also needs to allocate another piece of urban land to them in exchange when they acquire their land for the construction of infrastructure.

### *Type $II_1$ – Acquisition of rural land by the municipal government for land leasing*

This is the most recent type of land transaction and is becoming increasingly popular. Rural land is acquired by the municipality and then leased to other users through the market mechanism of negotiation, tender and auction. Because the municipality monopolizes the right of supply of this type of land, it can acquire rural land at a monopoly price from farmers and sell them to developers at market price. Considerable profit can be made by the municipality because of the great difference in land acquisition and land lease price. As land obtained from land leasing is more expensive than the previous three types of land, most users try to avoid acquiring land through land leasing as far as possible. If foreign users need land for a factory, they can seek various ways to use the existing land that is administratively allocated, for example, building a factory as a joint venture with a Chinese business partner who obtains the land from the administrative allocation methods (Types $I_1$ and $I_2$). The current customers of this type of land transaction are mainly foreign investors. The purpose of development is mainly for commercial housing, offices, hotels and industrial buildings. However, it has been reported that municipalities are facing increasing resistance from farmers who sometimes even resort to violence when their land is acquired by the municipality which will later lease it out in the land market. They are demanding higher compensation because the municipality can obtain a higher price in the land market through land leasing.

### *Type $II_2$ – Acquisition of existing administratively allocated urban land by municipal government for land leasing*

Municipalities acquire the previously administratively allocated urban land occupied by work-units and then lease the land to users who can pay market land price. Unlike type $I_3$ in which land is mainly used for the construction of urban infrastructures by the municipal government, that is, for non-profit use, this type of transaction which is similar to type $II_1$ is quite profitable. Compensation to the original users is higher than type $I_3$ because it is difficult for the municipal government to justify paying the same price as land acquired for state construction. As for land acquired for state construction, the socialist principle is that "local interests should obey national interests and individual interests should obey community interests." Under this principle, users were often asked to sacrifice for the construction of state projects and community infrastructures. However, it is not applicable to this type of transaction. Negotiations have to be carried out between the original land users and the municipality until a compromise of profit sharing is reached. It is not surprising that some city governments have to lease out their own compounds that are usually located at the city center and move their offices to other places. Surely, it is much more expensive and difficult for the municipal government to assemble this type of land than type $II_1$. This explains why leased land mainly comes from the acquisition of rural land and unoccupied urban land and why Economic and Technological Development Zones (ETDZs) which need large tracts of land are located at the periphery of the urban area.

Type $II_1$ and $II_2$ transactions are referred as the *primary land market* in China. Municipalities acquire land from the rural areas (type $II_1$) or existing administratively allocated urban land occupied by various work-units (type $II_2$) and then transfer the land use rights to users through land leasing by negotiation, tender or auction. Land leasing through negotiation is not much different from the traditional negotiation and compromise. For tender, the conditions of the parcel and the development constraints are announced and tendering is invited. Land auction is an open competition in which the highest bidding will get the land use right.

### *Type III – Exchange of administratively allocated land among work-units*

Work-units can exchange their administratively allocated land through negotiation. For example, one side may offer the other side some housing as a term for exchange. The exchange, however, does not follow market prices. Because of the lack of information and standard practice, such deals are based on previous practices. Exchange of land through this way is very time-consuming, difficult to balance the demand and supply, and to make compensation.

### *Type IV – Transaction of leased land among land users*

Only land that is acquired from the primary market can be transacted, i.e. leased land. This is referred as the *secondary land market* in China. In the newspapers and government documents, this type of transaction is sometimes further divided into transaction of land use rights among real-estate companies and transactions among ordinary users. The latter is referred to as the *tertiary land market* (Zhou 1992).

The process of decision-making is still largely based on the administrative system which decides the location and amount of land that will be put into the market. The market mechanism only operates on leased land and does not affect the majority of land that is administratively allocated. Such a feature is significantly different from that of Western cities where most of the land is privately owned under a freehold system. In Western cities, private initiatives and market mechanism are the basic processes which decide resource and land allocation. Government intervention which tries to correct market failures through urban planning is secondary to these processes (Yeh and Wu 1999).

## Urban development and land value under dual land market

Prior to the introduction of the paid land use system, very often land users would try to acquire more land than needed because land was allocated free. It is hoped that through the new system, land can be more efficiently used through the sale of the land use rights and the collection of land use taxes. Apart from economizing the use of land, this will also help to strengthen the tax base for constructing and maintaining the urban infrastructure.

The new paid land use system has provided the city government a new and substantial revenue source which was not available in the past when land was allocated free of charge. Lands are leased to foreign investors and local developers for industrial, residential and commercial uses. For example, in Hainan Province, money obtained from the property market can be as high as one-quarter of the total revenue at the highest peak of its property boom (Zou 1993). Revenue obtained from land is used to improve urban infrastructure which in turn can improve accessibility and thus open up new land for development. This will increase land value and thus increase government revenue which will provide further capital for building more infrastructure. This is often being referred to as "land breeding land development" *(yi di yang di)*. Many cities, for example, Guangzhou, are using the paid transfer of land use rights to finance their underground railway system. Developers are allowed to develop residential and commercial properties on top of the underground station and the money obtained from such development is used to finance part of the construction cost of the underground railway. As underground railway station is a convenient transport node, its land value is very high and in high demand. As land now has value, the city government is willing to invest in infrastructure to improve its environment and accessibility so as to further increase land value. In the past, the so-called "five connections and one leveling" (*wutong yiping*) – connecting roads, telecommunication, water, electricity and port and leveling of sites – are the main methods to provide a good investment environment for developing the special economic zones (Yeh 1985). These principles are actively applied in land development in the cities of China. With better accessibility and land formation, land can command a higher price in the market. The provisions of roads, water, electricity and telecommunications and leveling of sites are labor intensive. As labor cost is relatively cheap in China, with the high land price that can be obtained from the market, the rate of return for the provision of roads, water, electricity, and telecommunication and site leveling is very good. Revenue from land can now be used to fund infrastructure projects such as roads and telecommunications that were not possible before the introduction of the land market. The introduction of the land market has thus provided the funding and opportunity for planners to implement their plans which could not be carried out in the past because of the lack of funding.

However, despite the introduction of land leasing, a large percentage of land is still allocated through the administrative allocation system (Figure 4.2). The coexistence of the administrative allocated land is to avoid direct challenge to the vested interests of the existing users and the government departments and enterprises which have been receiving land free of charge. This is to ensure economic and political stability (Zhang 1997).

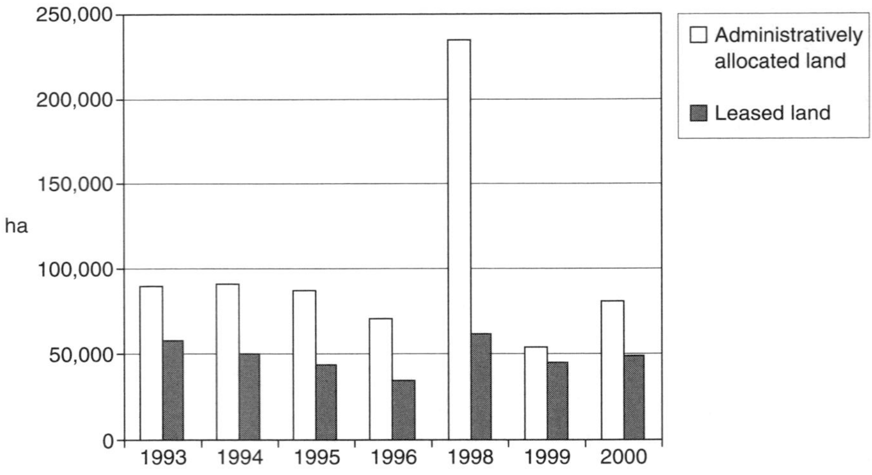


*Figure 4.2* The amount of leased and administratively allocated land in 1993–2000 (source: editorial department of *Yearbook of the Land of China*, 1995, 1996, 1997 and 2001).

Unregulated land use changes are occurring very rapidly in Chinese cities because of the existence of the black land market. Ever since the adoption of the paid land use system, land is no longer merely a space for living and production. It acquires commercial value and becomes a means to make profit. Land could be put into the production process as a productive factor (capital) and could bring economic return to the land owner. Theoretically, land is *de jure* owned by the state. However, in reality, there is no clear understanding about who has the property rights of an administratively allocated land. Thus, it is *de facto* owned by the land occupier. The land occupier can "sell" the land to other users for other uses in the black market to capture the rising land price as a result of the introduction of land leasing.

The dual land market has created a black market in China which led to a loss of land revenue for Chinese cities. There are two main forms of black markets – urban and rural black markets (Figure 4.1). The *de facto* owner of urban administratively allocated land – the state work-units or individuals who own private housing – can lease their land to other users through negotiation. There are several forms of black market of allocated land. First is the direct leasing of buildings to other users who are not the original users, such as the leasing of shops. Second is through joint venture in which the original user contributes the administratively allocated land and the investor, who does not have the land but has the money, pays for the construction costs of buildings. Because the transactions are illegal, the users cannot obtain a land ownership certificate from the land administration authority. Usually, the land is only used by the users themselves because they do not own the land and cannot build housing for sale in the market. The land occupier can use land as a form of investment in joint ventures to exchange land for housing or other benefits. According to the legislation, land use right cannot be transacted with other users in the market unless a premium is paid to the municipal government who is the *de jure* owner of the administratively allocated land. However, exchanging land for housing or even cash is very common in the central city. In Beijing, the land occupier can obtain 60 percent of housing constructed under such black-market arrangement.

Black markets also exist in the rural areas at the city suburbs. Farmers rent their lands to other users as a form of investment in joint ventures. In the urban fringe where rural land is facing immediate urban expansion, farmers prefer renting their lands to either foreign or domestic investors directly, because the price is higher than the compensation paid by the municipal government in transaction types $I_1$, $I_2$, and $II_1$. As a consequence, development in the urban fringe is chaotic.

As leased land is leased to users at a market price through negotiation, tender, or auction and administratively allocated land is allocated to users at an allocation price which is minimal compared to the market price, the difference between the prices of leased and allocated land has led to the development of a black market. The *de jure* system allocates land to state enterprises and institutions at a very low allocation price with the understanding that they are the sole users of the allocated land. However, because of the confusion in land ownership and ambiguity in property rights (Zhu 2002), very often land occupiers of administratively allocated land who are the *de facto* owners can sublet their land to other users under various forms of disguise to capture the rising land price as a result of the introduction of land leasing. Administratively allocated land is transacted in the black market which charges a price that is lower than the market price of leased land but high enough for the land occupiers to make a high profit. Revenues from such land leasing do not go to the city government but to the land occupiers of the administratively allocated land. Land leasing can potentially generate high amounts of revenue that is badly needed by the municipal government to improve the urban infrastructure, but only very little revenue has been generated because most of the revenues from land transactions have gone to the work-units through the black market.

Furthermore, the black market has also lowered the price of the leased land, diminishing government revenue that could be obtained from the leased land. The impact of the black market on land price and government land revenue can be illustrated by Figure 4.3. D is the land demand curve in the market and S is the supply curve of leased land. $E_0$ is the equilibrium point of the land demand and supply curve when the black market does not

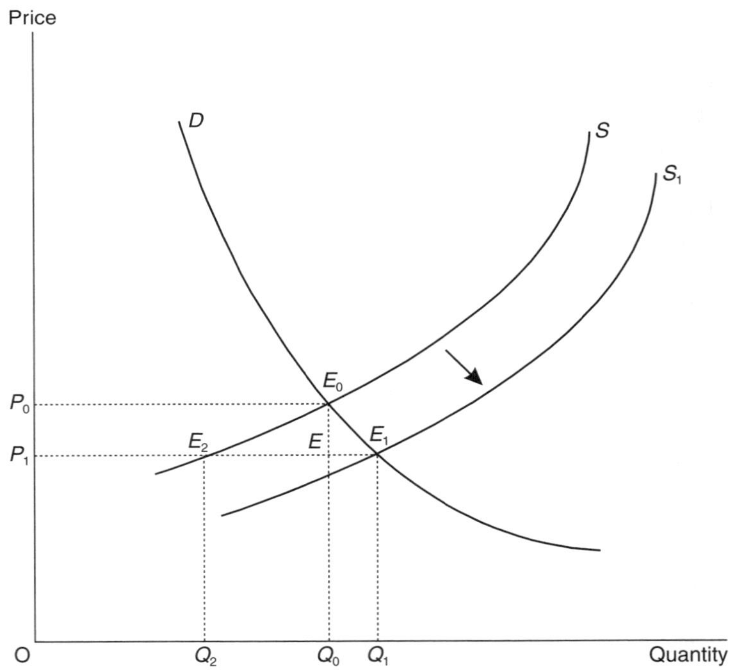


*Figure 4.3* Impact of the black market on land price and government land revenue.

exist. The land price will be $P_0$ and total land revenue that the government can get will be $P_0E_0Q_0O$. However, with the existence of the black market, the land supply curve will move to $S_1$ as the land price lowers to $P_1$ and the amount of land will increase from $Q_0$ to $Q_1$. As the amount of leased land is fixed by the government, curve S still represents the land supply in the leased land market. The quantity of land supply from the leased land will be $Q_2$ and the supply from the black market will be $Q_1 - Q_2$. Revenue loss from land leasing will be the area $P_0E_0Q_0O - P_1E_2Q_2O$. $E_2E_1Q_1Q_2$ will be the amount lost to the land occupiers who have leased their land to others in the black market. However, the above analysis is only from the perspective of a free-market economy. In fact, city government in China has to make annual land supply plan which means the quantity of leased land supply is fixed in each year. If the government decides to supply a quantity of $Q_0$ leased land according to the estimated market price, then because of the existence of the black market, the government will have a revenue loss from land leasing of $P_0E_0EP_1$ and the amount lost to the black market will be $EE_1Q_1Q_0$. It is estimated that the state lost at least 10 billion RMB (US$1.16 billion) annually through revenue loss from land leasing and to the black market (Tian 1994).

The black market hinders the development of a competitive land market system (Li 1997). The ambiguous property rights of the administratively allocated land enable them to compete with the leased land. As found by Zhu (2002), this led to an inefficient allocation of development resources and over supply of property in Shanghai. High vacancy rate is not a concern of land obtained through administratively allocated land but it affects the property value of development from leased land.

Furthermore, unlike land leasing in which government can obtain substantial planning gains, the black market also produces huge benefits that are exclusively captured by the users while the municipal government has to bear the cost of infrastructure provision for the increased intensity of land uses, such as roads and sewers. The relatively cheap land obtained from the black market also enables users to operate businesses in locations which might not be able to survive in a fair competitive land market system. Such a spontaneous conversion of land uses creates the problem of incompatible land uses in Chinese cities. There is an urgent need to regulate land transactions in order to prevent losses of government revenues to the black market and to ensure planning gain is controlled by the state and used for the public interest.

## Unregulated land use changes and urban planning under dual land market

Unregulated land use changes largely occur on administratively allocated land. The introduction of economic incentives is the basic motivation for the users to change their land uses in order to gain more "profit." The deficiency in development control makes it possible for the users to convert the use of land without seeking planning permission. The user discretion, a legacy from the project-specific mode of land development, is unlikely to be changed in a short period of time. As mentioned before, the administratively allocated land, mostly for industries and public organizations, was usually developed through the project-specific mode of land development (Yeh and Wu 1996). The users have gained the maximum control over the land since the compensation of land has been paid by the users and actual acquisition has been carried out through direct negotiation between the users and the farmers. In the past, because of absence of markets, there was no point for the urban-planning system to undertake the function of development control. The function of urban planning is different from that under a market economy where the private initiatives are the dominant source of investment and where there is a necessity of government intervention because of externalities. Under the planned economy, however, development was managed on project basis through economic planning. Thus, the land use of a project was not specified by urban plans but rather was left to users to decide the development intensity, layout and subdivision. The lack of management on land did not lead to a problem of chaotic land uses because projects usually only fulfill the assignment from the government.

The freedom to change land uses becomes problematic under the dual land system. Land uses are converted spontaneously regardless of compatible uses and environmental consequences. For example, the ground floor of some government buildings in Shanghai has been changed into shops. In Guangzhou, within the compound of Guangdong Science Academy, an exhibition of motor cars and lorries is held. On land allocated for military uses, hotels, restaurants and bars have been built. On the one hand, the change of land uses does enhance the efficiency of land uses which was often neglected under the old system. On the other hand, unregulated land use changes lead to mixed uses and incompatible uses, chaotic layout, inadequate infrastructure provision, obstacles to future urban development, and inconvenience to local people. For example, in Guangzhou, a huge furniture complex has been built by *de facto* users in the site of a proposed Guangzhou city highway junction. The construction does not apply for a planning permit from the city planning bureau which is required by City Planning Act 1989. The consequence is that either the building has to be demolished or the highway has to change its path which will raise the cost of the project substantially. During the unregulated land use changes, the government is unable to capture the benefit from land under the existing system. Unlike land leasing in which government can obtain substantial planning gains, the spontaneous land use changes produce huge benefits that are exclusively captured by the users, while the government has to bear the cost of infrastructure provision for the increased intensity of land uses. The loophole also brings negative effects on the establishment of a competitive market system. The cheap land enables some users to operate some business which might not be able to survive in a competitive commodity market.

The new land development process also has great impact on urban planning practice. The two-tier structure of urban planning (master planning and detailed layout planning) cannot control developments effectively (Yeh and Wu 1999). This is because the master planning is not an independent component of planning control. It has to resort to other measures that are provided by the planned economy, such as the registration of projects and investment monitoring. But, because of the devolution of power in China from the central government to local governments, these measures are relaxed and abandoned. Thus, the implementation of the masterplan becomes a serious problem. Unregulated land use changes are happening in China due to the lack of development control. Uncontrolled land use changes did not occur on the leased land, but on administratively allocated land developed through project-specific methods. Development of leased land can be controlled by the lease conditions although they are currently quite minimal. The main problem is with the administratively allocated land that is allocated to work-units in project-specific development. They are undergoing unregulated land use changes in the black market.

In the past, the internal structure of Chinese cities was strongly influenced by urban planning and state investment. However, they were of diminishing importance when there were more and more firms and factories owned by individual enterprises and foreign investments. Although the "City Planning Act" was enacted in December 1989, it was not effective in controlling land development. It mainly required the city government to prepare a masterplan. But the land use zones of the masterplans are too broad in controlling site-specific development, leaving too much discretionary decision to the building administration and local district governments. Disputes may occur between the applicant and the authority which grants planning permissions. It is difficult to reject a building with the existing broad land use zones. Because the existing land use zones are so broad, the actual location, type and intensity of development may not be what the planners intend to achieve in the master and detailed layout plans. For example, a site zoned for public building may be used to build tall office buildings, disregarding whether it is a suitable site for office building and depriving the area without cultural and recreational buildings which was the original intention of the zoning plan. In the past, most of the offices, shops and commercial activities were owned and operated by government departments. All non-residential activities and industrial lands were considered as public building land. This is different from the concept of public building land in Western free-market economy where most land that is considered to be "public building" land in China is considered to be office and commercial land.

There are now more private and foreign investments in the cities because of the post-1978 economic reform and open economic policy. Because of this, land zoned for public buildings may no longer be under the control of the city government. They may be developed into commercial offices for higher profits by government departments and state enterprises which want to make more money, leaving an inadequate amount of land for other public buildings to meet the demand of the community for sports, cultural and recreational facilities. Because of the broad land use zones, it is difficult for the city government to make sure that certain types of land will be available at the right location or will be available at all. For example, land zoned for public building may be used to build offices and hotels disregarding whether they are located at the right location. This is the reason why office buildings and hotels seem to be erected randomly in the city. If all the public building land were developed into offices, there would not be any land left for public buildings for sports, culture and recreation.

It is difficult for urban planning to control unregulated changes because Chinese cities do not have zoning legislation. Most land use changes are from residential and industrial to commercial use. They can be easily observed along the main streets of most cities where offices, hotels and restaurants have been erected in place of residential housing since economic reform in 1978. In Shanghai, small companies redeveloped old two-to-three-storey-high flats that were previously perceived as high-class detached private housing into offices. Along Zuopu Road in Shanghai, pubs and inns are quickly taking over ordinary housing after Deng Xiaoping's tour in the southern cities in 1992 when he urged that the pace of economic reform be quickened. It is now difficult to judge the use of buildings by their appearance. Land uses in cities are now becoming more chaotic and mixed. Unregulated land use changes are happening at different levels of the city – in buildings, street blocks and districts.

To deal with the problem of uncontrolled land use, urban planners in China are actively proposing the use of detailed development control plans (*kongzhixing xiangxi guihua*) which is very similar to a zoning plan but with more detailed control, such as plot ratios and site coverage. Attempts are made to refine the land use categories in the masterplan. The first national land use standard for urban planning in China, the National Standard of Urban Land Use Classification and Planning Standards (National Standard No. GVB137) was promulgated in 1990 and became effective on March 1, 1991. The Standard classifies land uses into ten main categories, forty-six sub-categories and seventy-three sub-types. Some large cities, such as Shanghai and Shenzhen, was considering preparing a zoning ordinance (Yeh and Wu 1999).

## Internal spatial structure of Chinese cities under dual land market

The dual land market has different influences on land use restructuring. Land use restructuring is normally taking place in a more orderly fashion and according to a development plan under the land-leasing system. The land-leasing system enables the government to capture the market land value. With the land-leasing system, city government can afford to resume land and then lease it to land use which can pay the highest price in the land market. In the past, without the land-leasing system, the city government did not have the financial resources to restructure the land use even if it wanted to do so. The introduction of the land market has thus provided the funding and opportunity for planners to implement their plans which could not be carried out in the past because of the lack of funding. Land leasing has led to the redevelopment of the city center to office buildings, forming central business districts which were absent in the Chinese cities in the past. Income from land leasing becomes a new source of income for urban development, helping them to have more income for improving their infrastructure. The use of land leasing and built, operate and transfer (BOT) are the current trend of fund raising for improving the urban infrastructure of cities. Cities such as Beijing and Shanghai are using them to attract foreign investment in urban renewal and the construction of the underground railway. The new paid land use system has provided the city government with a new and substantial revenue source which was not available in the past when land was allocated free of charge.

Land use restructuring is taking place very rapidly in administratively allocated land which is becoming one of the major headaches of the city government. Because of the problems of development control that has been discussed above, owners of administratively allocated land can find ways to change wholly or partly of their land for land use which yields higher return than the intended land use when land was administratively allocated to them. By subletting their land to other users through joint venture or other methods to avoiding giving the land back to the municipal government, they can capture some of the differences between the market land value and the value of land that they acquired almost free of charge through administrative allocation. Because of high demand and scarcity of land in the city center, they can get a good price in letting their land to other users. Land use changes in this type of land, although generally responding to market demand, are spontaneous and often creating urban traffic and environmental problems. Private businessmen who find residential areas that have potentials to open restaurants and nightclubs are willing to buy commercial housing in the suburbs to relocate the existing residents there. Residents could get compensation of up to 10,000 to 20,000 RMB (US$1,160 to $2,320). Some streets are evolving towards specialized functions, such as fashion shops in Huaihai Road and clubs and restaurants in Zuopu Road in Shanghai. In the past, the government had great difficulties in relocating economic activities and residents from the city center, but now the market forces are able to do so. Those who have visited Chinese cities a decade ago and go there nowadays will find that the urban landscapes, particularly those near the city center, have changed dramatically because of land reform (Yeh and Wu 1995; Wu and Yeh 1997, 1999).

The dual land market has led to the development of a new urban form through the urban restructuring process discussed above. As a result of economic reform which introduced land values and markets to its urban areas, most cities in China are changing from a compact city to a dispersed metropolis which is characterized by urban redevelopment in the city center, suburban expansion towards new suburban subcenters, and leapfrog urban sprawl in the urban fringe (Figure 4.4) (Wu and Yeh 1999; Zhang 2002). On top of this overall spatial structure which is mainly the result of the interactions of economic reform and urban planning is sporadic development of the administratively allocated land. The urban form of Chinese cities after land reform in 1988 can be considered as the composite of land use pattern which is the result of the better planned leased land and the sporadic development of administratively allocated land (Figure 4.5). Leased land is often put to the market for leasing after some form of urban planning. Land development in these lands is normally more orderly and less chaotic. Because of the black market and ambiguity in land use control, administratively allocated land is often converted into other land use which is not intended under the original land allocation. This often leads to sporadic and chaotic development. Such urban form which consisted of both a more well-planned development and sporadic development can be found in every city in China. This spatial structure is most prominent in the city proper where the land value is high, providing great attraction for the administratively allocated land to enter into the black market.

Because it takes time to resume land in the city center for redevelopment and large parcels of land may not be available for development, such as large-scale housing projects and economic and technological development zones, many new developments are decentralized to the urban fringe of the city proper through the acquisition of land from the farmers (transaction types $I_1$ and $II_1$), leading to urban sprawl (Zhang 2002). Economic and technological development zones (ETDZs) have been used as a mechanism for attracting foreign investment since the adoption of economic reform in 1978. The development of ETDZs was low before the land reform in 1987. But after the adoption of the paid transferred of land use rights when land suddenly had a value, more and more ETDZs have been set up because municipalities want to obtain more revenue through land leasing. In 1990, there were 1,874 ETDZs (Zou 1993). The number reached 2,700 in 1992 (Liang 1993). In 1992, the total area of ETDZs established in China reached 15,000 sq. km, which exceeded the total area of the built-up areas of the existing cities (Song 1995). Quite a large portion of ETDZs did not experience substantial development, and their land is left idle. The rapid increase of ETDZs wasted valuable agricultural land and caused ecological problems by changing the land cover. Only a few ETDZs in the large cities have achieved satisfactory results. Examples of these are Minhang, Hongqiao and Caohejing in Shanghai. In June 1993, many ETDZs that did not attract any development projects were ordered to close down by the central government. Apart from the proliferation of ETDZs in the urban fringe, there is also rapid development of large housing projects. Most of the land in the ETDZs is land leased out by the city government, but for some housing

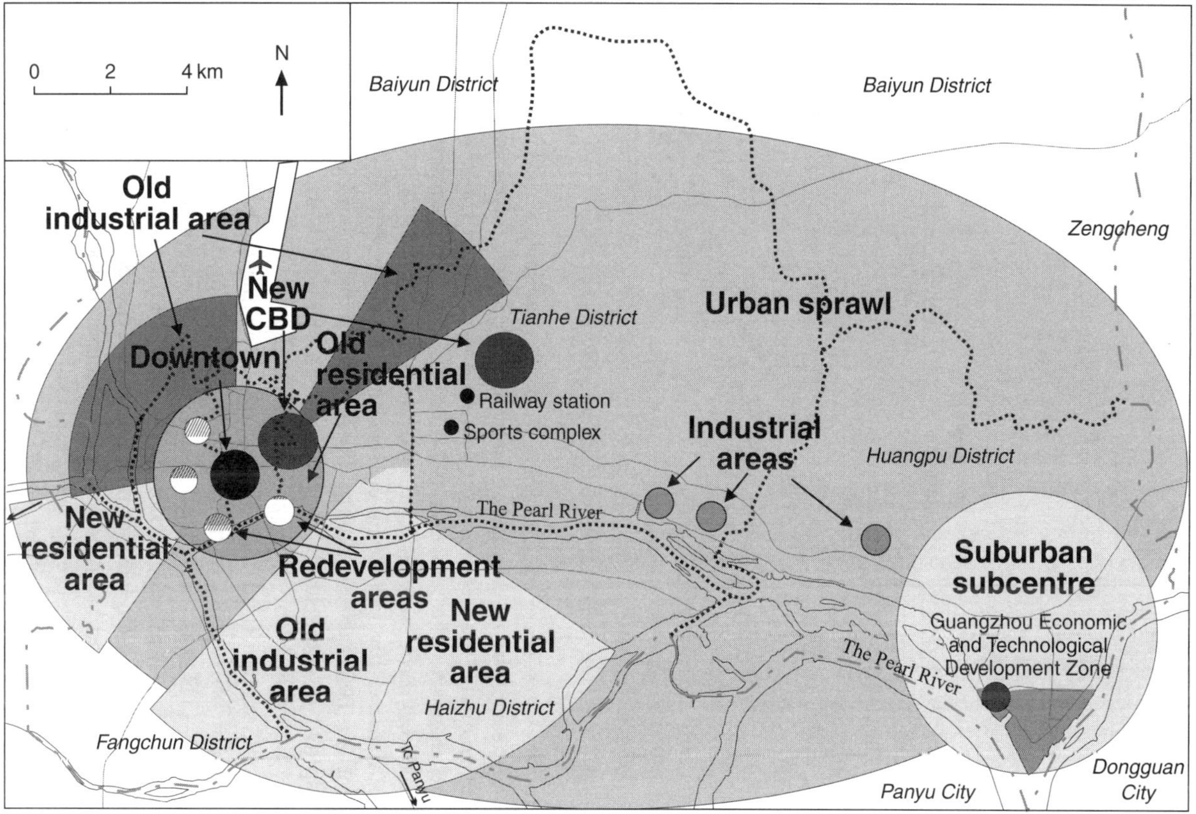


*Figure 4.4* Emerging new urban spatial structure of Chinese cities – an example of Guangzhou (source: Wu and Yeh 1999).

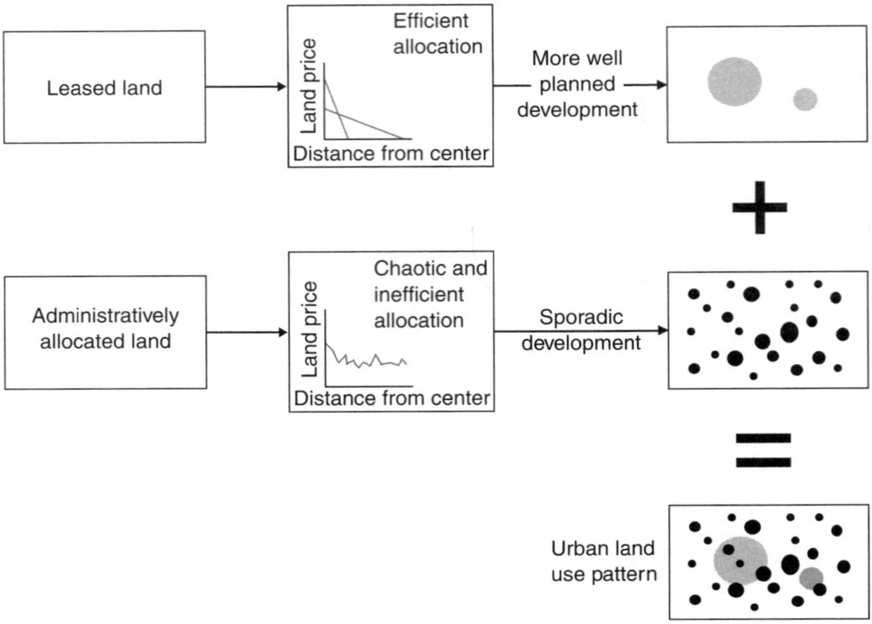


*Figure 4.5* Dual land market and urban form in Chinese cities.

projects, they are leased to developers through the black market. In the rural areas, farmers prefer to rent land to developers directly, often at a price that is lower than the market price for leased land, because the price that they obtained from the black market is higher than the compensation that they can receive from the state enterprises (transaction type $II_1$).

Urban sprawl has led to the encroachment on valuable agricultural land (Yeh and Li 1997, 1999, 2000). There was excessive land conversion and urban sprawl in the Pearl River Delta, north of Hong Kong. As a result of property boom in Hong Kong and the Pearl River Delta, rapid land development and agricultural land losses are taking place in Dongguan where 23.7 percent of the total area had undergone changes in 1988–1993 (Yeh and Li 1997). This is much higher than the 3.2 percent land use change in Hong Kong in a similar period in 1987–1995 (Yeh and Chan 1996).

Urban sprawl which does not take into consideration urban forms and valuable agricultural land has produced severe impacts on agricultural production and sustainable urban development, especially the consumption of energy in transportation. It is estimated that instead of urban sprawl, an alternative compact development, if used, can save as high as 35.7 percent of land development and infrastructure costs and 34 percent of gasoline consumption in Dongguan (Yeh and Li 2000). Also, as discussed before, the unregulated acquisition of land from black market for projects developed at urban fringe is problematic because very often their land use has not been approved by the government. It may not be provided with water supply and sewage, leading to environmental problems at the urban fringe. Land uses in administratively allocated land in the urban fringe have been converted spontaneously in the black market regardless of compatible uses and environmental consequences.

## Conclusion – challenges of the dual land market system

Despite the introduction of land leasing, a large proportion of land in China is still being allocated through administrative methods, leading to a rise in dual land market system. The difference in land prices in the dual land market can be enormous which often led to the development of a black land market. There is constant conflict among the municipalities, developers, existing land users and farmers who are attempting to capture the differential land value.

The dual land market system may lead to conflict with urban planning. The administratively allocated land is no doubt subject to urban plans that can be attuned to the public interest of the time. The leased land, however, is restricted by the lease condition. The land lease is a contract that is legally protected by the law of contract of the country. The experience of Hong Kong shows that under the leasehold land system there are potential conflicts between land leasing and zoning (Yeh 1994). The lease condition is not a good method of development control because of its rigidity. In order to use lease condition as a method to control development, a carefully prepared urban plan has to exist. The current land leasing in China specifies very clearly the use and development intensity. In the future, if the government wants to change the land use in accordance with a new urban plan, huge compensation may be incurred.

The dual land market and the resulting black market are vulnerable to corruption. Government officials can be bribed to approve development projects or to change the urban plan. Although land auction may capture the best land rent from the market, only a small proportion of land leasing in China is from land auction (Figure 4.6). The majority is from negotiation which may be susceptible to corruption. There is an urgent need to change the method of land leasing in China from negotiation to auction and to regulate land transactions and land use changes in order to ensure that the land value gained is under the control of the government and not directed to the work-units or state enterprises which obtain the administratively allocated land free of charge. The administratively allocated land is causing loss of land revenue, unregulated land use changes and urban sprawl. In the long run, the dual land market may have to be abolished. All land, except that for community purposes, should be leased land that operates on a land market principle. Existing administratively allocated land should be changed to leased land by paying premiums to the city governments. This can be started with the commodification of existing administratively allocated land for commercial and industrial uses (Xie *et al.* 2002).

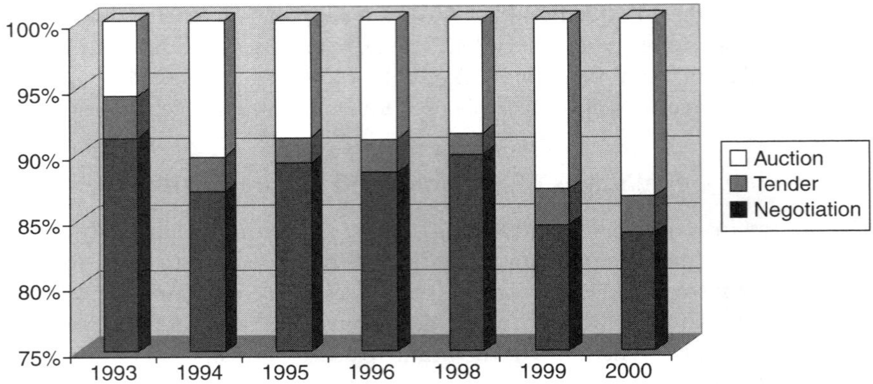


*Figure 4.6* Land-leasing methods in China in 1993–2000 (source: editorial department of *Yearbook of the Land of China*, 1995, 1996, 1997 and 2001).

## References


Bourne, L. S. (ed.) (1982) *Internal Structure of the City: Readings on Urban Form, Growth, and Policy*, New York: Oxford University Press.

Dowall, D. E. (1993) "Establishing urban land markets in the People's Republic of China," *Journal of the American Planning Association*, 59(2): 182–192.

Editorial Department of Yearbook of the Land of China (1995, 1996, 1997, 2001) *Yearbook of the Land of China*, Beijing: China Land Press.

Fung, K. I. (1981) "Urban sprawl in China: some causative factors," in Ma, Laurence J. C. and Hanten, Edward W. (eds), *Urban Development in Modern China*, Boulder, CO: Westview Press, 194–220.

Healey, P. (1991) "Models of the development process: a review," *Journal of Property Research*, 8: 219–238.

—— and Barrett, S. M. (1990) "Structure and agency in land and property development processes: some ideas for research," *Urban Studies*, 27: 89–103.

Knox, P. and Pinch, S. (2000) *Urban Social Geography: An Introduction*, Harlow: Prentice Hall.

Li, L. H. (1997) "The political economy of the privatization of the land market in Shanghai," *Urban Studies*, 34(2): 321–335.

Liang, C. (1993) "Seven cities chosen as pilot for real estate," *China Daily*, May 5, 1993.

Song, C. H. (1995) "Real estate development and urban planning control," *City Planning Review*, 2: 3–6 (in Chinese).

Tang, Y. B. (1989) "Urban land use in China," *Land Use Policy*, 6(1): 53–63.

Tian, F. (1994) "Review of the problems of real estate business in China," *Urban Problems*, 2: 51–54 (in Chinese).

Walker, A. (1991) *Land, Property and Construction in the People's Republic of China*, Hong Kong: University of Hong Kong Press.

Wu, F. L. (2001) "China's recent urban development in the process of land and housing marketisation and economic globalisation," *Habitat International*, 25: 273–289.

—— and Yeh, A. G. O. (1997) "Changing spatial distribution and determinants of land development in Chinese cities in the transition from a centrally planned economy to a socialist market economy – a case study of Guangzhou," *Urban Studies*, 34(11): 1851–1879.

—— (1999) "Urban spatial structure in a transitional economy: the case of Guangzhou," *Journal of the American Planning Association*, 65(4): 377–394.

Xie, Q. S., Parsa, A. R. G. and Redding, B. (2002) "The emergence of the urban land market in China: evolution, structure, constraints and perspectives," *Urban Studies*, 39(8): 1375–1398.

Yeh, A. G. O. (1985) "Physical planning," in Wong, K. Y. and Chu, D. K. Y. (eds), *Modernization in China: the Case of the Shenzhen Special Economic Zone*, Hong Kong: Oxford University Press, 108–130.

—— (1994) "Land leasing and urban planning – lessons from Hong Kong," *Regional Development Dialogue*, 15(2): 3–21.

—— and Chan, J. C. W. (1996) "Territorial development strategy and land use changes in Hong Kong," in Au, K. N. and Lulla, M. (eds), *Hong Kong and the Pearl River Delta as Seen from Space*, Hong Kong: GeoCarto International, 63–74.

—— and Li, X. (1997) "An integrated remote sensing and GIS approach in the monitoring and evaluation of rapid urban growth for sustainable development in the Pearl River Delta, China," *International Planning Studies*, 2(2): 195–222.

—— (1999) "Economic development and agricultural land loss in the Pearl River Delta, China," *Habitat International*, 23(3): 373–390.

—— (2000) "The need for compact development in fast growing areas of China: the Pearl River Delta," in Jenks, M. and Burgess, R. (eds), *Compact Cities: Sustainable Urban Forms for Developing Countries*, London: E&FN Spon Press, 73–90.

—— and Wu, F. L. (1995) "Internal structure of Chinese cities in the midst of economic reform," *Urban Geography*, 16(6): 521–554.

—— (1996) "The new land development process and urban development in Chinese cities," *International Journal of Urban and Regional Research*, 20(2): 330–353.

—— (1999) "The transformation of the urban planning system in China from a centrally-planned to transitional economy," *Progress in Planning*, 51(3): 167–252.

Zhang, T. W. (2000) "Land market forces and government's role in sprawl," *Cities*, 17(2): 123–135.

Zhang, X. Q. (1997) "Urban land reform in China," *Land Use Policy*, 14(3): 187–199.

Zhou, Y. P. (1992) *Perspectives on China's Real Estate Business (Zhongguo Fangdichan Toushi)*, Hong Kong: Joint Publishing (in Chinese).

Zhu, J. M. (2002) "Urban development under ambiguous property rights: a case of China's transition economy," *International Journal of Urban and Regional Research*, 26(1): 41–57.

Zou, D. (1993) "An impact by the leasing system on urban land upon urban development," *China City Planning Review*, 9(1): 27–30.

# 5 Irregular trajectories

## Illegal building in mainland China and Hong Kong

*Alan Smart and Wing-Shing Tang*

### Introduction

The People's Republic of China was unusual among low-income countries in that it did not develop large, uncontrolled illegal settlements. In part this was because of restrictions on urbanization, so that the urban population stayed relatively low as a proportion of the total, and partly due to the greater capacity of the communist state to control activities compared to the capitalist Lesser Developed Countries (LDCs). Since economic reforms were adopted in 1979, illegal building has increased substantially, with illegal land use accounting for about 10 percent of the total changes in land use between 1993 and 1999 (Tang and Chung 2002: 47). However, the nature of this illegal use has very distinctive features related to the unusual characteristics of reform China. In this chapter, we draw on comparisons with another territory with a Chinese population, Hong Kong, to clarify the specific features of the illegal building situation in contemporary China. The comparison will also be used to provide some insights into the political economy of irregular forms of urbanization.

Certain kinds of illegality are difficult to control because of their invisibility (contraband drugs) or because of the absence of complaining victims (vice). Neither condition applies to illegal settlements, by their very nature fixed in place and difficult to hide. Thus explanations for illegal building must differ from other persistent illegal activities, and generally relate to limitations on or divisions of regulatory power (Smart 2000). Spaces of illegality must be seen within the context of broader political processes, on which in turn they provide new perspectives. Fernandes and Varley (1998: 3) estimate that 40 percent of the population of the major cities of Asia, Africa and Latin America "are living in illegal conditions" in terms of land tenure, infrastructure requirements and building standards. Illegality pervades the world's poorer cities, but is important everywhere. Persistence of illegal construction provides us with a lens to examine some of the basic processes through which law, and the failure to observe it, impact upon the urban environment (Karst *et al.* 1973; Perez Perdomo and Nikken 1980; Heyman and Smart 1999, 2001; Yonder 1998). Since spatially fixed and visible illegalities cannot rely on difficulty of detection, as can much of the informal sector, explanations for their survival concern more directly the nature of governmental interventions/non-interventions and the actions and self-organization of illegal builders. The two are not necessarily mutually exclusive, since government officials can be actively involved in promoting or profiting from illegal building, as the discussion of the Chinese case below will demonstrate.

Another crucial question concerns the nature of the illegality. It is logically possible that irregular structures can be illegal from, so to speak, top to bottom. That is, occupants have no rights to the land on which the structure sits, have no permission to build, build with no concern about regulations on how construction takes place, and structures are transferred between occupants without reference to the artefacts of legal culture (agreements, contracts, etc.). Research suggests, however, that more commonly some legal rights pertain at some point in the processes of housing provision and occupation, and that even where governments acknowledge no legal rights at all in these spaces and structures, private law in the form of commodification and contracts for transfers develops informally (Smart 1986). Similarly, continued occupation might depend only on the ability of occupants to mobilize sufficiently credible threat of force to deter their removal, but in practice accommodations are often reached to allow continued residence. One of the major dimensions of the nature of illegal occupation is the extent to which it is bureaucratically or judicially routinized, as in Hong Kong, or remains a variable and negotiated outcome of local political contestation, as seems generally to be the case in China.

The term "ownership" can be usefully seen as referring to the "greatest possible interest in a thing which a mature system of law recognizes" (Honore 1961, cited in Hann 1998: 6). Illegal structures veer far away from this level, and yet in both Hong Kong and China, they have had sufficient stability to be seen as worthy of profit-oriented investment, even to the extent of commercial loans in China (Tang and Chung 2002: 56). Hann (1998) follows a long anthropological tradition in insisting on property as embedded in social relations that support, legitimate and limit it. By exploring how illegal building takes place in China and Hong Kong, we attempt to delineate these social relations that allow fixed and very visible (after all, they are often referred to as "eyesores") but considerably illegal products to persist. In doing so, we also hope to shed some light on the broader social and political contexts in which illegal buildings are embedded.

## The political economy of illegal building

The combination of rapid urbanization, poverty and ineffective governmental control over space produced the rapid growth of settlements that veered to a lesser or greater extent from fully legal development in most of the world's low-income countries. Illegal settlements have accounted for large proportions of all housing produced in these countries, although this varies considerably from a few percent of the total up to as much as 80 percent. Almost all governments attempt to prevent the illegal occupation of land, but in most cases illegal settlements are nonetheless constructed and often survive. How these areas are established and developed varies greatly. Governments often eventually accept the presence of illegal settlements, and may even regularize their irregular tenure situation and initiate programs to upgrade their facilities, such as streets, sewers, water supplies and electricity. Expectations about the future trajectory of an irregular settlement have as much (or more) influence on how they develop as do the formal rules.

There is an immense literature on irregular housing, which we cannot attempt to summarize here. Much of this literature disagrees even about such fundamentals as the labeling of the issue. Slums, shantytowns, squatter settlements, informal housing, spontaneous communities, self-help housing, unconventional development, and a plethora of adapted local terminology such as *favelas* or *bustees*: all of these cover some of the conceptual space that we are concerned with here. As relatively neutral labels, although ones that come with their own share of ethnocentric baggage, we opt here to begin by talking about "irregular" settlements or structures or urbanization processes to cover the broader range of specific phenomena, with our own analysis focusing on the issue of "illegal building." Reference to irregularity avoids questionable assumptions about self-building or spontaneity, while recognizing that the settlement fails to meet some of the standards for legally constructed dwellings. The specific forms of irregularity determine whether a structure should be considered to be a squatter dwelling, in contravention of building or zoning ordinances, or a legal structure that has been illegally occupied by persons other than the owners, to name only three permutations. The disadvantage of the terms is that they carry with them the assumption that what is "regular" construction or development is clear and consistent, and that the definition of "illegality" is unambiguous and easily definable. In many contexts these assumptions must be used with considerable care. In China and Mexico, for example, much, perhaps most, illegal building is done with the knowledge or active collusion of local officials, and even agencies of the central state have been active in the process (Leaf 2002; Varley 2002). In such contexts, "illegality" is often an *ex post facto* political conclusion, rather than something that can be determined in advance. In both Hong Kong and China, at least some portion of the illegal construction was illegal because rules had been changed to remove rights to build that had previously been held by local residents.

Two main factors lead to the proliferation of irregular structures. First is the inability of the political economy, however organized, to provide sufficient affordable legal housing when poor cities grow rapidly. The second factor is inability of governments to control informal solutions to these failures, either by individual households or by entrepreneurs producing accommodation for profit (the two are not mutually exclusive, as many who produced their own housing also rented or sold additional space). Since often strenuous efforts to encourage migrants to return to the countryside were generally complete failures (one of the key differences between China and capitalist LDCs is that mobility controls were much more effective in China at least until 1979 when unrationed food became available), irregular spaces proliferated where government had little control over the new housing (and sometimes factories, stores, schools, and other facilities).

However, if governments were incapable of eradicating irregular uses of space, this does not mean that these spaces operated outside the influence of government practices. In the first place, it was not uncommon for politicians and even ruling parties to organize "invasions" of land (usually plots controlled by their rivals or the supporters of their opponents), sometimes in support of social justice but more often in order to gain blocks of reliable supporters (Collier 1976). In the second, state policies and practices in other arenas have a considerable impact on how even illegal occupation is organized (McAuslan 1998).

Gilbert and Ward (1985: 128) argued that the process of land allocation is the most crucial variable involved in determining the forms of low-income housing that predominate in a particular city. In otherwise similar cities, there can be great variation. For example, in Valencia, Venezuela, invasions of vacant land accounted for 45 percent of the housing stock, and most were "owner"-occupied, whereas in Bogota (where peripheral agricultural land was fertile and valuable), land invasions were rare, and 59 percent of the population acquired their dwellings through what was locally known as "pirate urbanization": subdivisions of agricultural land that were illegal because they do not conform with local planning regulations, are not serviced, and may not be properly registered. Gilbert and Ward (1985: 24) concluded that the "proportions of the urban population owning, renting, and sharing accommodation is largely explicable" in terms of how land is allocated and alienated, which in turn was determined by:

> a combination of factors including the general price level of land, the pattern of landownership and the nature of the state involved. Whether land is invaded, community land is alienated, or land is subdivided illegally depends upon the local conjunction of these factors.

They were also critical of the informal–formal housing distinctions, demonstrating that the two markets were closely linked and mutually influencing.

Illegal forms of access serve to facilitate access for the very poor, and serve to stabilize the system "by offering the poor some stake in the property system" (Gilbert and Ward 1985: 49). The nature of the linkage between the two markets, however, is strongly influenced by the probable ultimate state of an irregular settlement: where regularization is likely, market forces will tend towards price convergence between informal and formal housing of similar amenities, investment in improvements is a better bet, and middle-class individuals are more likely to move in. If eradication is recognized as the inevitable fate, though, as in Hong Kong, it is the conditions for compensation, if any, for dislocated residents that will have the greatest effect on the illegal housing market (Smart 1986). Where the outcome is "in the balance" or dependent on political developments, collective mobilization to influence decision-makers may be more likely. Trajectories for illegal settlements in China are very uncertain, politically contingent, and vary considerably from one locality to another. However, prevailing conditions make collective action particularly dangerous, with the result that reliance on local officials or power brokers (see Zhang 2001) is more common than are the organized invasions or social movements often found in Latin American informal housing processes, although the situation during military governments may have more similarities given the absent utility of low-income votes and electoral cannon fodder.

## Hong Kong

Housing has been a problem in Hong Kong since establishment as a British colony. Only after the Japanese Occupation ended in 1945 did illegal building become widespread. Destruction of the housing stock combined with the influx of refugees to generate huge demand for affordable housing, but the re-established colonial government initially failed to meet these needs. Squatting filled part of the void, along with intensified use of existing private structures. The number of squatters grew to 300,000 by 1949 and numbers eventually peaked in 1981 with over 750,000 (Wong 1978: 208; Smart 1992).

Legal development was hobbled by the central role played by government in the development process. Disorganized after the Second World War, it could not get significant amounts of building land through its varied controls and restrictions. A policy of allowing the private sector to build (and price) as it wished might have produced much more housing, although perhaps at great cost to the poor. The private sector did in practice produce a great deal of housing for lower-income groups, but illegally, as squatter dwellings for rent and purchase (Smart 1986, 1992). Much of the "squatting" was only illegal because of the administrative distinction between agricultural land and building land: villagers could not legally take advantage of their fields to meet the demand for accommodation. In addition, building and related ordinances multiplied the varieties of illegality. Contrary to the consensus in the literature that the private sector was unable to cope with the demand for housing so that the government was eventually forced to intervene (in the form of public housing, which now provides accommodation for half the population), the state was already centrally involved in the land development process. This involvement was the source of much of the problem. State intervention limited prior rights of indigenous villagers to build as they wished on their own land (Chun 2000). Restrictions were particularly apparent in New Kowloon, where the majority of the urban squatter structures were built (Smart 2003).

In 1898, the Hong Kong government leased the New Territories from China for ninety-nine years. What is now commonly referred to as the New Territories excludes part of the treaty area: the land between the ceded portions of Kowloon peninsula, and the rugged Kowloon mountains (New Kowloon). These mountains reduced access to the rest of the New Territories. Only land with buildings standing in 1901 was classified by the Land Court as building land; the remainder was registered as agricultural land, and had to be converted to building land prior to development. The label "squatter" is not restricted in Hong Kong to those who encroach on Crown or other people's land, but is also applied to those who build illegally on their own private agricultural land.

The policy to distinguish between building and agricultural land was subject to initial disagreement. The president of the Land Court believed that there should be two different types of lease, agricultural and building, whereas the land officer argued for a system with only one type of lease, but where improvements would be assessed higher taxes. The main argument for the single type of lease system was that free substitution of use would promote development. The land officer thought that the sooner land was taken up for building purposes the better for the colony, as well as for the state's revenues. However, the single lease system required two collection methods. The decision was made in favor of two forms of lease. Agricultural leases would have to be exchanged for building leases at a reassessed rent and premium.

In 1908, land conversion was far from easy and afterward became much more difficult. To acquire building leases, owners of agricultural land needed to comply with layout (town-planning) schemes. If the plan was incomplete, they were allowed only temporary building permits. The assistant land officer felt that plans for future development usually made it inexpedient to allow conversion. The result was that development in New Kowloon was subject to considerable administrative restrictions, delay and required high premiums.

Indigenous villagers lost the opportunity to profit from the urbanization of New Kowloon. Legal development primarily took the form of larger commercial schemes and required extensive and expensive consultation and cooperation with government departments. Self-help, even in the form of building a dwelling on one's own land, was precluded. At the same time, the government was ineffective at making land available for private development. Between 1945 and 1958, no more than 200 acres of land were made available for housing in the urban areas of Hong Kong (Smart 1992: 39). The centrality of a government in postwar disarray in land development posed a serious obstacle to private development. At the same time, the proliferation of illegal settlements occupied much of the space on the urban fringe that would have been most easily built on.

The bottleneck was finally broken after 1954 by the squatter resettlement program, which was not simply a public housing scheme but also a way to clear land for private and public construction. While simply demolishing their dwellings would have freed up even more land, the combination of the squatters' potential for militant resistance and the diplomatic instability of Hong Kong's colonial situation after the victory of the Chinese Communist Party limited the viability of this repressive option (Smart 1989). Squatter resettlement completely transformed property relations in the illegal building sector. Previously, the government attempted to prevent new squatting, eradicated those areas it needed to without compensation for evictees, and tolerated squatters on land that it did not need for the moment. Between 1954 and 1984, the Hong Kong Government followed a policy of toleration of squatting combined with refusal to extend ownership rights to irregular structures. Those living in registered squatter dwellings when they were scheduled for demolition were eligible for rehousing according to their household characteristics and current policies. To avoid appearing to acknowledge property rights, it was the structures that were registered and tolerated, rather than occupants. In this context, squatter dwellings had considerable value, but without property rights being ascribed or officially enforceable. After 1954, the illegal housing market was primarily for sales of dwellings rather than rentals (since the occupant rather than the "owner" would receive compensation in a clearance), while initial illegal occupation depended on either local connections or raw force, particularly through the triads and other criminal gangs (Smart 1985).

Governmental regulation of illegal building, and hence the system of property relations, changed fundamentally after a territory-wide registration in 1984 and 1985 of all occupants of squatter dwellings. Only these registered occupants would be eligible for resettlement in permanent public housing. In practice, unregistered residents were in most cases resettled in temporary housing areas (Smart 2003). Despite this change, an illegal market for squatter dwellings continues to exist. Continued demand for squatter dwellings results from lack of alternatives for many. The property boom that took off with the signing of the Sino-British agreement and ran until 1997 (followed by a bursting of the bubble, reducing prices to half their peak) made access to private housing even less affordable than it had previously been. Obtaining any kind of affordable accommodation, no matter how bad and insecure the conditions, had a substantial market value to many people.

The market for illegal structures was formed and reformed in response to shifting policies and the broader context. A period of repression eventually gave way to routinized squatter control, clearance and resettlement, which since 1984 has changed to a regime of effective control over new building combined with relocation policies that tended to exclude ever greater proportions of the existing squatter population (Smart 2002).

In the next section, we turn to a very different administrative landscape.

We suggest that the situation in China resembles most closely illegal building in Hong Kong before 1954. However, in many other ways the property relations of illegal building are radically divergent, and are likely to produce new developmental trajectories. Initially, it seems likely that the pattern will be closer to Latin American experience than to the Hong Kong case. There are clear similarities in the underlying systems of land use rights, though (state monopoly of ownership combined with a system of long-term leaseholds), suggesting that the landed property system by itself does not determine the development and treatment of illegal building. There are also differences that relate to the dissimilar natures of the respective state forms: reform communism in post-1979 China versus a diverse set of regimes focused around often unstable democracies and military governments in Latin America.

## China

Illegal building in China is important for both practical and academic reasons. Although the statistics are disputed, the loss of arable land (already low in proportion to the population) due to urbanization has been very substantial, with a fall of 4.5 percent in total arable land between 1978 and 1995. The State Council was concerned enough about the situation to place a moratorium in 1997 on arable land conversion which was in place until 1999. Illegal building played a partial but significant role in this process: about 10 percent of all land use conversions between 1993 and 1998 were illegal, accounting for 1,209,225 hectares. Illegal land use, in these calculations, refers to "illegitimate occupation and use, illegitimate application approval, buying and selling of land and illegal transfers" and thus does not necessarily involve illegal building since transferred land may remain vacant (Tang and Chung 2002: 46). The black market for illegal land transfers limits the efficacy of urban planning and farmland conservation.

Illegal building provides a useful vantage point for examining China's transformational dynamics created by the shift of increasing proportions of the society and economy beyond the traditional loci of social control: the rural collective and the urban "work-unit" (*danwei*) (F. L. Wu 2002). Changes in land allocation policy were among the early demands of foreign investors, and alliances between investors and local governments often provided "gray" equivalents of land markets long before market transfers of land use rights were formally accepted (Smart 1998). Illegal building must be seen in the context of contestation over the classification of "gray" practices (Zhu 2002), and expectations about whether they will eventually become "white" as economic responses to reforms consistently "pushed the envelope" of acceptable practice (Smart 2000). Some practices, such as secondary market exchange of allocated land, eventually became legal and regulated, while others eventually became more clearly unacceptable in the form of the "black market." The system of property rights in land has gradually become more routinized, particularly since the implementation of the revised Land Administration Law in 1999 (Xie *et al.* 2002; Cartier 2002).

Any system of landed property rights is more complicated than can be summarized briefly, and a society being gradually transformed from central planning to "market socialism" adds the difficulty of having to describe both the "old system," the new practices and the interaction between them (Zhu 2002; Ho and Lin 2004; Tang 1998, 1994). Prior to the onset of reforms beginning in 1979, the Chinese state owned all urban land, while most rural land was controlled by local collectives. As Ho and Lin (2004) summarize "only the state has all the rights of ownership. The collective can possess, use, and benefit from the ownership of land, but does not have the right to dispose of the land."[1] In the cities, land was allocated (without time limit and for a nominal charge) to state enterprises and urban collectives, who had no rights of transfer (although a system of barter and black markets emerged regardless). Gradually, a commercial system of land use rights emerged. While formal ownership status has not changed, the old system of land allocation (*huabo*) has been supplemented with a system of "conveyance" (*churang*) where commercial users obtain the additional rights of transfer to other users and the ability to use their use rights as collateral for loans (Ho and Lin 2004). However, allocation still accounted for about 70 percent of land distributed between 1993 and 1998 (Ho and Lin 2004). Even for the portion under "conveyance," administrative intervention is by no means absent. In Shenzhen Special Economic Zone, for example, 87.76 percent of the land by area from 1987 to 1994 was transferred by negotiation between the government and land users. If one adds the amount with price waivers – another hidden administrative measure – the figure jumps to 96.12 percent (Tang 1998: 326). Fees can be paid to move land from state use to commercial use, thus bringing secondary market influence into the state allocation system. The potential profits in shifting land from one track to the other prompts rent-seeking behavior at all levels of the system (Zhu 2002).

Similarly, laws, ordinances and administrative procedures on urban planning and administration have contributed to the proliferation of illegal land uses and illegal constructions. In 1984, when the City Planning Ordinance was implemented, it demanded that every city should prepare a masterplan, which supposedly took command within a definite spatial boundary. Every development should conform to the use specified in that plan. In other words, all development proposals must obtain from the relevant planning authority the development land permit and building construction permit before development could actually take place. The definitions of illegal land uses and illegal constructions could then be deduced from these clauses. They were explicitly categorized later in, for example, the City Planning and Administration Methods in Guangzhou in 1986. As a result, a variety of land occupancies were classified as illegal and subject to regulation (for a classification scheme, see Ng and Xu 2002). Over time, as more cities have completed the formulation of masterplan (and also detail plans), activities within a larger spatial reach (from the urban to the rural and from one city district to another) were subject to the classification of being illegal. As a finer definition and classification of illegality was introduced under successive promulgation of relevant laws and ordinances (including the City Planning Act in 1989 and the Urgent Notices Concerning the Curbing of Illegal Land Uses and Illegal Constructions in 1994 by Guangzhou), a wider spectrum of land occupancies has been rendered illegal over time. In short, it was the state's attempt to put more development activities, partly initiated by foreign investors, under increasing control that led to the proliferation of illegal land uses and illegal constructions (Tang and Chung 2002).

It is crucial to recognize that not all resort to the black market or gray practices results from individual venality. The greatest amount of illegal land use resulted from the activities of rural collectives and state organizations, and although much of this may benefit individuals, often it is primarily intended for the economic development of the local unit. Furthermore, many of the regulations put in place to control land use have been seen by rural officials, particularly, as stripping them of the prerogatives that they *de facto* previously held, such as making land or a building available to foreign investors. The "illegalization" of previously accepted actions is particularly clear in relation to the construction of new housing by villagers (Tang and Chung 2002). Cultivated land converted to rural housing accounted for 33 percent of all farmland conversion from 1985–1988 but dropped to 8 percent by 1993 (Ho and Lin 2004: 19).

The revised Land Administration Law, adopted in 1999, has been partially effective in reducing the loss of arable land, in part because the use of remote-sensing technology has reduced the reliance of the central state on information provided by local authorities (Cartier 2002). Carolyn Cartier (2001) argues that previous legislation actually encouraged the conversion of arable land due to contradictions between policy and practice.

The trend seems to be that peasants and rural collectives attempt to increase their share of the profits from commercialization by not going through the required procedures (Zhu 2002). The expropriation procedures that are part of the official track for transferring rural land to building land status generally give the villagers insufficient resources to compensate for the loss of their livelihood, and often little or nothing from the fees paid goes to the actual users of the land. The result is illegality at another level, as those "who have lost their cultivated land attempt to make use of their residential land – their only resource – to support themselves. This induces them to construct illegally" (Tang and Chung 2002: 49). This strategy is particularly lucrative in the peri-urban areas around rapidly growing cities and towns. Demand for accommodation is fueled by the influx of migrant workers without access to formal sector housing (Smart and Smart 2001; Taubmann 2002; W. P. Wu 2002). Zhang (2001) provides a detailed ethnographic account of how one migrant village developed in Beijing, the consequences for social organization, and the ways in which self-organization to resolve pressing problems (such as the need for children's education) was seen by officials as more threatening than individual suffering or criminal activities and thus prompted repression. Since local arrangements and alliances influence the extent and specific forms of illegal construction, local diversity of patterns is considerable. The following case study should not be seen as generalizable but as illustrative of some of the patterns that have developed.

In a study of Tianhe Village, near a new commercial core being developed on the fringes of Guangzhou, the capital of Guangdong province, Tang and Chung (2002) found that 95 percent of the village's houses had been expanded illegally. Although rural housing in the area is limited to four stories, many are higher, some even with seven stories. Why has such visible flaunting of the rules proliferated in this area? Kinship relations between local cadres and villagers combine with the fact that the officials have themselves undertaken illegal expansions to encourage toleration of infractions, particularly since the practices have considerably contributed to the local economy. However, the "safety net" provided with the complicity and sympathy of local officials is a fragile guarantee which will "disappear at any time when governments at higher levels decided to erase illegal constructions" (Tang and Chung 2002: 57). While this may be the outcome in Tianhe Village, it would seem that in many cases, illegal building either persists indefinitely or eventually becomes legalized through the payment of various fees. What this means is that the future trajectory of illegal settlements is profoundly uncertain, and subject to interplay between local and wider politics. The outcomes are thus likely to vary considerably from one locality to the next.

The situation in China is much closer to the Latin American experience than to the bureaucratic process found in Hong Kong, where after the disorder of the first postwar decade, the administrative situation of the squatter areas became clearly defined, even if politics led to various adjustments in the eligibility requirements for the various categories. In China what we see is not a situation of white and temporarily tolerated black, but a spectrum of varying shades of gray. Ambiguous property rights and political changes and internal conflicts mean that there is even considerable uncertainty about the shade of particular practices. Similarly, in Mexico, Ann Varley (2002) discovered that similar housing developments in Puebla and Guadalajara were considered to be respectively illegal and legal, because in the latter irregular developers were closely connected to government officials while in Puebla they were not. As she notes, the:

> characteristics of the urban environment in subdivisions in these two cities were therefore essentially the same. Yet the Guadalajara subdivisions were classified as legal, while those in Puebla were not. This questions the opposition of legality and illegality.
>
> (p. 453)

If we see the spectrum of conformity to legal norms as a gray scale, it is in practice more like a photograph of one that is in the process of being developed, so that only relative comparisons allow us to guess whether the outcome will be light gray or dark black.

Similarities between the Chinese and Latin American cases should not, however, blind us to the differences, from which we can also draw insights. While in both cases spaces are opened up for the proliferation of illegal building by political divisions, in Latin America these divisions have typically been between contenders for power (opposition parties, social movements, etc.), while in China they are between different levels of the (more or less) unified administrative hierarchy. As yet, spaces in which the state is almost completely absent or ineffective have not yet spread to the extent that they have in Latin America (O'Donnell 1993). Another crucial difference has been that in Latin America, once established, illegal settlements have often been left to their own purposes in the provision of services and the formation of internal systems of governance (Leeds 1996; Mangin 1967), but when migrant communities have attempted to do the same in China, this has been likely to provoke repressive intervention, even demolition, on the part of authorities who see such self-organization as more threatening than anomie (Zhang 2001).

## The cases compared

In this section we compare illegal building in Hong Kong and China. We concentrate on, first, describing variance in the particular forms of illegal building in each context, its accomplishment and contestation. Second, we address the ways in which political, economic and social contexts influence these processes and their outcomes.

Despite great differences between the political economies of Hong Kong and the rest of China, there are also some considerable similarities between them in relation to land. First, in both all land (at least in the fullest sense) is owned by the state, and private rights consist at most of long leaseholds, or rural collective limited rights over agricultural land. Second, land revenues are of considerable significance to governmental budgets. Third, substantial amounts of land have been built upon illegally and have persisted in their illicit state for considerable periods of time. Fourth, in both cases large amounts of illegal building continued even after governments devoted considerable efforts to control it.

The forms taken by illegal building in China are very different from those found in Hong Kong after 1954, although there are more commonalities with Hong Kong's pre-resettlement era. Since the regime of illegal building was almost completely transformed from the mid-1950s in association with the implementation of a regime of squatter resettlement (Smart 2001), our comparison is really tripartite rather than dual.[2] The administration of illegal occupation became increasingly standardized and the prospects generally understood by those who moved into the sector.[3] This does not mean that the conditions for particular settlements were immune from modification: pressures from uncontrolled migration and high housing costs meant that new structures were repeatedly amnestied and provided with similar prerogatives of toleration and rehousing until the land was required. The regulation of squatter settlements became a bureaucratic, much more than a political or judicial, process (Smart 2002). By contrast, the disposition of the vast quantity and diverse characteristics of illegal land use in China is only gradually moving from the arena of political influence and complicit self-interest on the part of the administrators towards a clearer "rule of law" in the land use regime. A clearer bureaucratic division of labor, with a single agency possessing sufficient capacity to regulate illegal building, would seem to be a prerequisite for moving towards the Hong Kong style of bureaucratic modality of regulation. Overlapping and contested spheres of authority in China suggests that a political mode for regulating illegal building is likely to continue for the foreseeable future. Whether or not the courts become more heavily involved, as is common in the West, remains to be seen. The impact of foreign investors may encourage a judicial approach, since contracts often mandate international arbitration for disputes, and new rights for them are being provided by the WTO accession. A judicial replacement for the political mode also seems more likely in the context of almost certainly continuing divisions of interest within the administrative hierarchy.

Since 1954, almost all the irregular structures have been illegal from top to bottom in Hong Kong: even where illegal structures were built by villagers on agricultural land (progressively less common after 1954), they generally lost control over these dwellings because of the unavailability of legal remedies when tenants refused to pay their rent. This thorough-going illegality has resulted in very clear visible differentiation between illegal and legal building sectors. No one observing an urban squatter settlement in Hong Kong would confuse it with legally constructed accommodation. There is one major exception to this obvious boundary between the sectors: illegal additions to legal private sector buildings. During the 1980s it was estimated that replacing the space that would be lost if all these additions were demolished would require the equivalent of public housing for 200,000 people if they were rehoused at the minimum space allocation in effect at the time. In China, although some illegal building is visually distinct, much of it looks like other low-quality legal construction, and the two are often mixed together in complex patterns. Furthermore, since some illegally built housing will eventually be regularized, patterns of investment do not create the same dynamic that ensures temporary construction in the Hong Kong case.

Variation in who controls the illegal units constitutes another major difference between post-1954 Hong Kong and China. Most accounts of Chinese illegal housing indicate that the dwellings remain the property of the "landowners," and that outsiders rent rather than purchase their accommodation (Chan *et al.* 2003). Prior to 1954, there is evidence that the rental of huts and cubicles in illegal settlements was common in Hong Kong as well, but this was replaced by a sales market, due to a regulatory system that compensated only occupants, not "owners," in the case of demolition (Smart 1988). This tenure change reflects the insecurity of ownership in Hong Kong irregular settlements; the prevalence of renting suggests at least the belief of owners in the security of control over units that have been rented out in illegal developed buildings. The availability of loans to support their development is an indication of how prevalent these attitudes are. Security of ownership has been demonstrated to encourage both the building of better quality housing and its improvement over time (Turner 1976; Varley 2002; Smart 2003). It is not just the legal status itself that determines the quality and character of a structure, or of its inhabitants, but expectations about how that legal status will influence (or fail to influence) the future of the structure and its surroundings. In certain contexts, authorities take these expectations into consideration in planning their regulatory policies, for example, by accepting their inability to prevent new illegal building and putting in place a framework by which its operations are influenced by enabling policies that channel it into particular forms and places. The trajectories of irregular settlements are thus subject to dialectical influence between residents, authorities, and the wider population of potential illegal builders. Conflicts of interest between different levels of government add to the complexity of the situation. Such divisions are much more apparent in the case of China than in either pre- or post-1954 Hong Kong. Their impact seems to make the future development of existing illegally built structures and the expansion of areas of illegal building much more uncertain and locally diverse than has been the Hong Kong experience.

## Conclusions

The complex reconfiguration of space that is taking place in China provides great challenges for our interpretive frameworks. Ma (2002) has suggested that excessive reliance on Western urban theories should be counteracted by greater efforts to compare China to the post-socialist cities of the Central and Eastern Europe and the former Soviet Union. We agree with this agenda, but also suggest that for some dimensions of the Chinese reform urban landscape, a comparison with capitalist LDC cities may offer more insights. The relevance of LDC experience particularly applies to those dimensions where central control has eroded, and irregular and illegal spaces have come into being and persisted despite various efforts to eradicate them or reassert control. The literature on LDC cities is full of concepts and frameworks that can help make sense of interplay between informal refusals to follow the plans, the rules, the accepted orders, and the interventions of power-holders in divided and exclusionary urban places. Scholars of urban

China have only rarely drawn upon this work to help make sense of the transformation of landscapes and social organization, even though the similarities have become ever more apparent. One risk in neglecting this important body of research is that effort will be wasted through attempts at reinventing the wheel, another is that policies will recreate mistakes that were learned from long ago in Brasilia or Kuala Lumpur.

In this chapter we have attempted to suggest some of the advantages of looking at Chinese cities with the assistance of a classic theme from the study of LDC cities: irregular settlements. Much of what is distinctive about LDC cities derives from the extent of the spaces of illegality and irregularity. Such spaces are emerging and are increasingly receiving detailed attention from scholars of urban China. For example, Zhang (2001) provides a brilliant account of how migrants are producing a new kind of space which is illegal and informally constructed but "nevertheless deeply intertwined with officialdom through informal patronage ties" (3). Yet, there are almost no references in her book to the Latin American literature, or other capitalist LDCs, where these processes have garnered decades of thoroughly researched and theoretically incisive analysis. When we are considering urban processes that exuberantly burst the boundaries of what was previously allowed by the state, these academic resources must be better utilized than they have been thus far. While no longer an LDC, Hong Kong still retains substantial spaces of illegally built structures, and given the similarities in landed property systems and culture, provides a useful mix of similarities and differences. We have tried to draw out the comparison, using it to highlight what is distinctive about the Chinese case, and how in some ways the dynamics of illegal building share more in common with the Latin American cases than with Hong Kong.

At the present time, expected trajectories for illegal settlements in China are very uncertain, politically contingent, and vary considerably from one locality to another. Regularization has been a common recommendation for illegal settlements. Hernando de Soto (2000) has received considerable attention to his argument that legalizing the extralegal in the LDCs can help them follow the successful path of the rich countries since "what the poor are missing are the legally integrated property systems that can convert their work and savings into capital" (227). Yet, Ann Varley (2002: 455) concludes that legalization may be a "less efficient engine of change than its supporters . . . suppose" since the difference between conditions before and after legalization is not necessarily as great as they assume. Both in Mexico and China, illegal property is reasonably secure given certain political guarantees, can already be used as collateral for loans, and offers trajectories out of poverty into reasonable security and comfort. Understanding de Soto's mystery of capital, then, may relate to the mysteries of local governmental support for reasonably secure future trajectories for illegal property, and the more crucial mystery may involve an understanding of how and why governments support or undermine either illegal or legal forms of property.

## Notes

1 This points out the limitations of the much-repeated statement that rural land is "owned" by the collectives: it is only true if one recognizes the "catch" in that it is ownership of a category that is fundamentally limited. "True" ownership involves control over essentially all rights: of transfer, of transformation of usage, etc. and these are clearly rights that are not formally distributed to the collectives but are monopolized by the central state.

2 One could argue that the changes associated with the 1997 Land Law are far-reaching enough to justify dividing the Chinese experience into two cases as well. However, the degree of continuities afterwards resulting from incomplete implementation results in sufficient continuities in the before and after dynamics that it seems possible to restrict the comparison to three cases for our purposes here.

3 Evidence for this can be seen, for example, in the way that advertisements for structures available for purchase in the early 1980s always included details about the registration status of the structure (the color of the official markings indicated the period during which it was registered: recent construction had no markings and had much less secure resettlement rights).

## References

Cartier, C. (2001) "'Zone Fever,' the arable land debate, and real estate speculation: China's emerging land use regime and its geographical contradictions," *Journal of Contemporary China*, 10(28): 445–469.

—— (2002) "Land development, regulation theory and the regional economy in South China," *Asian Geographer*, 21(1–2): 33–52.

Chan, R. C. K., Yao, Y. M. and Zhao, S. X. B. (2003) "Self-help housing strategy for temporary population in Guangzhou, China," *Habitat International*, 27: 19–35.

Chun, A. (2000) *Unstructuring Chinese Society*, Amsterdam: Harwood Academic Publishers.

Collier, D. (1976) *Squatters and Oligarchs*, Baltimore: Johns Hopkins University Press.

De Soto, H. (2000) *The Mystery of Capital: Why Capitalism Triumphs in the West and Fails Everywhere Else*, London: Bantam Press.

Fernandes, E. and Varley, A. (1998) "Law, the city and citizenship in developing countries: an introduction," in Fernandes, E. and Varley, A. (eds), *Illegal Cities: Law and Urban Change in Developing Countries*, London: Zed Books, 3–17.

Gilbert, A. and Ward, P. (1985) *Housing, the State and the Poor*, New York: Cambridge University Press.

Hann, C. M. (1998) "Introduction: the embeddedness of property," in Hann, C. M. (ed.), *Property Relations: Renewing the Anthropological Tradition*, Cambridge: Cambridge University Press, 1–47.

Heyman, J. M. and Smart, A. (1999) "States and illegal practices: an overview," in Heyman, J. M. (ed.), *States and Illegal Practices*, Oxford: Berg, 1–24.

Ho, S. P. S. and Lin, G. C. S. (2001) "China's evolving land system," Center for Chinese Research Discussion Paper 01-1, Vancouver, BC: Centre for Chinese Research.

—— (2004) "Converting land to nonagricultural use in China's coastal provinces," *Modern China*, 30(1): 81–112.

Karst, K., Schwartz, M. and Schwartz, A. (1973) *The Evolution of Law in the Barrios of Caracas*, Los Angeles: Latin American Centre.

Leaf, M. (2002) "A tale of two villages: globalization and peri-urban changes in China and Vietnam," *Cities*, 19(1): 23–31.

Leeds, E. (1996) "Cocaine and parallel polities in the Brazilian urban periphery: constraints on local-level democratization," *Latin American Research Review*, 31(3): 47–83.

Ma, L. J. C. (2002) "Urban transformation in China, 1949–2000: a review and research agenda," *Environment and Planning A*, 34: 1545–1569.

Mangin, W. (1967) "Latin American squatter settlements: a problem and a solution," *Latin American Research Review*, 2: 65–99.

McAuslan, P. (1998) "Urbanization, law and development: a record of research," in Fernandes, E. and Varley, A. (eds), *Illegal Cities: Law and Urban Change in Developing Countries*, London: Zed Books.

Ng, M. K. and Xu, J. (2002) "Development control in post-reform China: the case of Liuhua Lake Park, Guangzhou," *Cities*, 17(6): 409–418.

O'Donnell, G. (1993) "The browning of Latin America," *New Perspectives Quarterly*, 10(4) 50–53.

Perez Perdomo, R. and Nikken, P. (1980) "Law and housing ownership in the barrios of Caracas," *Urban Law and Policy*, 3: 365–402.

Smart, A. (1985) "The squatter property market in Hong Kong," *Critique of Anthropology*, 5(3): 23–40.

—— (1986) "Invisible real estate: investigations into the squatter property market," *International Journal of Urban and Regional Research*, 10(1): 29–45.

—— (1988) "Old huts and new regulations: changes in the Hong Kong squatter property market," *International Journal of Urban and Regional Research*, 12: 303-307.

—— (1989) "Forgotten obstacles, neglected forces: explaining the origins of Hong Kong public housing," *Environment and Planning D Society and Space*, 7: 179–196.

—— (1992) "Making room: squatter clearance in Hong Kong," Hong Kong: University of Hong Kong, Centre of Asian Studies.

—— (1998) "Economic transformation in China: property regimes and social relations," in Pickles, J. and Smith, A. (eds), *Theorising Transition: The Political Economy of Post-Communist Transformations*, London: Routledge, 428–449.

—— (2000) "The emergence of local capitalisms in China: overseas Chinese investment and patterns of development," in Li, S. M. and Tang, W. S. (eds), *China's Regional Development*, Hong Kong: Chinese University Press, 65–95.

—— (2001) "Unruly places: urban governance and the persistence of illegality in Hong Kong's urban squatter areas," *American Anthropologist*, 103(1): 30–44.

—— (2002) "Agents of eviction: the squatter control and clearance division of Hong Kong's housing department," *Singapore Journal of Tropical Geography*, 23(3): 333-347.

—— (2003) "Impeded self-help: toleration and the proscription of housing consolidation in Hong Kong's squatter areas," *Habitat International*, 27: 205–225.

—— and Smart, J. (2001) "Local citizenship: welfare reform, urban/rural status, and exclusion in China," *Environment and Planning A*, 33(10): 1853–1869.

Tang, W. S. (1998) "An inquiry into urban land system reforms in Shenzhen," in Yeung, Y. M. (ed.), *Urban Development in Asia: Retrospect and Prospect*, Hong Kong: Hong Kong Institute of Asia-Pacific Studies, 307–340.

—— (1994) "Urban land development under socialism: China between 1949–1977," *International Journal of Urban and Regional Research*, 18(3): 392–415.

—— and Him Chung (2002) "Rural–urban transition in China: illegal land use and construction," *Asia Pacific Viewpoint*, 43(1): 43–62.

Taubmann, W. (2002) "Urban administration, urban development and migrant enclaves: the case of Guangzhou," in Wong, K. Y. and Shen, J. F. (eds), *Resource Management, Urbanization and Governance in Hong Kong and the Zhujiang Delta*, Hong Kong: Chinese University Press, 79–96.

Turner, J. (1976) *Housing by People*, London: Pantheon Books.

Varley, A. (2002) "Private or public: debating the meaning of tenure legalization," *International Journal of Urban and Regional Research*, 26(3): 449–462.

Wong, L. S. K. (1978) "The squatter problem," in Wong, L. S. K. (ed.), *Housing in Hong Kong*, Hong Kong: Heinnemann Educational Books, 204–232.

Wu, F. L. (2002) "China's changing urban governance in the transition towards a more market-oriented economy," *Urban Studies*, 39(7): 1071–1093.

Wu, W. P. (2002) "Migrant housing in urban China: choices and constraints," *Urban Affairs Review*, 38(1): 90–119.

Xie, Q., Parsa, A. R. G. and Redding, B. (2002) "The emergence of the urban land market in China: evolution, structure, constraints and perspectives," *Urban Studies*, 39(8): 1375–1398.

Yonder, A. (1998) "Implications of double standards in housing policy: development of informal settlements in Istanbul, Turkey," in Fernandes, E. and Varley, A. (eds), *Illegal Cities: Law and Urban Change in Developing* Countries, London: Zed Books, 55–68.

Zhang, L. (2001) *Strangers in the City*, Stanford: Stanford University Press.

Zhu, J. M. (2002) "Urban development under ambiguous property rights: a case of China's transition economy," *International Journal of Urban and Regional Research*, 26(1): 41–57.

# 6 Globalization and the development of new central business districts in Beijing, Shanghai and Guangzhou

*Piper Gaubatz*

## Introduction

One of the most visually striking aspects of China's contemporary urban transformation is the rise of ultra-modern skyscrapers which tower above both the low- and mid-rise factory and apartment compounds of the pre-reform era and the one- and two-storey courtyard houses of traditional China. These new urban spaces embody the new interactions between China and the world in the reform era. The location and design of new high-rise "central business districts" (CBDs) in China is a direct result of the impacts of globalization on the organization of planning authority, the planning process and changing demands for new types of urban space. This chapter begins with a discussion of the reconceptualization of Chinese cities from locally oriented to globally oriented entities, the devolution of planning authority during the reform era from the central government to municipal governments, and the increasing involvement of private contractors – especially multinational firms – in urban design. It then presents case studies of the development and redevelopment of financial districts/central business districts in Beijing, Shanghai and Guangzhou, and reflects on the implications of these trends for urban form.

## Changing conceptualization of Chinese cities

Chinese cities have undergone fundamental transformations since the announcement of China's economic reforms at the end of 1978. These transformations have taken place not only in the actual form of the cities, but also in the fundamental conceptualization of the cities embodied in their plans. In previous publications, I have addressed many of the basic aspects of contemporary Chinese urban form as it has been reshaped during China's drive toward globalization (Gaubatz 1995a, 1995b, 1999a, 1999b and 2003). The most recent form of the "ideal" Chinese city, which I call the "Great International City," is characterized by: (a) the development of distinct functional areas such as residential districts, commercial districts, and industrial and economic development zones; (b) the development of new transportation and communication arteries; and (c) a new vertical dimension with the introduction of high-rise structures, elevated expressways and monumental bridges (Gaubatz 1999b).

Although there are a number of distinct and interrelated motivations for redevelopment of large cities such as Beijing, Shanghai and Guangzhou, none is more striking than the apparent desire to create cities which resemble the world's financial capitals. This arises from a significantly changed conceptualization of the function of China's large cities in the domestic and global economies. China's cities were substantially altered after 1949 in an effort to transform them from cities of consumption to cities of production. This production was oriented toward domestic markets, and indeed, during some eras, primarily toward local and regional markets. The resulting urban form was relatively undifferentiated in the sense that it was devoid of specialized districts such as CBDs. In the early years of the post-1978 reform era, the function of large cities continued to be presented as one of production for domestic markets, although perhaps with a greater orientation toward national marketing and distribution strategies. Production oriented toward the global economy was limited initially to specially designated economic development zones.

During the 1980s, however, participation in the global economy increased greatly in all large cities, especially in the east. Urban plans prepared in the early 1990s presented a reconceptualization of large cities as globally oriented entities. While still maintaining a commitment to domestic functions, these new plans advocated aggressive development toward full participation in the global economy and the world system of cities. For example, Beijing's functional orientation was redefined from a political and cultural center to a center for politics, culture and international affairs and finance (Zhang 2003a). This reconceptualization was intensified during the late 1990s with the campaign for entry into the World Trade Organization, bids for hosting international sports events and conferences, and the beginning of city "brand marketing" campaigns (F. L. Wu 2000b; Chen and Pan 2003; Hong Kong 2003). The ultimate result of this reconceptualization is the aspiration of some of China's cities to "world city" status).[1] During the 1990s, "nearly 50 cities set their development goal as becoming an 'international city'" (Zhou 2002: 60; Xu 1995). Of these, Beijing, Shanghai and Guangzhou (and their metropolitan regions) are three of the four most likely to become world cities (Zhou 2002: 73). In discussing Shanghai's potential to become a world city, Yusuf and Wu (2002: 1229) observe that "cities at or near the apex of the international urban hierarchy . . . have acquired large economic, cultural, and symbolic roles." The development of new high-rise "central" business districts designed to house the global finance industry is a key strategy in promoting this conception of a world city by increasing the city's economic and symbolic roles in the global urban hierarchy and conforming to the expectation that major "international" cities should have a "downtown" (see, for example, DesignArchitecture 2001).

This new conceptualization of the city, while produced and reproduced to some extent by a broad web of planning professionals, the media, the business community and the public, is officially articulated by municipal planners through the preparation of comprehensive planning documents. The production of urban plans in post-Mao China falls into two broad categories: municipal comprehensive plans and detailed plans for specific sites.[2] Comprehensive plans prepared by municipalities are broad scale documents which provide the fundamental rhetoric for the conceptualization of the city's image and functions, spell out the city's goals in areas ranging from land use to population planning, and provide guidelines for the spatial distribution of land use, economic and social activities within the city. Although these goals vary between cities, most large cities and many medium-sized cities have shared a common set of urban planning and development goals during the era of globalization which have a significant impact on urban form, which I summarize in Figure 6.1 (and see Gaubatz 1995a, 1995b, 1996, 1999a, 1999b, 2002 and 2003).

These new urban visions have evolved against the backdrop of China's increasing "globalization." Of Short and Kim's (1999: 3) three different aspects of globalization: "economic globalization (a global economy), cultural globalization (a global culture) and political globalization (a global polity)," the first two have the most relevance for contemporary Chinese urban development. There are three interrelated ways to perceive "globalization" as it relates to changes in Chinese urban form: (a) the globalization engendered by the changing economics and organization of production (Storper 1997), such as the privatization of China's state-owned businesses and housing; (b) the globalization engendered by the utilization of foreign capital in local development (Harvey 1985a, 1985b), such as the establishment of economic development zones; and (c) the globalization engendered by new approaches to the construction of knowledge and the information economy (Castells 2000), such as China's recent trend toward diversifying the planning process and "outsourcing" planning projects to foreign firms.

One of the most common strategies municipal planners are now using to

1 The creation of a distinct and marketable image for the city.
2 The creation of specialized land use areas, including:
   (a) The redevelopment/reintroduction of commercial activities in the urban core;
   (b) The relocation of housing (and residents) from the urban core to the urban periphery;
   (c) The relocation of industry to the periphery.
3 Infrastructure improvement and development ranging from transportation to water and power improvements.
4 The "modernization" and "internationalization" of urban landscapes, particularly through the development of high-rise structures and grade separated expressways.
5 Other aesthetic improvements such as increasing the area of "green space."

*Figure 6.1* Common planning goals for twenty-first century Chinese cities.

transform locally oriented cities into internationally oriented cities is the designation of CBDs. There has been a concerted effort to insert high-rise business districts into the existing urban fabric of many large and medium-sized cities. In Beijing, Shanghai and Guangzhou, the motivation seems not only to accommodate the growing business sector, but also to create signature districts which can develop as internationally recognizable symbols for the cities, on a par with New York's Wall Street.

The rhetoric of urban image-building and place promotion (Wu 2000b) in China thus commonly includes references to such signature places. For example, Shanghai's Nanjing Road has been compared to the Champs Elysées in Paris and Fifth Avenue in New York City (Xinhua 11–22–01) and the designers of the Beijing CBD project equate their work with Hausmann's redesign of Paris (DesignArchitecture 2001). There is a tendency to promote a single district above the others as a "central" business district to create a signature center, even though in practice the growing international business sector is accommodated through multi-nuclear plans. While the municipality provides significant financial and political support for these signature centers, the responsibility for their actual planning and design increasingly falls to local district governments[3] and domestic and foreign consultants.

## Decentralization of planning authority

> Some say that the present planning system in Beijing is still a product of a planned economy, basically involving a "top-down" process. It is in fact not so. Urban planning and management have actually long been weak, and the municipal government lacks the means for providing necessary policy guidance, macro-regulations, and methods of cross district coordination.
>
> (L. Y. Wu 1999: 207)

Municipal-level planning was weak during the Maoist period to the extent that its main role was to facilitate the economic priorities of the central state (Hsu 1996: 896; Ma 1979, 2002; L. Y. Wu 1999). Although authority over urban planning and development has devolved from the central government to the municipalities, nonetheless the municipalities are losing control over many aspects of planning and development as responsibility for these activities continues to devolve toward smaller units of government. During the 1980s, the central government granted municipalities the right to prepare their own urban plans, regulate development and to issue land use permits. In turn, increasing decentralization within municipalities has led to increased involvement in planning by urban district governments and other "territorial organisations"[4] (Ning 1998; Abramson *et al.* 2002; F. L. Wu 2002a, 2002b). As development control has devolved to district governments, many development projects have come under their direct supervision. For example, in 1994, the Beijing municipal government granted district governments the power to grant project rights for housing renewal, thus giving them power over housing construction and real-estate development (L. Y. Wu 1999: 205). F. L. Wu (2002) has observed that whereas the role of the urban district was peripheral in the era of state socialism, it has become central today.

The rapid changes in urban form and infrastructure since 1949 have thus been accompanied by rapid changes in planning and development processes. This has proven a challenge for effective planning. As Yusuf and Wu (1997: 43) observed, "[t]here was no body of rules on how to proceed." Since that time, urban management, particularly the regulation of growth and development, has become a common topic in the Chinese-language urban literature (see, for example, Zhang 2003b). The implementation and enforcement of urban planning has been complicated by shifting balances of power between the government, represented by policy-makers and planners, and economic interests which might include government bureaus, industries and foreign ventures. Although China's urban planners now utilize internationally recognized planning practices, their efforts are often stymied by conflicting regulations, by ill-defined enforcement procedures and a power structure which permits numerous concessions to high-profile developments (Gaubatz 1999a, 1999b). As Yeh and Wu (1995: 544) have noted, "[t]he actual location, type and intensity of development may not be what the planners intended in the master and detailed plans, but rejecting a proposed building is difficult." Moreover, as planning is increasingly linked with urban economic development, different urban jurisdictions, particularly district governments, have begun to compete for investment and development funds, at times to the detriment of the effective execution of comprehensive planning.

Thus, from the highly centralized state socialism of the Maoist period in which the central government maintained ultimate authority over even the most mundane aspects of urban development, Chinese cities have now arrived at an extraordinary new situation in which increasingly small units of territorial authority negotiate directly with multinational corporations over the production and reproduction of urban space.

## The rise of design competitions and the contract-bidding system

> Perhaps the salient point in the current controversy [over Albert Speer's 36 square mile Beijing Olympic Boulevard Plan] is that Beijing – once isolationist and xenophobic – has chosen a foreigner to revamp its capital.
>
> (August 2003: 22)

While the involvement of multinational firms in planning and development for large cities is commonplace in many parts of the world, it is extraordinary in the Chinese context, where the process was largely closed to outsiders from the departure of Soviet advisers in the late 1950s until the beginning of internationally focused planning in the late 1980s. Whereas detailed urban designs – masterplans for specific sub-units of the district territory which lay out land use functions, building mass, open space and landscape guidelines – were once prepared by large government bureaus or "sub-contracted" to university departments, today domestic and foreign contractors are increasingly responsible for the details of urban plans, their development and execution.[5] Concomitantly, the role of the urban planning bureaus is changing from the preparation of detailed plans to the preparation of overall schemes and regulations designed to facilitate detailed planning by others. If this transition continues, planning activities carried out by Chinese urban planning bureaus will increasingly resemble those of large global cities elsewhere.

The recent adoption of foreign approaches to planning, such as Guangzhou's recent utilization of conceptual planning techniques developed in Great Britain (Wang *et al.* 2001: 6), and the employment of foreign planning consultants (see below) represents a recognition of the need for planning "international" spaces. Nowhere is the globalization of the planning process more evident than in the contract-bidding system. The earliest example of high-profile foreign firms competing for a planning project may well be the Lujiazui Zone of Shanghai Pudong New Area, which in the early 1990s invited high-profile foreign firms to prepare alternative plans (Olds 1997). Under the heading "First-Class Urban Planning," the Lujiazui website stated in 1997:

> With a view to building Lujiazui into a world-class urban area and the CBD of Shanghai in the early 21st century, the Shanghai Municipality invited world-renowned experts and domestic architects to work out the masterplan, the traffic plan, and the urban design of the financial center.

This "invited" bidding system characterized several early collaborations between the Chinese and foreign architects and planners, but has recently been replaced by a competitive bidding system. Although China utilized competitive bidding for World Bank-financed projects during the 1980s and experimented with competitive bidding for government purchases in the mid-1990s, the country's first Competitive Bidding Law was not adopted until 1999. The law requires competitive bidding on large-scale projects (including engineering, construction and infrastructure projects). The introduction of design competitions is part of China's efforts to align economic practices with the World Trade Organization (Lardy 2002: 59, 197). In this context, there has been a marked change in the process of urban development as planners have focused more on developing project guidelines for bids than on the preparation of the plans themselves, and as numerous foreign and domestic firms have become involved in Chinese urban planning. The foreign firms range from international giants, such as Skidmore Owings and Merrill (SOM) (US), Sasaki (US), Johnson Fain (US) and Albert Speer (Germany), to smaller firms such as TOWNS: Urban Development Consultants (Canada) and Urbis (Australia). The domestic firms are often recently privatized government units that must now contend with international competition.

The participation of multinational design firms has become a key component of the globalization of Chinese cities, as the Chinese evidently hope to ensure the production of globally oriented spaces which will attract global capital. Thus "the Shanghai municipal government hired Skidmore Owings Merrill in the mid-1990s to help realize its dreams of becoming a world city again" (Armstrong 2003). In open, international competitions for large-scale projects, high-profile foreign firms dominate. Table 6.1 illustrates this through a sample of competitions for large-scale master planning projects. There are high stakes here both for the firms and for the cities. The fees paid to those firms chosen as finalists in the competitions can be as high as the US$200,000, for example, awarded to finalists in the competition for the planning of the "Guangzhou Pearl Riverside." The winner of each competition receives considerably more.

Yet while such competitions are the accepted norm throughout the developed world, they are sometimes controversial. Noted architectural historian Witold Rybcynski (2002: 139), for example, argues that design competitions tend to push the designers toward "flamboyance rather than careful thought." Others argue that publicity connected with a high-profile design competition tends to infuse the project with monumentalism that becomes "an integral part of the project's extraordinary symbolic essence" and leads to fundamental transformations of the identity of the place (Larson 1994: 478; Olds 1997: 120). One critic observes that:

> everyone bidding for the new contracts [for Chinese architecture and planning projects] asserts that they are deeply inspired by Chinese aesthetics. The buildings may be huge and monolithic, like the new headquarters of the Bank of China, but the architects, I M Pei's sons, believe the design evokes the courtyard houses it replaced . . . it is hard to see how an atrium, big enough to house a football match and 2,000 people, resembles an intimate courtyard.
>
> (Becker 2003: 13)

Thus although bids often include references to urban traditions, in practical terms the plans tend more toward the global than the local, minimizing references to Chinese culture in favor of landmark towers and postmodern references. Chinese architects and planners have criticized not only the architecture itself, but the urban designs as well. For example, the Sasaki–Huahui plan for the Beijing 2008 Olympic Green has been criticized for failing to comply with traditional principles of Chinese urban design by

*Table 6.1* Selected master planning projects carried out through international design competitions in China

| *Year winner announced* | *Project and location* | *Nationalities and/or names of competitor firms (winner(s) in* **bold***)* |
|---|---|---|
| 1993 | Shanghai Pudong Lujiazui Central Finance District Master Plan | 1 **Great Britain: Richard Rogers**<br>2 Italy: Massimiliano Fuksas<br>3 Japan: Toyo Ito<br>4 France: Dominique Perrault |
| 1996 | Beijing Finance Street | US: **SOM (Skidmore, Owings and Merrill)** |
| 1999 | Shanghai Luwan Master Plan | 1 **US: SOM** |
| 2000 | Guangzhou Pearl Riverside | 1 **Japan: Kisho Kurokawa**<br>2 US: Sasaki<br>3 US<br>4 Great Britain |
| 2001 | Zhengzhou, Henan. Zhengdong New Town Master Plan | 1 **Japan: Kisho Kurokawa**<br>2 US: Sasaki<br>3 Australia<br>4 France<br>5 Singapore<br>6 China: Institute of Urban Planning and Design |
| 2001 | Beijing CBD | 1 **US: Johnson Fain**<br>2 US: SOM<br>3 Germany: GMP<br>4 Netherlands: Kuiper Compagnons<br>5 Japan: Toshi<br>6 US: NBBJ<br>7 China<br>8 China |
| 2001 | Beijing World Exhibition and Sports Center | 1 **US: RTKL**<br>16 other proposals |
| 2002 | Beijing 2008 Olympic Green | 1 **US: Sasaki (partnered with China: Tianjin Huahui)**<br>54 other proposals |
| 2003 *expected in late August | Guangzhou Development Area Urban Design | 1 US: Johnson Fain<br>2 US: KDG + SWA<br>3 Great Britain: ATKINS<br>4 China (Hong Kong): EDAW<br>5 China: Suzhou Jurong |

*continued*

*Table 6.1* Continued

| *Year winner announced* | *Project and location* | *Nationalities and/or names of competitor firms (winner(s) in* **bold***)* |
|---|---|---|
| 2003 | Shanghai Expo 2010 Plan | **Germany: Albert Speer** |
| 2003 | Yiwu: Central Business District | 1 **Australia: Urbis**<br>2 US<br>3 Hong Kong |
| 2003 | Tianjin World Class CBD | 1 **France: AAUPC**<br>2 Italy: PCMR<br>3 US: RNL<br>4 Germany: SBA<br>5 Japan: Urban Design Institute |
| 2003 | Hainan Downtown Waterside (The Hainan "Bund" Project) | 1 **Spain: BBJ & FTE (Federico & Tatania Echevarria)**<br>2 Netherlands: Kuiper Compagnons<br>3 US: Duany Plater-Zyberd & Co.<br>4 Australia: Urbis Keys Young |
| 2003 | Beijing Olympic Boulevard Plan | **Germany: Albert Speer** |

downplaying the monumentality of Beijing's central axis (Li 2002; Liu 2002). Given the success of non-traditional design bids it could be argued that international firms have gauged current Chinese ambitions, although others have argued that international architects have difficulty understanding the cultural context and mistakenly assume the Chinese are fascinated by (and satisfied with) Western-style designs (Yager 2003). In regard to the controversial French design for the now nearly-complete Chinese National Theater just West of Tian'anmen Square, Chinese-Canadian architect Peigen Peng argued that "to believe blindly in Western 'futurism' showed a lack of understanding and confidence in Chinese culture and Chinese people" (O'Neill 2000: 17). Debates centered in globalization have also arisen in competitions in which the majority of the entries have been from Chinese firms, such as the controversy over following European or Chinese proportions and other design principles for the new People's Square in Taizhou (Luo *et al.* 2002).

## Financial districts

The production of new CBDs/financial districts in Beijing, Shanghai and Guangzhou illustrates the effects of the decentralization of planning authority and the disengagement of urban governments with the details of urban design. Interest in the creation of new financial districts, in part a result of

WTO requirements for opening financial markets, is a startling new direction. Throughout the reform era, there have been significant operational and geographic controls on international participation in China's banking sector. In 1997–1998 some licenses were granted to foreign banks to accept domestic currency deposits, with the geographic restriction that they must be located in Shanghai Pudong New Area or in the Shenzhen Special Economic Zone. But all geographic restrictions on foreign banks will end by 2005, and all operational restrictions will end by 2010. Geographic restrictions will end, as well, for foreign insurance firms and other professional services, such as accounting, law, urban planning, architecture and engineering along a similar timetable (Lardy 2002: 67–74). These changes have made China a new frontier for global firms. Administrators of China's large eastern cities have interpreted this as an opportunity to bring new foreign investment. Many cities have developed plans for a business district designed to appeal to the expected flood of foreign firms. Where once there was concern over the overstock of new commercial space, now the most internationally oriented cities, especially Beijing, Shanghai and Guangzhou, expect high occupancy rates in the near future. Shanghai's Lujiazui Finance and Trade Zone is already home to more than twenty foreign financial institutions, and Guangzhou's Tianhe District is working to attract finance-oriented firms. In Beijing, however, the establishment of a new, internationally oriented financial district has been more controversial.

### *Beijing*

Beijing has been self-consciously planned to be a powerful symbol of China's prevailing ideological currents. The recent desire to transform Beijing into a "world-class" capital city reflects China's new outward orientation. This, in turn, seems to have generated a desire to create new types of urban space, such as a monumental high-rise CBD. There is so much interest in creating a CBD that there are now at least three different districts – Chaoyang, Xicheng and Haidian – developing internationally oriented CBD/financial districts (Figure 6.2).

Plans for a pre-eminent business district first became central to the city's development strategy in the early 1990s. The 1991–2010 Comprehensive Plan placed the new financial district in eastern Beijing's Chaoyang District, beyond the Second Ring Road, in order to maintain the low-rise character of central Beijing, with the Forbidden City at its heart. This location had long housed internationally oriented businesses and embassies and thus seemed a logical choice (Gaubatz 1995a and 2003; BMCUP 2002b; Zhou 1998; Dai 2003).

A second financial district was launched, however, when the Bank of China constructed a new high-rise headquarters in western Beijing's Xicheng District, and Xicheng District planners responded by establishing a 1.7 km long "Financial Street" district just within the Second Ring Road

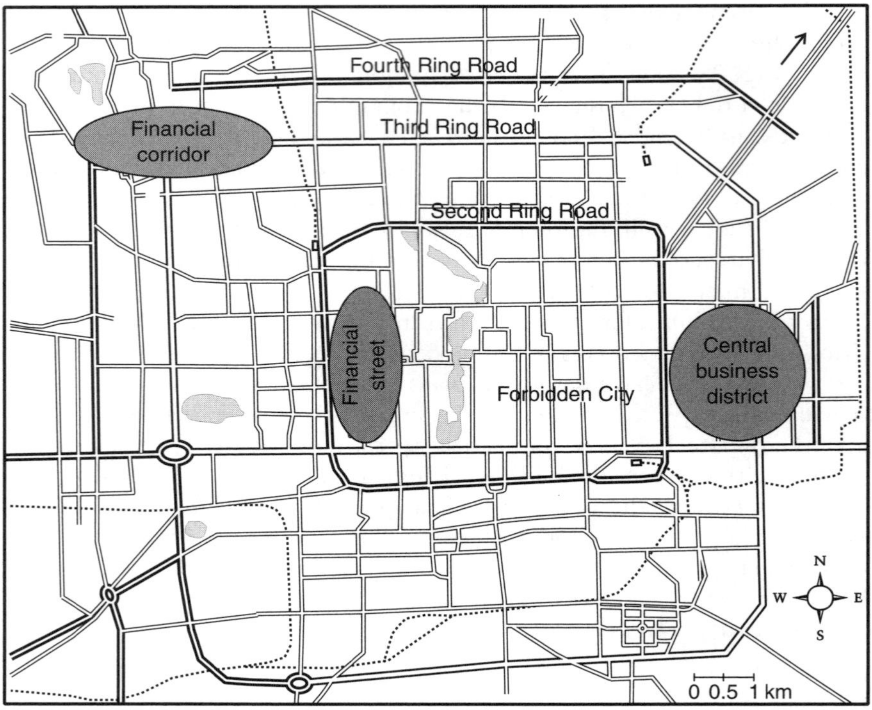


*Figure 6.2* Three competing financial districts in Beijing.

adjacent to the Bank of China. The 1991–2010 Beijing comprehensive plan characterized the Bank's move as part of an effort to consolidate state-owned financial institutions and ministries, thus concentrating domestically oriented activities in western Beijing and internationally oriented activities in eastern Beijing. The Xicheng District plan, however, set out to attract foreign firms to western Beijing. Although this proposal contravened the 1991–2010 comprehensive plan, intense lobbying efforts ultimately gained Xicheng the support of both the State Council and the municipality. As the city's largest single redevelopment project in the 1990s, the project relocated nearly 4,800 families and more than 100 enterprises in order to accommodate about 3,000,000 $m^2$ of newly constructed space in mid- and high-rise structures, including twin 30-storey towers – the tallest structures within Beijing's Second Ring Road (Gaubatz 2003; Fang 2000: 66–68; BMCUP 2003). These structures are centered on a paved plaza containing an obelisk surmounted by a globe (Figure 6.3). Critics such as Liangyong Wu, Qinghua University's senior architect and urban planner, have argued that it will adversely impact traffic patterns and will provide too much competition

*Figure 6.3* Financial street project, Beijing.

for the eastern Beijing financial center. Moreover, the Xicheng District plan has prompted other districts to attract international financial firms. For example, Xuanwu district planned the "International Financial Center" and in 2002 Haidian District announced its intention to create a 4-km long "Financial Corridor" (Wu 1999: 205–207). The Xicheng Financial Street initially developed slowly, attracting relatively little foreign investment. However, the Beijing Urban Planning and Design Commission held a competition for a new design in 2001, and selected an American firm, Skidmore, Owings and Merrill, to complete the project by 2005. SOM planners are clear that they are designing a new CBD, though they recognize that it will not be the sole CBD for the city (Yeh 2003). The project has rapidly evolved from a "maverick" venture to one of the key components of the city's "central business district" identified in the plan for the 2008 Olympics (BMCUP 2002a; BOC 2002).

Meanwhile plans for the Chaoyang District financial center have continued with the State Council's approval of a revised Chaoyang District plan entitled "Beijing CBD." A study carried out by the Beijing City Planning Commission cites the area as the focus of the development of the city's "new economy" (Lu 2002). In the 1980s to 1990s a number of internationally oriented high-rise office, hotel and apartment complexes were constructed in the Jianguomen area of Chaoyang District. In 2000, the State Council announced support for plans for a new "CBD" in Chaoyang District to house financial, insurance, commercial and cultural ventures (CD, 7/02/2000). And the office of the Mayor, in preparation for China's Tenth Five-Year Plan, supports the accelerated development of a Chaoyang central business district (BMCUP 2002b).

The "Beijing Central Business District" plan calls for retaining high-rise commercial structures, embassies and historic buildings, but demolishing forty-two factory compounds built in the 1950s that cover more than 60 percent of the project's 4 $km^2$ land area (BMCUP 2002b). In 2000 the city accepted bids from eight firms (two domestic and six foreign) for the design, and awarded the contract to Johnson Fain Partners (United States). The Johnson Fain plan, which includes 55 skyscrapers, centers on a 140-storey building and a concentration of financial services and office structures comparable to the Financial Street development (Chen and Gan 2001). The project will also benefit from its designation as the main locus for the development of new hotels in preparation for the 2008 Olympics.

### *Shanghai*

Shanghai is extraordinary to the extent that it, arguably more than any other of the former Treaty Ports, retains its pre-1949 CBD/financial district despite the city's loss of its position as an international financial center during the Maoist period (Tang 1998). Plans are under way to rebuild and expand the CBD while attracting foreign investment, relocating population out of the central city, developing a tertiary economy in the city center, and shifting secondary activities to outlying areas (Ning 1995: 50–51). Twenty-first century Shanghai will be formed around a high-rise central business district spanning both banks of the Huangpu River, with retail and services on the West bank in the old commercial core, and a financial center on the east bank. Beyond this core, the city will take a polynucleated form (Figure 6.4).

In 1990, the city announced the opening of the 522 $km^2$ Pudong New Area, designed to double the size of central Shanghai. Located across the Huangpu River from Shanghai's famous European-style downtown, Pudong encompasses nine development zones reserved for finance, manufacturing, free trade and recreational areas. These zones are interspersed with agricultural, residential, commercial and green areas (SPNAA 1993; Yeh 1996).

Pudong's Lujiazui finance and trade zone has created a new signature skyline immediately opposite the imposing early twentieth-century European bank structures of Shanghai's pre-1949 financial district (Figure 6.5). Lujiazui is rapidly developing under the combined forces of unprecedented support from the central and municipal governments, international and

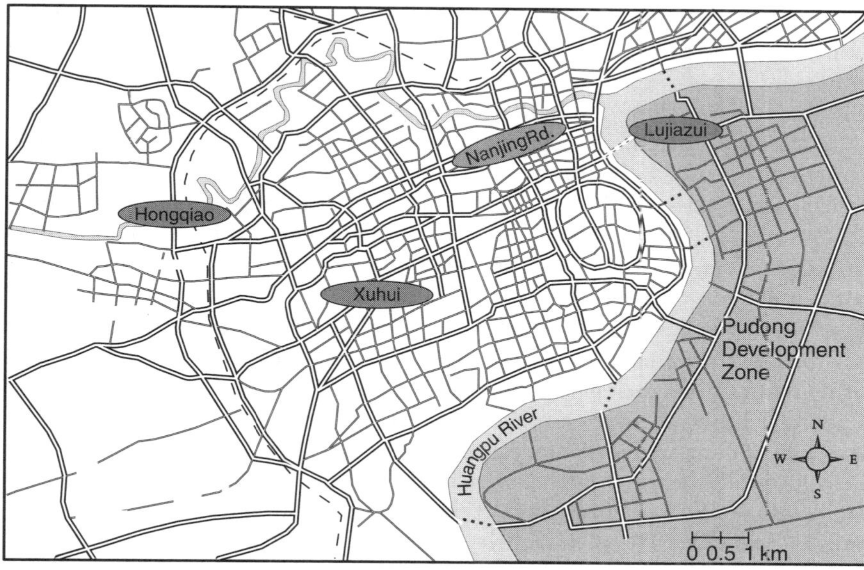


*Figure 6.4* Planned layout of Shanghai (base map adapted from: Shanghai Economic Zone Tourism Volume. Academy Press, Shanghai, 1990 and Shanghai Pudong New Area. Shanghai China Cartographic Studies Press, 1992).

*Figure 6.5* New development in Lujiazui, Pudong New Area, Shanghai.

domestic investment, and the prospects of the development of the financial sector with China's accession to the WTO. Located on a 28 $km^2$ promontory, Lujiazui's cluster of high-rises creates a spectacular profile defined by a cluster of skyscrapers, among them the 420 m, 88-storey SOM-designed Jinmao Tower – China's tallest building and the fourth highest in the world in 2003, and Shanghai's new icon, the 468 m Oriental Pearl Tower. Construction resumed in February 2003 on the 492 m, 101-storey, Kohn, Pederson, Fox (US)-designed Global Financial Center which was intended to be the world's tallest building (Cushman and Wakefield 2003). Foreign firms participated in a highly visible way in the development of the Lujiazui Master Plan, as Olds (1997) has outlined. In the early 1990s, after a visit to Paris' renowned "La Defense" area – then a newly built high-rise business district – then-Shanghai Mayor Zhu Rongji invited four architectural firms with "international reputations" to prepare plans for redesigning Pudong's Lujiazui zone, which had been planned originally in 1991 by the East China Architecture and Design Institute. A competition was held for which representatives of British, French, Italian and Japanese firms and a team of Shanghai architects prepared their plans after a week's orientation in Shanghai. The plan of British architect Richard Rogers was awarded first place, following which a team of Chinese planners prepared three alternative scenarios: one based on Rogers' plan, one based on the Shanghai team's plan, and a third which made only minor modifications to the existing 1991 plan. Ironically, the Pudong New Area administration chose the third option, with the result that most of the input of the foreign consultants was not utilized. Nonetheless, as Olds concludes, the high-profile process generated a heightened national and international interest in the Pudong New Area.

Shanghai, like Beijing, is experiencing lack of development coordination and growing competition between urban districts (Zhao 1994). Lujiazui and the Huangpu District on the opposite bank of the river (areas of which have been revitalized with designs by Skidmore, Owings and Merrill and other international firms) are planned to complement each other and to constitute a single functional commercial district. Yet Huangpu District proclaims pre-eminence with statements such as "Huangpu District is the symbol of Shanghai's prosperity," and has re-emerged as a financial center with sixty-three domestic and foreign banks, twenty insurance companies, and fifteen other financial organizations (Huangpu District 2003). Several of Shanghai's other districts, such as Xuhui and Hongqiao, have become loci of significant alternative development. Efforts to focus attention on Lujiazui continue. Wu (2000b: 351–352) has noted the "strong state support" for Pudong as a strategic point in national economic planning and its impact on place promotion for Pudong. Important support for Lujiazui and Pudong as a whole was provided by the geographic limitation on foreign banks accepting domestic currency to Pudong and Shenzhen in the late 1990s and by relocation of Shanghai's international airport to Pudong in 1999.

### *Guangzhou*

The city of Guangzhou, the traditional economic and political center of southern China, has also planned for financial district/CBD development. It is also experiencing competition between its urban districts for such investment despite efforts to promote a single signature CBD development. Unlike Beijing and Shanghai, Guangzhou faces competition for international investment from other municipalities within the Pearl River Delta region, particularly Hong Kong, Shenzhen, Zhuhai and Foshan. Guangzhou's position in the regional economy is gradually declining (Lin 2002: 250). Intra-regional competition has become so great that the provincial government has begun a new initiative for city integration and coordinated regional planning in the Pearl River Delta (Xinhua, 4–12–2003).

The conceptualization of Guangzhou's urban planning changed markedly in the reform era with quite different plans adopted in the 1980s and 1990s. The 1984 plan divided the city into ten zones with distinctive, specialized functions in order to create a polynuclear form, such as cultural and communications activities for the old central city district, and science, research and education activities for the newly developing Tianhe District. In contrast, the 1991–2010 plan divided the city into only three zones and channeled internationally oriented development especially toward Tianhe District, which was now designated as the city's new CBD (Dai 1994; Fan 1994; Li *et al.* 1994; Wu and Yeh 1999) (Figure 6.6). Tianhe District is centered on a

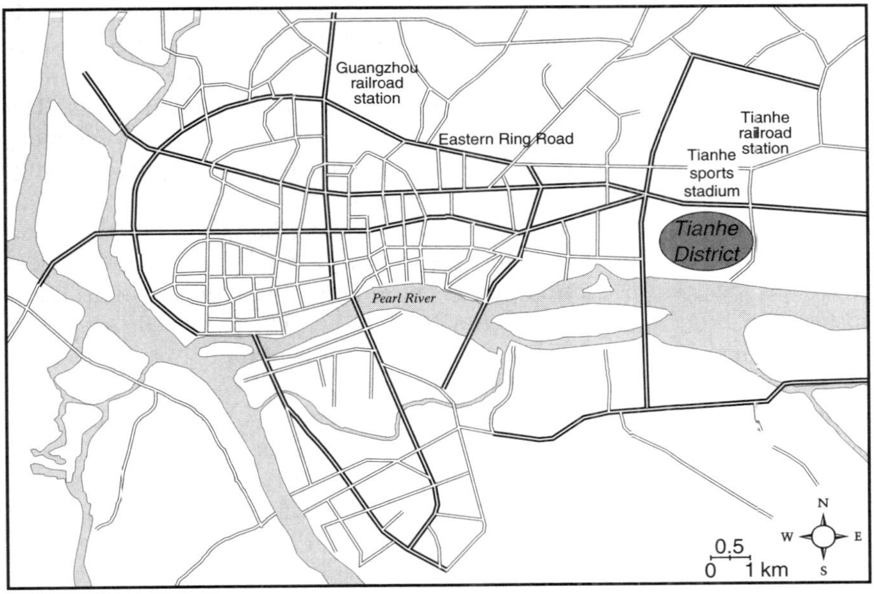


*Figure 6.6* Planned layout of Guangzhou.

large open site that had been used as an airport in the 1950s and 1960s, and was transformed in 1984 into a large sports complex dominated by the Tianhe Stadium. The immediate area around the stadium was developed in the late 1980s and 1990s for recreation and office/commercial space and the surrounding area was developed with high-rise housing, a railroad station connecting Guangzhou to Shenzhen and Hong Kong, and commercial structures. In 1991 new development designed to attract foreign investment began in earnest with the establishment of the Tianhe New and High Technology Zone, a development zone which offered financial incentives for investment in computer and other high-technology industries (Dai 1994).

The designation of Tianhe as a new CBD in the 1991 plan was, as in the case of Shanghai's development of Lujiazui, an attempt to develop a new district on land previously used primarily for farming and relatively low-density housing. Moreover, like Lujiazui, the development of Tianhe serves to extend the city toward its new export-oriented industrial facilities, in this case the Guangzhou Economic and Technological Development Zone 31 kilometers east of the city center. The Tianhe plan was prepared in 1992 with the aid of an American planning firm, Thomas Associates, and Hong Kong developer Li Ka-shing. The plan included the creation of a new axis running north–south from the Tianhe train station to the Qigang Huaqiao New Village (Fan 1994; Li *et al.* 1994).

Today Tianhe describes itself as "a modern downtown" (Thnet 2003) and boasts a concentration of nearly twenty impressive, glass-walled skyscrapers (more than any other Guangzhou district) as well as newly developed parks and broad boulevards. Tianhe's 391 m, 80-storey CITIC Plaza building is the sixth tallest skyscraper in the world. Tianhe has yet to develop as a pre-eminent CBD, however, and Guangzhou's internationally oriented commercial development continues to take place in several different districts. The differences between the degree of recent CBD/financial district development in Guangzhou and that in Beijing and Shanghai remain quite striking. Although Guangdong Province receives by far the largest share of foreign direct investment of any province in China, Guangzhou lags behind Beijing and Shanghai in terms of the numbers of foreign-funded enterprises, offices of foreign financial institutions, offices of foreign telecommunications institutions and other economic indices (Zhou 2002). The pace of development in Guangzhou was further slowed by the 1999 bankruptcy of the Guangdong International Trust and Investment Corp (GITIC), one of the main sources of domestic investment funds in the region. As a result, Tianhe was notable for having some of the most spectacular examples of "*lanweilou*" ("rotten tail buildings"), unfinished structures whose development had been abandoned in 1999. The city of Guangzhou itself paid to have the glass facade installed on the twin towers of China City Plaza in Tianhe to improve the urban landscape for the 2001 Ninth National Games. Work is now resuming on many of these structures (Mitchell 2002; Li 2003). While Tianhe has attracted some foreign investment, it continues to compete for internationally oriented development with several of Guangzhou's older districts, especially Yuexiu and Dongshan where there is a significant concentration of new internationally oriented development near the Guangzhou Railroad Station and along the eastern Ring Road. Moreover, Guangzhou is likely to experience much slower financial district growth than nearby Shenzhen and Hong Kong, both of which, unlike Guangzhou, are the sites of stock exchanges, giving those two cities an advantage in the anticipated increase in international financial transactions in China. Indeed, Tianhe District recently has turned increasingly toward the promotion of high-technology industries, rather than financial services, to attract foreign investment.

## Conclusion

This chapter has addressed three interrelated aspects of the impacts of globalization on Chinese urban form: the reconceptualization of cities, the devolution of urban planning authority, and the privatization and internationalization of detailed urban design. The reconceptualization of large cities from local to global entities has contributed to the restructuring of the functional organization and layout of urban space, from the relatively undifferentiated space of the Maoist city to the highly specialized land use designations developed during the reform era. Moreover, the devolution of planning and development authority from the metropolitan government to district government has created a situation in which it is possible for district governments to make some independent decisions about their territories' functional specialization. As financial structures change, district governments see potential in both high-profile and high-profit land uses. Furthermore, the introduction of the contract bidding system, directly a result of China's adoption of international economic and regulatory practices, allows local district governments to carry out urban design projects without needing urban design departments. This system has tended to award high-profile projects to large-scale foreign firms whose facility with generating internationally oriented spaces is well established. This favors development which appeals to multinational firms seeking easily adaptable spaces and favorable business locales for their global subsidiaries. The process does little, however, to preserve or reproduce the local character of the cities in which they are situated.

These shifts in planning and development point toward a significant weakening of the power of comprehensive planning at the municipal level at a time when municipal governments have become increasingly interested in developing their cities toward specific goals for image and form. This has been one important factor in the sometimes prominent contradictions between overall city planning and current development. In Beijing, serious competition is developing between financial districts just as the support of both the central and municipal governments seems to have coalesced behind a single project (Beijing CBD). And in Shanghai, districts on the West bank of the Huangpu River continue to carry out high-profile internationally oriented redevelopment even as the Pudong area on the east bank has consistently been put forward as the city's focal point for globalization. Thus there is every likelihood that by 2010 the form, functional layout and landscapes of these cities will bear little resemblance to the urban visions set forth in the 1991–2010 plans.

Changes in the planning process such as devolution of planning responsibility to urban districts, the increasing prevalence of competitive bidding, and design by foreign firms, together with local efforts to attract foreign investment through signature developments, new financial districts and CBDs, are dramatically increasing competition between territorial units within and between the cities and creating potential for greater uneven urban and regional development. Guangzhou's Tianhe District has already experienced the embarrassment of abandoned construction projects, the future of Shanghai's Hongqiao District is uncertain now that the city's international airport functions have been relocated from Hongqiao to Pudong, and Beijing's three financial districts are unlikely to be equally successful.

These processes have important consequences for urban residents. Globalization has often been identified with increasing social polarization in cities (Sassen 1991, 1994, 1995; Short and Kim 1999). The development of internationally oriented commercial districts in China has tended both to transform everyday urban life and economic opportunities for many residents and to destroy long-established neighborhoods and villages by relocating residents and businesses on a massive scale (Gaubatz 1995a and b; 1999a and b; Fu 2002; Gu and Liu 2002; Wu 1999).

Current controversies such as competition between districts for premier place within the context of polynuclear development are evidence of how far globalization has already reworked earlier Chinese assumptions about urban design and planning. So too is the use of high-profile international design firms for urban planning, a practice which is commonplace in the world's financial capitals, but which is unprecedented in China. In recent years as urban policy-makers and planners have worked toward the production of "international cities," Chinese urban planning, form and landscapes have become more similar to those in other parts of the world at the expense of the distinctive urban forms and landscapes of both traditional and Maoist Chinese cities.

## Notes

1 John Friedmann has published some of the most widely accepted definitions of world cities. He argues that world cities are a class of cities that play a leading role in the spatial articulation of the global economic system (Friedmann 1998; Friedmann and Wolff 1982).

2 See Yeh and Wu (1999) for a detailed analysis, but also note that the process has undergone significant changes since that time, not only through the increasing

privatization of the housing and real-estate markets (Wu 1998) and the devolution of planning to district governments (Wu 2002a), but also through the development of the contract-bidding system for major urban planning and development projects (Lardy 2002).

3 China's large cities usually consist of municipalities which comprise urban and suburban districts (*chengqu* and *jiaoqu*) and counties (*jiaoxian*), which in turn are divided into urban subdistricts (*jiedao*) and rural townships.

4 Wu (2002) identifies "territorial organizations" as bodies having jurisdiction over specific areas, such as the municipality as a whole, urban districts, street committees and residents' committees.

5 The contract-bidding system has been applied to other types of planning as well, especially transportation and basic utilities planning and development.

## References

Abramson, D., Leaf, M. and Ying, T. (2002) "Social research and the localization of Chinese urban planning practice: some ideas from Quanzhou, Fujian," in Logan, J. R. (ed.), *The New Chinese City*, Oxford: Blackwell, 167–180.

Armstrong, D. (2003) "S.F. architects helping to reshape Shanghai," *San Francisco Chronicle*, Sunday January 26 (online edition).

August, O. (2003) "Beijing signs up son of Hitler's architect," *The Times* (London), February 14: 22.

Becker, J. (2003) "Empire building," *The Independent*, March 24: 13.

Beijing Municipal Commission for Urban Planning (BMCUP) (2002a) "Beijingshi 'shiwu' shiqi jianzhuye fazhan guihua (Plans for the development of the architectural profession in Beijing under the Tenth Five-Year Plan)," January 15.

—— (2002b) "Zhongyang shangwuqu – Beijing CBD" (Central business district – Beijing CBD), Report of the Committee, July 30.

—— (2003) "Beijing jinrongjie (Beijing financial street)," Report of the Committee.

Beijing Olympic Committee (BOC) (2002) "Beijing aoyun xingdong guihua (Plan for the Beijing Olympics)," July 25.

Castells, M. (2000) *End of Millennium*, Oxford, Malden: Blackwell.

Chen, G. and Gan, J. Z. (2001) "Beijing shangwu zhongxinqu guihua Fangsu Zhengji de Zhutiao" (Competition for the planning of the Beijing CBD) *Chengshi Guihua* (*Urban Planning Review*), 25(6): 64–67.

Chen, Y. and Pan, D. (2003) "Chengshi xingxiang wangzaozhong de jingxiaoxue linian (Moulding city images with the guidance of marketing principles)," *Xin jianzhu (New Architecture)*, 2.

*China Daily* (CD) (2000) "Chaoyang official plans central business district," July 2, 2000.

Cushman and Wakefield Co. (2003) "Research: Office snapshot Shanghai." Online, available at: www.cushwakeasia.com/research/resrch_doc/snapshot_sh/ SH%200103.pdf (accessed January 2003).

Dai, C. (2003) "Chongtu yu xietiao: Beijing CBD zongshu" (Conflict and harmony: analysis of the Beijing CBD, *Shidai Jianzhu* (*Modern Architecture*), No. 2: 26–28.

Dai, F. (1994) "Kuashiji de Guangzhou chengshi zongti guihua (Turn-of-the-century Guangzhou master urban plan)," *Guangzhou chengshi guihua (Guangzhou City Planning)*, January: 5–8.

DesignArchitecture (2001) "Beijing gives nod to Johnson Fain Partners for vision of new city center." Online, available at: www.designarchitecture.com (accessed May 9, 2001).

Fan, R. L. (1994) "The planning and construction of the new CBD," *Guangzhou chengshi guihua (Guangzhou City Planning)*, January: 9–12.

Fang, K. (2000) *Dangdai Beijing jiucheng bianxin: Diaocha, yanjiu, shensu* (*Contemporary Conservation in the Inner City of Beijing: Survey, Analysis, and Investigation*), Beijing: Zhongguo jianzhu gongye chubanshe.

Friedmann, J. (1998) "World city futures: the role of urban and regional planning policies in the Asia-Pacific region," in Yeung, Y. M. (ed.), *Urban Development in Asia: Retrospect and Prospect*, Hong Kong: Chinese University of Hong Kong Press, 22–54.

—— and Wolff, G. (1982) "World city formation: an agenda for research and action," *International Journal of Urban and Regional Research*, 6(3): 304–344.

Fu, Z. J. (2002) "The state, capital, and urban restructuring in post-reform Shanghai," in Logan, J. R. (ed.), *The New Chinese City*, Oxford: Blackwell, 106–120.

Gaubatz, P. (1995a) "Changing Beijing," *Geographical Review*, 85(1): 79–96.

—— (1995b) "Urban transformation in post-Mao China: impacts of the reform era on China's urban form," in Davis, D., Kraus, R., Naughton, B. and Perry, E. (eds), *Urban Spaces in Contemporary China: The Potential for Autonomy and Community in Post-Mao China*, Woodrow Wilson Center Press Series, Cambridge: Cambridge University Press, 28–60.

—— (1996) *Beyond the Great Wall: Urban Form and Transformation on the Chinese Frontiers*, Stanford: Stanford University Press.

—— (1999a) "China's urban transformation: patterns and processes of morphological change in Beijing, Shanghai, and Guangzhou," *Urban Studies*, 36(9): 1495–1521.

—— (1999b) "Understanding Chinese urban form: contexts for interpreting continuity and change," *Built Environment*, 24(4): 251–270.

—— (2002) "Looking west toward Mecca: Islamic enclaves in Chinese frontier cities," *Built Environment*, 28(3): 231–248.

—— (2003) "Planning the Chinese city: vision, implementation and form," in Petruccioli, A., Michele, S. and Giuseppe, S. (eds), *The Planned City?* Vol. 3 Bari, Italy: Uniongrafica Corcelli Editrice, 805–810.

Gu, C. L. and Liu, H. Y. (2002) "Social polarization and segregation in Beijing," in Logan, J. R. (ed.), *The New Chinese City*, Oxford: Blackwell, 198–211.

Harvey, D. (1985a) *Consciousness and the Urban Experience: Studies in the History and Theory of Capitalist Urbanization*, Baltimore: Johns Hopkins University Press.

—— (1985b) *The Urbanization of Capital: Studies in the History and Theory of Capitalist Urbanization*, Baltimore: Johns Hopkins University Press.

Hong Kong (2003) "Brand Hong Kong," official website of the city of Hong Kong. Online, available at: www.brandhk.gov.hk/brandhk/eindex.htm.

Hsu, M. L. (1996) "China's urban development: a case study of Luoyang and Guiyang," *Urban Studies*, 33(6): 895–910.

Huangpu District (2003) "Shanghai Huangpu," Online, available at: www.en.njilzs.com.

Lardy, N. (2002) *Integrating China into the Global Economy*, Washington, DC: Brookings Institution Press.

Larson, M. (1994) "Architectural competitions as discursive events," *Theory and Society*, 23: 469–504.

Li, H. (2002) "Beijing decides on 2008 Olympic venue construction plan," *People's Daily Online*, July 29.

Li, P. P., Tang, L. M., Song, Z. Y. and Huang, X. Z. (1994) "Zhujiang xincheng guihua zonghe fangsu jieshao (An introduction to the plan of the new CBD along the Pearl River)," *Guangzhou Chengshi Guihua (Guangzhou City Planning)*, January: 13–17.

Li, S. (2003) "Revived GDI resumes work on Guangzhou development," *South China Morning Post*, June 20: 3.

Lin, G. C. S. (2002) "Region-based urbanization in post-reform China," in Logan, R. J. (ed.), *The New Chinese City*, Oxford, Blackwell, 245–257.

Liu, X. (2002) "Winning design caught in public dispute," *China Daily*. Online, available at: www1.chinadaily.com.cn/ (accessed July 23).

Lu, H. J. (2002) "Xinjingji shidai de zhongyang shangwuqu – Dui Beijing CBD dingwei he fazhan silu de jianyi (The CBD of the new economy – the development of the Beijing CBD)," Beijing Municipal Commission for Urban Planning report, July 5, 2002.

Luo, Q. P., Shen, M. G. and Qiu, F. (2002) "Sheji, pengzhuang, qishi: yi Taizhoushi shimin guangchang sheji wei li (Design, conflict, enlightenment: the example of the design of People's Square in Taizhou)," *Jianzhu sheji (Architectural Design)*, July 26, 2002.

Ma, L. J. C. (1979) "The Chinese approach to city planning: policy, administration, and action," *Asian Survey*, 19: 838–855.

—— (2002) "Urban transformation in China, 1949–2000: a review and research agenda," *Environment and Planning A*, 34(9): 1545–1569.

Mitchell, T. (2002) "Statistical paper trail throws veil over reality of economic trend," *South China Morning Post*, July 23, 2002: 3.

Ning, Y. M. (1995) "Case study of Shanghai, China," *Metropolitan Planning and Management in the Developing World: Shanghai and Guangzhou, China* (Nairobi: Habitat).

—— (1998) "'Xin chengshihua jincheng – 90 niandai Zhongguo chengshihua dongli jizhi he tedian tantao' (New processes of urbanization – Dynamics and features of urbanization in China in the 1990s)," *Dili xuebao (Acta Geographica Sinica)* 53(5): 470–477.

Olds, K. (1997) "Globalizing Shanghai: the 'Global Intelligence Corp' and the building of Pudong," *Cities*, 14(2): 109–123.

O'Neill, M. (2000) "To build or not to build: work has ceased on Beijing's national theatre as a decision is awaited on its controversial design," *South China Morning Post*, September 21, 2000: 17.

Rybcynski, W. (2002) "The Bilbao effect: public competitions for architectural commissions don't necessarily produce the best buildings," *Atlantic Monthly*, 290(2): 138–142.

Sassen, S. (1991) *The Global City: New York, London, Tokyo*, Princeton, NJ: Princeton University Press.

—— (1994) *Cities in a World Economy*, Thousand Oaks, CA: Fine Forge Press.

—— (1995) "On concentration and centrality in the global city," in Knox, P. L. and Taylor, P. J. (eds), *World Cities in a World System*, Cambridge: Cambridge University Press, 63–75.

Shanghai Pudong New Area Administration (SPNAA) (1993) *Shanghai Pudong New Area Handbook*, Shanghai: Shanghai Far East Publishers.

Short, J. R. and Kim, Y. H. (1999) *Globalization and the City*, New York: Addison Wesley Longman.

Storper, M. (1997) *The Regional World: Territorial Development in a Global Economy*, New York: Guilford.

Tang, J. Z. (1998) "The evolution and functional readjustment of Shanghai's central business district," in Foster, H., Lai, D. and Zhou, N. S. (eds), *The Dragon's Head: Shanghai, China's Emerging Megacity*, Canadian Western Geographical Series Volume 34, Victoria, BC: Western Geographical Press, 247–258.

Thnet (2003) "Urban construction in Tianhe." Online, available at: www.thnet.gov.cn.

Wang, M. H., Duan, X. F. and Tian, L. (2001) "Guangzhou chengshi zongti fazhan gaijin guihua de shensu yu shizhao (The practice of conceptual planning for the city of Guangzhou)," *Chengshi Guihua (City Planning)*, 25(3): 5–10.

Wu, F. L. (1998) The new structure of building provision and the transformation of the urban landscape in metropolitan Guangzhou," *Urban Studies*, 35(2): 259–284.

—— (2000a) "Modeling intrametropolitan location of foreign investment firms in a Chinese city," *Urban Studies*, 37(13): 2441–2464.

—— (2000b) "Place promotion in Shanghai, PRC," *Cities*, 17(5): 349–361.

—— (2002a) "China local governance and city planning in transition," Final Report for the British Council China Studies Grant, October 2002.

—— (2002b) "Real estate development and the transformation of urban space in China's transitional economy, with special reference to Shanghai," in Logan, J. R. (ed.), *The New Chinese City*, Oxford: Blackwell, 153–166.

—— and Yeh, A. G. O. (1999) "Urban spatial structure in a transitional economy: the case of Guangzhou, China," *Journal of the American Planning Association*, 65(4): 377–387.

Wu, L. Y. (1999) *Rehabilitating the Old City of Beijing: A Project in the Ju'er Hutong Neighborhood*, Vancouver: University of British Columbia Press.

Xinhua News Agency (11–22–2001) "Shanghai to build top-class street."

—— (4–12–2003) "Guangdong to lift competitiveness by city integration."

Xu, J. Z. (1995) "About the development space of China's international cities," *City Planning Review*, 19(3): 23–25.

Yager, G. (2003) "Care with China/great cities being reinvented," *Building Design*, February 21, 2003: 19.

Yeh, A. G. O. (1996) "Pudong: Remaking Shanghai as a world city," in Yeung, Y. M. and Sung, Y. W. (eds), *Shanghai: Transformation and Modernization Under China's Open Policy*, Hong Kong: Chinese University Press.

—— and Wu, F. L. (1995) "Internal structure of Chinese cities in the midst of economic reform," *Urban Geography*, 16(1): 521–554.

—— (1999) "The transformation of the urban planning system in China from a centrally-planned to transitional economy," *Progress in Planning*, 51(3): 167–252.

Yeh, P. (2003) Project Manager, Beijing Finance Street, Skidmore, Owings and Merrill, San Francisco office, personal communication.

Yusuf, S. and Wu, W. P. (1997) *The Dynamics of Urban Growth in Three Chinese Cities*, New York: Oxford University Press.

—— (2002) "Pathways to a world city: Shanghai rising in an era of globalisation," *Urban Studies*, 39(7): 1213–1241.

Zhang, J. G. (2003a) "Ba Beijing jiancheng – liude xiandaihua guojichengshi

(Build Beijing into a first-class modern and internationalized city)," *Chengshi Wenti (City Problems)*, 2003, No. 2.

—— (2003b) "Beijing chengshi jingying yu chengshi guihua (City management and city planning in Beijing)," *Journal of Beijing Union University*, January 2003.

Zhao, M. (1994) Interview, August, 1994. Professor Zhao is associate professor and vice head of the department of urban planning at Tongji University, Shanghai.

Zhou, Y. (1998) "Beijing and the development of dual central business districts," *Geographical Review*, 88(3): 429–436.

Zhou, Y. X. (2002) "The prospect of international cities in China," in Logan, J. R. (ed.), *The New Chinese City*, Oxford: Blackwell, 59–73.

# 7 Historical memory, community-building and place-making in neighborhood Shanghai

*Tianshu Pan*

## Introduction

From the mid-1980s to the late 1990s, urban China witnessed the birth and growth of an industry of nostalgia in the realm of popular culture. A shared history of Maoist China dominated by successive political movements provided a space for the construction of a culture of nostalgia. Shanghai, China's most cosmopolitan city, was affected by a similar culture of nostalgia. Shanghai nostalgia, however, was *not* for its recent revolutionary past (such as "Mao nostalgia" in other parts of China), but for its colonial heritage. Moreover, Shanghai's nostalgia has been actively promoted by the city's social and economic elites as well as government officials who assumed a pivotal role in the ongoing privatization, stratification and liberalization in many spheres of everyday urban life. This is rather striking because until very recently, Shanghai's colonial past, as represented in the official textbooks, was nothing but a "century of humiliation," beginning with China's defeat in the Opium War (1840–1842).

Under state socialism, colonial Shanghai was often referred to as a site of imperialist domination over the Chinese people prior to the 1949 revolution. The sign that reputedly read "No Dogs or Chinese" at the gate of a public park in the Bund became a symbol of colonial oppression as if it summed up the total history of pre-revolutionary Shanghai. The death of Chairman Mao in 1976 and the return of Deng Xiaoping to the center of the Party leadership led to a significant change of China's political life in the late 1970s. Local historians lost no time in capturing the opportunity to rewrite their colonial past. Hundreds of books on Shanghai society under colonial rule were published in the name of facilitating academic research and meeting the needs of the reader who wanted to know more about old Shanghai. The easy access granted by the Municipal Government to the newly renovated Shanghai Archives enabled Western scholars to amass a sizable body of historical data from all the pre-1949 files.

Colonial Shanghai or "old Shanghai," depicted in the imaginative world of historical accounts, scholarly papers, novels, and on countless websites, has become a "saturated symbol" for the local social actors deeply involved in the multifaceted cultural process of manipulating collective memories for the rediscovery, reevaluation and reinvention of Shanghai's past as they keep repositioning themselves and reasserting their identities within a rapidly restratified society. Based on my field research, I argue that the ongoing historical reconstruction and spatial reconfiguration in arguably China's most cosmopolitan city has answered the strategic need of the local elites in their bid for a more internationally oriented millennium. More importantly "Shanghai nostalgia," the nostalgia for a selectively remembered and reimagined pre-Communist colonial past, serves the social and political agenda for Shanghai's transition toward a global city.

Under the conceptual umbrella of "Shanghai nostalgia," I examine in this chapter the impact of the collective memory of the city's neighborhoods, dichotomized as the "upper and lower quarters (*shangzhi jiao, xiazhi jiao*)," on community-building and place-making processes in various localities. Echoing anthropologist Arjun Appadurai's (1988) call to "put hierarchy in its place," I attempt to go beyond textual representations (e.g., Lee 1999; Zhang 2000) and locate the cultural and political symbols of "Shanghai nostalgia" in *actual sites*. Mapping "Shanghai nostalgia" in time and space would thus make the contradictions and tensions more visible in the urban processes within which upper quarters and lower quarters as imagined communities come to be "attached to imagined places" (Gupta and Ferguson 1997: 39).

The earth-shattering changes that affected every corner of the city since the early 1990s made it easy for us to assume that territoriality has become less of an issue than it was in the recent past. Over the course of my research, however, I came to realize that the age-old dichotomy between the lower quarter (low end) and upper quarter (high end) had hardly been blurred by the profound social transformations of the last one hundred and fifty years. I found, rather, that the lower/upper quarter dichotomy has remained a linguistic device strategically appropriated by the local residents to map out neighborhood Shanghai and their perception of the social and economic reality of where they belonged. From the colonial past to the late socialist present, both the lower quarters and upper quarters were among the most meaningful categories for the people of Shanghai to articulate one's status and position in society. Such a binary opposition between the high and low, as I will show in this chapter, had not lost its currency in the everyday discourse. The persistence of the dichotomy points to the limits of the socialist social engineering attempts at eliminating inequality and disparity within and between residential quarters. It was my point of entry to Bay Bridge, a territorial subdivision of Shanghai's Luwan District where I conducted field research on community building practices intermittently between 1998 and 2002.

## The dialectics of the "upper quarters" and the "lower quarters"

In their mental universe, generations of Shanghai residents created spatial terrains corresponding to the particular socioeconomic echelons in which they situated themselves. One could sense a logical relationship between the social and the spatial in the Shanghainese use of the term *jiao*, which can be literally translated as "locality" or "corner." The notion of *jiao* conveys two layers of meanings of place in John Agnew's terms: *locale*, the settings in which social relations are constituted; *location*, the geographical area encompassing the settings for social interaction as defined by social and economic processes operating at a wider scale; and sense of place, the "local structure of feeling" that defines the emotive sensibilities people develop about the different spatial settings of their lives (Agnew 1987: 28). Such an acute sense of place embedded in the local notion of *jiao* allows us to see the ways in which the Shanghainese adjust themselves to everyday life in their social worlds.

Using *jiao* as a trope to suit their needs and express their values, the Shanghainese mapped out city neighborhoods in terms of a dichotomy between the "upper quarter" or "higher corner" and the "lower quarter" or "lower corner" more than a century ago. In the local dialect, the location of one's residence or workplace in the "upper quarter," the so-called *shangzhi jiao*, could be viewed as the equivalent of the English terms "uptown" or "the right side of the tracks." Similarly the term "lower quarter," or *xiazhi jiao*, is the equivalent of "downtown" or "wrong side of the tracks." In many ways, the dichotomy between the "lower quarter" and the "upper quarter" in Shanghai is the Chinese equivalent of such metaphors of "upscale sections" and "ghetto" that encode the race and class relations in many American cities.

The products of urbanization and industrialization, the spatial dichotomy indexes two social echelons representing radically different lifestyles, individual and family histories, native place identities, and living environment for the past one hundred and fifty years during which Shanghai was transformed from a rural county seat into a cosmopolitan metropolis. Social historians never failed to take note of the general social bias or snobbery based on "quarters" or "corners" (Honig 1992; Lu 1999: 15, 376). As reflections of both historical imagination and social reality, the two "quarters" have remained the key terms used self-consciously by the local residents, municipal officials and real-estate agents as a strategic device to position themselves in the web of social relationships. In short the divide between the upper quarters and lower quarters reflects the widening disparity between the upper and the lower echelons as neighborhood Shanghai becomes an increasingly stratified society.

As an emerging cultural industry, "Shanghai nostalgia" is largely concerned with those whose past experience was an integral part of the "upper quarter" and is hardly relevant to the everyday life of the ordinary people, especially the underclass in the "lower quarter." Within academe scholars have paid scant attention to the city's peripheral area, the working-class neighborhoods that formed the "lower quarters" of the city – home to hundreds of thousands of migrant laborers and refugees in pre-1949 Shanghai. The works of a handful of scholars such as Emily Honig (1992), Lu Hanchao (1999) and Elizabeth J. Perry (1993 and 1997) best represent the systematic research focus in the social life of Shanghai "beyond the neon lights" (Lu 1999). Because of their efforts, Shanghai is demystified and stripped of its stereotypical images as the "Paris of the Orient," the "Whore of Asia" and the "Paradise of Adventurers." Their insights have effectively directed our attention to the part of Shanghai that is not always associated with the tales of magnates, gangsters, writers, artists and prostitutes as I proceed to examine the largely under-researched "lower quarters" in this chapter.

## Upper quarters and lower quarters as "imagined communities"

Historically the "upper quarters" were the neighborhoods with enclaves of foreign populations – the French, the British, the American, the Russian and East Europeans of Jewish origins, and the Japanese. In 1843 the Western powers started to divide and rule their own spheres of influence within Shanghai, a treaty port forcefully opened to foreign trade after China lost the Opium War (1842). For example, the British victors became the rulers of the International Settlement, which was later jointly governed with the Americans. Within the boundaries of the International Settlement (created in 1863) was the old downtown area built on both sides of Nanjing Road, the nation's busiest shopping street. To its east was the waterfront financial center known as the "Bund." One could hardly lose sight of the prideful display of bank buildings and hotels with distinctive styles. As a contrast to the hustle and bustle of the International Settlement, the French Concession occupied a substantial section of the "upper quarter" as a residential area with a poetic flavor as indicated by the buildings of distinctive French architectural design and tree-shaded boulevards. The French Concession, established in 1849, was home to the colonial officers and Chinese compradors. Under socialism the Western style buildings located in the Bund continued to perform their former functions as banks, trading companies, customs office, municipal court and other government institutions for over fifty years. Within the "upper quarters," a beautiful house with a garden and backyard well protected by the iron-gate and thick walls is often the residence for a top government official. Yet in the same neighborhood an apartment building of colonial style could be occupied by more than a dozen families who moved in after its landlord fled Shanghai on the eve of Communist takeover in 1949.

Owing to the historical legacy of Maoist urban planning scheme aimed at both limiting population and controlling residence (Ma and Hanten 1981; Whyte and Parish 1984), inner-city neighborhoods in Shanghai have retained their "upper quarter" status despite the profound changes of Shanghai's city landscape brought forth by the construction boom during the 1990s. It is important to note that for most of the ordinary residents living in the former International Settlement and the French Concession, their sense of superiority is derived from the very location (*jiao*) of their home and not necessarily their actual housing conditions. The old Chinese city, officially referred to as Nanshi ("the Southern District") which occupies the middle ground between the "upper quarters" and the "lower quarters," is a good case in point. Now part of Huangpu District (previously the International Settlement) as a result of recent redistricting, Nanshi has been turned into a site of tourist attraction, exhibiting a reinvented local culture with an origin that could be traced to more than seven hundred years ago. While feeling proud of the Temple of City Gods, a garden and tea-house, and the Confucian Temple restored to its original site, the local residents are rather embarrassed by their living environment. Until very recently, Nanshi District had been the most densely populated residential area in the city. Age-old wooden houses and alleyway houses have been the major forms of housing. Despite the inconvenience of having to put up with communal kitchens and public toilets, many would choose to remain in the District for both nostalgic and practical reasons. On the one hand, their sense of belonging originates from being the residents of a unique place rich in history. Living within walking distance of the "upper quarter," on the other hand, makes them feel protected from the bad influences from the "lower quarter" only minuets away to the west of Nanshi.

In local terms, the "lower quarter" is essentially a synonym for shantytown housing or simply shacks (*penghu*) and has always been associated with stereotypical images of narrow lanes inhabited by the Subei people, the descendants of migrants and refugees from northern Jiangsu speaking a dialect distinctively different from the Shanghainese. The derogatory term "Subei" might be just a term that was conceived by those residents elsewhere in the city and not necessarily an objective description of their place of origin, as Honig (1992: 28–35) has rightly argued. Yet the "lower quarter," the very source of prejudice against the Subei people in Shanghai, has remained a material reality and a mental category for decades. Like the popular misconceptions about shacks and squatter settlements, Shanghai's *penghu* (lower quarter) is perceived as the armpit of the city, and stereotyped as where one could expect to see vicious circle of urban poverty, illegal housing, family breakdown and social disorganization. If the "upper quarter" stands for modernity and civilization,[1] then the "lower quarter" is no more than a symbol of backwardness and underdevelopment.

As a set of metropolitan knowledge and individuated experiences, the dichotomy between "lower" and "upper" quarters has been a key reference point for the city administrators as they proceeded to identify the social and economic characteristics of a particular neighborhood and mark out the boundaries of residential enclaves. After 1949 the area within Shanghai Municipality was reorganized into administrative districts (*qu*), each with several sub-districts referred to as "streets and avenues" or *jiedao* in Chinese. Each sub-district formed a constituency governed by a Street Office headed by a chief appointed by the district government. Because of Communist city-planners' desire to eliminate if not minimize the inequalities in income and housing conditions among districts, the goal of redistricting was to combine administrative areas that fell into the pre-1949 categories of the "lower quarters" *and* "upper quarters." The boundaries that separated poor districts from the rich ones disappeared on the city map of new Shanghai after 1949. Yet, within each newly configured district, the boundaries that used to separate the "lower quarters" from "upper quarters" in the pre-1949 past continue to exist.

While the demarcation lines between the "quarters" such as walls, fences and paths were gradually becoming less visible, the establishment of sub-district street offices served to reify the difference between the social and economic status of those inhabiting neighborhoods that represented two totally different social worlds. In everyday bureaucratic practices, as Lu Hanchao notes, the leaders of the Street Office usually acknowledged the existing differences by establishing residents' committees based on the types of neighborhood, the living conditions, and even the native-place origins of the inhabitants (Lu 1999: 316). The networks of alleyways and lanes that each street office formed as well as the allocation of space to particular uses and sizes of buildings therefore became an overt expression of the total gamut of behavior characteristic of a certain residential quarter within its jurisdiction. As if to rid the city of its colonial past and to reflect the changes brought by the founding of New China, the English and French names of the streets within the "upper quarters" were changed into Chinese ones. Ford Lane became Fujian Road and Route Lafayette, Fuxing Road (meaning "the street of revitalization"). The street names within the "lower quarters" for the most part remained unchanged with the same connotations for their place origins and status.

Therefore one should not be much surprised if "revolutionary leaderships, consciously or unconsciously, come to play the lord of manor," as Benedict Anderson (1991: 160) has cautioned us. Anderson makes it clear that it was the leadership not the ordinary people who came to "inherit the old switchboards and palaces" (*ibid*: 161). As Chairman Mao Zedong and his marshals took residence in Beijing in the imperial mansions in the vicinity of the "Forbidden City" of China's last Dynasty, his comrades in Shanghai quietly moved into the colonial style villas and gardens located in the section of the former French Concession that had always been off-limits to the ordinary people.[2] The marble mansion that was the headquarters of the Hong Kong and Shanghai Banking Corporation in the former International Settlement remained one of the most imposing office buildings on the Bund and served the administrative needs of the Municipal Government until 1999. Such privileges the Communist victors enjoyed made a mockery of the socialist ideal of egalitarianism and illustrated what Bahro refers to as "social stratification under actually existing socialism" (Bahro 1978: 163–182; see also J. Watson 1984: 1–15; R. Watson 1994).

The socialist citizenship was not only defined by class affiliation but also reflected in the official spatializing strategies in neighborhood Shanghai. Redrawing administrative boundaries between the urban districts and renaming streets and lanes after the Communist takeover in 1949 never blurred the "upper–lower quarter" dichotomy which was not only reflected on the social and political maps but also materialized on the ground level. Luwan District, the primary field site for this study, is one such example in which one could see reclaimed gentility in its "upper quarter" and the debris of socialism down in its "lower quarter."

After 1950 Luwan had within its jurisdiction four subdistricts under the control of four street offices. Three of the four subdistricts were located in the former French Concession (1842–1945). Within these subdistricts Avenue Joffre was renamed Huaihai Road commemorating a military campaign that led to the revolutionary victory in the Civil War (1947–1949). On the one hand, the "upper quarter" remained the upscale section of the residential areas of Luwan where the District leaders lived and worked. On the other hand, Bay Bridge, the fourth subdistrict added to Luwan after 1949, like their counterparts in Shanghai, had retained its "lower quarter" status inhabited by the descendants of migrants, refugees and farmers who were not able to speak the Shanghai dialect properly. Moreover, as I describe in the next section, Bay Bridge seemed to share all the defining characteristics of a "lower quarter" in terms of the geographical location, native-place origin, and the social composition of the local population.

## Bay Bridge: the social life of a "lower quarter"

Having grown up in a mixed neighborhood in Huangpu District (a borderland between the pre-1945 International Settlement and French Concession) that was separated from the northern section of Luwan District by one street, I myself had never heard of Bay Bridge as a teenager even though I did go to a dental hospital on its east end (belonging to another District). When I appeared rather embarrassed at such an oversight if not ignorance of the actual existence of real communities, my cadre friends and residents assured me that there was no reason why one should know about Bay Bridge in the first place because it was a "lower quarter after all."

The "lower-quarter" status of Bay Bridge was reified, as I gradually learned over the years, because of the additional sources of stigma attached to the entire locality. First, a well-developed funeral service prior to the 1949 Communist takeover was believed to have disrupted the system of geomancy that regulated the forces of "wind" and "water" (*fengshui*). The locality was further "polluted" in the aftermath of the notorious Japanese bombing of Shanghai in 1937 that had effectively turned Bay Bridge into one of the city's biggest graveyards where thousands of dead bodies were discarded without proper burial. During the Civil War (1947–1949), the "ghost land" of Bay Bridge became a haven for both the bandits evolved from the defeated Nationalist Army (KMT) and the landlords who had fled their home villages in the aftermath of the Communist Land Reform (1945–1950). Even during the present construction boom, the inauspicious indications of an unspoken *and* unspeakable past often disturbed those living in the present as human bones and skeletons were unearthed in virtually all the sites upon which high-rise office and apartment buildings would be built.

The pre-1949 Bay Bridge had more than enough attributes to qualify as the lowest of the city's "lower quarters" – graves and garbage, dirty ponds and stagnant creeks, squatter settlements, beggars and tramps, mosquitoes and flies. The historical connection with death, funerals, filth as well as the starving beggars became a major source of stigma that reinforced the marginalization of the area even after 1949. The lack of a decent school confirmed its social and economic status. As "the people without history," to paraphrase Eric Wolf (1982), the people of Bay Bridge could never expect to be treated equally by the legitimate Shanghainese in the "upper quarters" even though they were legally registered as the permanent residents of Shanghai under the *hukou* system. Small wonder that the local cadres dispatched from the district government (which was, ironically, located in the "upper quarter," the former French Concession) referred to Bay Bridge as "Luwan's Siberia," a "heart-broken island." In the eyes of these civilizing agents, Bay Bridge was a special reserve for the disenfranchised people who had no culture, no traditions, no history (Wolf 1982).

What set Bay Bridge apart from other comparable underclass neighborhoods in the city's periphery was, however, its proximity to the northern section of Luwan District, which attained its "upper quarter" status by virtue of being the former French Concession. To the best knowledge of the urban planners I interviewed, Bay Bridge was among the very "lower quarters" in the city that was in the vicinity of historical landmarks, the architectural representations of "monumental time" (Herzfeld 1991) of modern Shanghai. To its north was the restored holy site of the "Birthplace of the Chinese Communist Party (CCP)" where the First Congress of the CCP was held in July 1921. To its south was Jiangnan Shipyard (which began as an arsenal during the later Imperial era), which was called the "Cradle of China's Proletarian Class" in the official historiography. To its east, the "City of Temple Gods," the old walled city for the Chinese residents in the colonial days, was showcase exhibiting the Shanghai local cultures and customs. Located to its far west was a Catholic cathedral known for its pivotal role in promoting the Chinese understanding of the West three centuries ago.

In Luwan's northern section (the "upper quarter"), colonial-style villas, hotels and clubs were well preserved and restored, while the red-brick terrace buildings that used to house the pre-1949 middle class were carefully renovated. Although within walking distance of all these historical landmarks that monumentized the city's glorious past, Bay Bridge remained marginalized. More importantly, it was never a part of the pre-1945 French Concession. Xujiahui Road (now a four-lane expressway) had always served as the demarcation line separating the "upper quarter" of Luwan District from the unfashionable south, where Bay Bridge was located.

As I walked from the northern Luwan to Bay Bridge, I could detect a gradual change in housing patterns, from the fancy little European style villas sandwiched between the postmodern high-rises, to the more traditional terraced houses which blended into rows of matchbox-shaped apartment buildings. Within Bay Bridge, the matchbox-shaped buildings (five or six storey walk-ups) and the century-old and poorly maintained wooden houses were the most representative form of housing for the local residents. Occasionally, one could even find traces of squatter settlement in Bay Bridge's oldest residents' community called "the commoners' village" (a homonym with "the paupers' village," all pronounced as "*pinmin*" in the Shanghai dialect).

Until it became an officially designated "Model Community" in 1995, Bay Bridge had rarely been considered as a place worth writing about. As the biggest territorial subdivision of Luwan occupying an area of 3.07 $km^2$ inhabiting 81,634 registered residents, Bay Bridge is given hardly any attention in the District Gazette of 244 pages – merely two pages of introduction to its population and neighborhood organization structure. For the District officials who set their visions high, Bay Bridge could not possibly match the standard of the prosperous Luwan District that has been often associated with images of colonial gentility and post-socialist prosperity.

The conspicuous absence of an official account of everyday life in the city's underclass neighborhood can be explained by the social and political context of post-Deng Shanghai's relentless bid for the status of a global city in the Asia Pacific region. Yet how could a "lower quarter" like Bay Bridge represent itself in an increasingly global city? In the summer of 2000, I presented a street officer with a report on the undocumented history of the pre-1949 past of Bay Bridge with the hope that they could use some of my findings for an upcoming exhibition on the past and present of the neighborhood. To my surprise and dismay, the street officer and his colleagues acknowledged the time and effort I took in "writing so much about so small a community" but showed very little interest in what I wrote about. They implicitly criticized me that I was too obsessed with the past of Bay Bridge, the past had practically nothing to do with their present, let alone the future they envisioned for themselves.

My cadre friends' lack of interest in my work could be conveniently explained by the fact that the newly acquired knowledge of Bay Bridge's past had allowed me to intrude not just into the private lives within the communities, but also the privacy of their collective space that Michael Herzfeld referred to as "cultural intimacy" (Herzfeld 1997). However, after several follow-up visits to Bay Bridge, I realized that aside from the potential damage that such intimate knowledge of a previously polluted locality could do to the image of Bay Bridge, there is serious economic interest at stake. The difference in the sales value of a housing unit located in the lower quarter and that of a unit in the upper quarter literally on the right side of the track (or toward the north end of Bay Bridge) could be as significant as 1,000 *yuan* (approximately $120) per square meter. Unfortunately the local cadres could not stop the *fengshui* (geomancy) masters from Hong Kong, Macao and Taiwan from collecting and disseminating such "dangerous" information about the fact that Bay Bridge was once a highly polluted locality haunted by "hungry ghosts."

Like their counterparts in charge of "lower quarters" in neighborhood Shanghai, the cadres of the Bay Bridge would seize every possible opportunity to improve the social and economic status of their neighborhood and develop various strategies of competing for community resource, media coverage and attention from the municipal officials. The community construction movement, beginning in the early 1950s, offered them opportunities to change the images of a "lower quarter" by erasing its unspeakable past and monumentalizing its present for the sole purpose of creating a community as the model for a civilized and scientific way of living in a modernist city. Each generation of street officers shared the same idea that as a typical "lower quarter," Bay Bridge was like a blank sheet of paper on which they could paint beautiful pictures. Some of the pictures, as I show in the remainder of this chapter, became faded as time went by while others turned into mere showcases if not figments of their imagination.

## Upscaling Bay Bridge: community-building in a gentrifying neighborhood

Since the early 1990s, large-scale infrastructure reconstruction and various urban revitalization projects have reconfigured Chinese social space in both the coastal and inland regions. The frequently updated city maps of Beijing and Shanghai are the most obvious manifestations of the gentrification process that is transforming everyday life of ordinary neighborhoods. The current transition toward a post-socialist civil society has thus changed the nature of urban experiences and reshaped power relationships between different localities in the Chinese cities.

China's urban landscape has been redefined not merely by the constant construction of roads and buildings, but more importantly by massive demographic shifts in the past decade. With the acceleration of post-Mao reform in the late 1980s, the Party-state has been taken over by a new generation of technocrats who readily abandoned conventional strategies of urban development that limited the size and population of the cities. Urban sprawl has given rise to unprecedented migration that is beyond the capacity of the local authorities to control and monitor the flow of the so-called "floating population" (Solinger 1999; Zhang 2001). Meanwhile the restructuring of the poorly managed or bankrupt state-owned enterprises (SOEs) has created millions of unemployed industrial workers through the layoff (*xiagang*) system.

As casualties of the structural reform, millions of unemployed industrial workers in China find themselves in a painful process of adjusting to a new way of life centering on their community rather than their workplace. Unemployment has drastically enlarged the marginal section of the Chinese society officially labeled as the "hardship population" which traditionally included the elderly, disabled and sick in the neighborhood. Despite the fact that Shanghai has arguably the most sophisticated social welfare scheme in China, the fight against urban poverty has been an uphill battle. City officials have identified unemployment as one of the greatest sources of social instability. This is largely because of Shanghai's historical position as the nation's industrial base, the locus of hundreds of state-owned textile factories, steel plants, shipyards and mechanical works. The local cadres have to devote considerable amount of time and energy to accommodating the needs of the unemployed and underemployed inhabiting the "lower quarters" because both parties are well aware of the tradition of rebellion in the birthplace of the proletarian class and the Chinese Communist Party (Perry 1993).

The gradual emergence of transnational and private firms and the death of state enterprises in late socialist Shanghai have made the city a place where different modes of living are displayed under and beyond the neon lights reminiscent of the city's glorious colonial past. Since the early 1990s the global flow of knowledge and ideas have had a profound impact on local bureaucratic practices in both the upper and lower quarters. Along with the introduction of Western concepts of efficient business management, city administrators have replaced outmoded socialist jargon with such terms as "social work," "sustainable development" and "community service." Such terms are used to authorize the gentrification and revitalization processes.

As an observable housing pattern in the postindustrial world, the term "gentrification" has often been interpreted as middle-class settlement or resettlement in older inner-city neighborhoods formerly occupied mostly by working-class and underclass residents. In this chapter, "gentrification" refers to a social and political process of urban change that must be understood in the historical context of socialist engineering and community development during the reform era (1978–present). As far as space reconfiguration in neighborhood Shanghai is concerned, we are able to discern at least two forms of gentrification in the "upper quarters" and "lower quarters." Within the former French Concession and the International Settlement, the ongoing gentrification process can be explained in its conventional terms except that inner-city neighborhoods in a socialist city were never a "downtown" despite the signs of urban decay in certain parts. Many such neighborhoods have retained the image and self-image of being in the "upper quarter" (uptown) since the 1949 revolution. Generally speaking gentrification efforts in the former colonial concessions are manifested by the restoration and renovation of old-style villas and mansions along with the construction of skyscrapers and modern apartment buildings.

The gentrification process in Shanghai's "lower quarter" that I describe in this chapter was initiated a decade ago after Deng Xiaoping's unexpected post-Tiananmen "tour of the south (*nanxun*)." As Shanghai engages in a relentless bid to regain its pre-1945 status as the financial center of Asia by renovating the colonial-style bank buildings and constructing high-rises for foreign companies, factories located in the city's peripheral areas where most of the "lower quarters" are located are approaching the end of their productive life. On the one hand, gated communities managed by local or transnational real-estate developers have made their appearance on the land acquired from the owners of bankrupt or poorly managed SOEs. In a sense the built forms of high-rise apartments in a typical "lower quarter" such as Bay Bridge were literally the sprouts of capitalism grown out of the debris of socialism. On the other hand, the advent of gated communities did not immediately result in the forceful relocation of the local residents. In fact, the process of gentrification was relatively slow especially in comparison with what went on in the "upper quarter." The "upper quarter" (the pre-1945 French Concession), which used to be part of a famous shopping street named after Joffre, a French General, is reliving its glorious past as overseas investors assume an active role in transforming its cultural and political landscape. More and more shopping malls (one of which was Sincere whose owner fled to Hong Kong on the eve of the 1949 liberation) have been put up. The historical building of the colonial French Police Station has been renovated and become home to cafes, restaurants and stores selling brand-name merchandise.

As recipients of a culture of conspicuous consumption, an integral part of "Shanghai nostalgia," the new generation of "yuppies" in Luwan's "upper quarter" are usually employees of joint ventures and foreign companies working in the office high-rises concentrated in the district's Pacific-Hong Kong Plaza, which is essentially a recast of its original in Hong Kong. As college postgraduates born in the 1970s, the yuppies are well educated, with good computer and English language skills. The beneficiaries of the economic reform that favors the young, the beautiful and the affluent, many of the yuppies have become local representatives of transnational corporations and earn an annual income equivalent to what their parents would have made as factory workers in at least ten years. The "upper quarter" of Luwan District allegedly has the highest concentration of "head hunting" companies serving the needs of "job hoppers" working in the adjacent buildings.[3]

The images of affluent "yuppies" that dominated the covers of fashion magazines unabashedly showcasing economic prosperity in the "upper quarter," stood in stark contrast to what went on in the "lower quarter" – Bay Bridge, which witnessed the emergence of a new generation of *xiagangers*. As more and more state-owned factories (mostly in the textile and steel industries) declared bankruptcy, hundreds of thousands of laid-off workers became members of the "urban poor" eligible for welfare assistance. A sad truth for the District bureaucrats to confront was that the increase of *xiagangers* has clearly outpaced the growth of yuppies in Luwan and elsewhere in Shanghai. While a yuppie is enjoying a cup of cappucino in Park'97 (a cafe modeled after its Hong Kong prototype) or recently opened Starbucks and planning to buy a second or third apartment in the old French Concessions, the *xiagangers* in Bay Bridge are worrying about how to pay their monthly bills. Widely publicized annual events such as the "Rose Wedding"[4] in October (organized by the District Tourist Bureau) starring one hundred yuppie couples in shiny limos parading in the "upper quarter" only heighten the sense of loss, betrayal and despair on the part of the gradually marginalized *xiagangers* in the "lower quarter."

In Bay Bridge, many recently unemployed workers are psychologically unprepared for the sudden loss of the "iron rice bowl" of secure lifelong employment. They are afraid of re-entering the labor market because they thought that they had long lost the ability to compete. Whenever they become nostalgic for the "good-old days" when the chimneys of their factories were still puffing out smoke, they feel they are being short-changed. They are overwhelmed with a sense of disorientation, betrayal, despair and misplacement because they could no longer wear the badge of honor as members of the vanguard class. For the ordinary residents of Bay Bridge, "Shanghai nostalgia," experienced by those living in the "upper quarter," represents nothing more than a lifestyle and mode of consumption they could not afford to partake.

The disappearance of state factories and plants in the city's depressed industrial areas coincided with the neighborhood gentrification of "lower quarters" such as Bay Bridge. By the 1990s, the state factories in the neighborhood were either shut down or on the brink of declaring bankruptcy. A gated community in Bay Bridge is often located in the midst of several "workers' new villages (*gongren xincun*)" made up of dozens of six-storey walk-ups in the typical form of government housing projects built during the 1970s. For the current generation of technocrats of the District Government and subdistrict street offices, the gated community, whose residents are total strangers in the neighborhood, represent the ideal form of a "model community" – a perfect showcase for cultural citizenship. The most active gentrifying agents at the local level were the young and well-educated street officers who formed an alliance with property and real-estate developers, local entrepreneurs, and urban planners with the same visions of high modernity and scientizing principles divorced from the social realities of a locality that had yet to live with its unspeakable past as a "lower quarter" of Luwan.

In the eyes of the young and forward-looking street officers, it would thus make perfect sense to showcase a garden-like "model community" (often known as "civilized community") supervised by an elected neighborhood council and or owner's association. In so doing they would help the local residents forget their unforgettable past, restore their confidence in the present, and envision the future of Bay Bridge as a transformed locality. However, balancing the needs of managing the debris of socialism and showcasing citizenship is no easy task. With mountainous unresolved problems such as unemployment and housing shortages, Bay Bridge does not seem to stand a chance in the city-wide model community contest and beautification campaigns. As a result, the street officials have to adopt a selective strategy by showcasing one or two newly established gated communities such as the "Volkswagen Town," home to the staff members of the German automobile company in Shanghai, to attract media attention and official recognition.

The new technocrats who have taken control of the entire neighborhood were pleased with the presence of "Volkswagen Town" and "Redbud Pavilion" because these gated residential communities were essential to a rapid process of gentrification that would "improve" the neighborhood by attracting more prestigious kinds of cultural citizenship. By choosing only a couple of new housing communities to represent a significantly improved "lower quarter," however, the technocrats have also reminded the remaining majority of those living in Bay Bridge of "what we used to be" *not* "what we are supposed to be." Seemingly the "urban villages" in Luwan's "lower quarter" are now airbrushed from the social and political map. Underneath the false facade of a harmonious, hygienic and crime-free environment, Bay Bridge continues to be treated as an atypical "lower quarter," or just another "dirt" (Douglas 1966) – a matter out of place in the everyday discourse of a rapidly stratified urban society.

## Conclusion

Using the intimate perspectives derived from ethnographic fieldwork, I have in this chapter probed the local grounding of "Shanghai nostalgia" in terms of the "lower–upper quarter" dichotomy, the age-old stereotype reminiscent of the colonial era and the apparent contradictions in community-building and place-making processes. Instead of focusing on deciphering the multiple layers of meanings in the "Shanghai nostalgia" literature, I have chosen to situate this study in the broad social and political contexts within which the historical memory of various imagined communities is reconstructed and new social spaces are (re)produced by an emerging class of social and economic elites. Making explicit links between local history, structural power and concrete places, I have thus explored the ways in which "Shanghai nostalgia" was reified and spatially manifested in dichotomized "upper and lower quarters" as a consequence of the conscious efforts by local residents and municipal officials.

In the futuristic visions of Shanghai's ambitious leaders, "Shanghai nostalgia" could be employed as a political strategy to accomplish the long-term goal of turning the increasingly global city into a leading commercial hub of East Asia. Community-building, while being translated into various schemes of revitalization and reconstruction, thus served the official interests of beautifying both the "upper quarters" and the "lower quarters" so that the Shanghainese could reclaim the prestigious kinds of cultural citizenship associated global standards of affluence and wealth that have always been associated with the popular imageries of "colonial Shanghai."

In everyday practice, revitalization schemes as a series of civilizing projects aimed to upscale the city's lower quarters have become the top priority of neighborhood organizations. Participation in model community contests in particular has captured the attention of the Party-state, the mass media and academics across the nation. There was great tension between the modernist visions of the young technocrats who began to dominate the leadership of the Street Office and the harsh conditions of lower quarter neighborhoods such as Bay Bridge. The publicity of such events as showcasing citizenship and model community contest has never altered the structure of inequality exemplified by the spatially dichotomized "upper quarters" and "lower quarters" and the strong sense of place attachment shared by generations of residents of China's most cosmopolitan city.

## Notes

1 Wasserstrom (2002) has questioned the prevalent views of prior writers that Shanghai's colonial settlement was a thoroughly modern place in which one would expect to see strong evidence of the development of modern citizenship. While acknowledging his historical insights and sharing his doubts about the uniqueness of the Foreign Concessions in Shanghai, in this chapter I place my emphasis on the impacts of the differences, real and perceived, between the two quarters and their symbolic meanings on community-building and the strategies of utilizing such a dichotomy between the high and low ends in place-making.

2 In the summer of 2003, I accompanied my former landlord, a Shanghai enthusiast who was fascinated with the colonial style houses for a sightseeing tour in the heart of the former French Concession. One morning as we began to take pictures of the quiet and well-preserved neighborhood of Kangping Road, we were immediately cornered by several policemen stationed at the crossroads and then taken to the local police station for interrogation. We should have realized that as the home to generations of top government officials (including the retired Chinese President, Jiang Zemin), the Kangping (meaning "healthy and peaceful") community became such a symbol of power that no visitors would dare even to appear curious as they passed through the neighborhood.

3 Most of the "head-hunting" companies had connections with large corporations in the US. For ethical reasons I cannot reveal their names because these "head-hunting" companies operated in a gray area and circumvented the control of the official Shanghai Foreign Enterprise Services Company (SFESC).

4 In 1999 the event took place in Hong Kong, Shanghai's twin city, because the young elite couples were not satisfied with the recreated romantic surroundings in the former French Concession.

## References

Agnew, J. A. (1987) *Place and Politics: The Geographical Mediation of State and Society*, Boston: Allen & Unwin.

Anderson, B. (1991) *Imagined Communities: Reflections on the Origin and Spread of Nationalism*, revised edition, London: Verso, 160–161.

Appadurai, A. (1988) "Putting hierarchy in its place," *Cultural Anthropology*, 3: 36–49.

Bahro, R. (1978) *The Alternatives in Eastern Europe*, London: Verso.

Douglas, M. (1966) *Purity and Danger: An Analysis of the Concepts of Pollution and Taboo*, London: Routledge and Kegan Paul.

Gupta, A. and Ferguson, J. (eds) (1997) *Culture, Power, Place: Explorations in Critical Anthropology*, Durham and London: Duke University Press.

Herzfeld, M. (1991) *A Place in History: Social and Monumental Time in a Cretan Town*, Princeton, NJ: Princeton University Press.

—— (1997) *Cultural Intimacy: Social Poetics in the Nation-State*, New York and London: Routledge.

Honig, E. (1992) *Creating Chinese Ethnicity: Subei People in Shanghai, 1850–1980*, New Haven: Yale University.

Lee, L. O. F. (1999) *Shanghai Modern: the Flowering of a New Urban Culture in China, 1930–1945*, Cambridge: Harvard University Press.

Lu, H. C. (1999) *Beyond the Neon Lights: Everyday Shanghai in the Early Twentieth Century*, Berkeley, CA: University of California Press.

Ma, L. J. C. and Hanten, E. W. (eds) (1981) *Urban Development in Modern China*, Boulder, CO: Westview Press.

Perry, E. J. (1993) *Shanghai on Strike: The Politics of Chinese Labor*, Stanford: Stanford University Press.

—— and Li, X. (1997) *Proletarian Power: Shanghai in the Cultural Revolution*, Boulder, CO: Westview Press.

Solinger, D. (1999) *Contesting Citizenship in Urban China: Peasant Migrants, the State, and the Logic of the Market*, Berkeley, CA: University of California Press.

Wasserstrom, J. N. (2002) "Questioning the modernity of the model settlement: Citizenship and exclusion in Old Shanghai," in Goldman, M. and Perry, E. J. (eds), *Changing Meanings of Citizenship in Modern China*, Cambridge and London: Harvard University Press, 110–132.

Watson, J. L. (ed.) (1984) *Class and Social Stratification in Post-Revolution China*, London and New York: Cambridge University Press.

Watson, R. S. (ed.) (1994) *Memory, History, and Opposition under State Socialism*, Santa Fe, NM: School of American Research Press.

Whyte, M. K. and Parish, W. L. (1984) *Urban Life in Contemporary China*, Chicago and London: University of Chicago Press.

Wolf, E. (1982) *Europe and the People without History*, Berkeley, CA: University of California Press.

Zhang, L. (2001) *Strangers in the City: Reconfiguration of Space, Power, and Social Networks within China's Floating Population*, Stanford: Stanford University Press.

Zhang, X. D. (2000) "Shanghai nostalgia: postrevolutionary allegories in Wang Anyi's literary production in the 1990s," *Positions: East Asia Cultures Critique*, 8(2): 348–387.

# 8 Uneven development among Shanghai's three urban districts

*Tingwei Zhang*

## Introduction

In explaining the differences in urban development patterns among cities and communities, many factors have been examined. These factors include public policy debates, the behavior of the market, and community features, especially the social (such as race) and economic (such as class) background of a community. It is commonly accepted that development differences are largely attributable to the interaction of these factors. However, since American urban communities differ so significantly in race and class compositions, it is hard to measure how and to what extent public policy integrated with other factors contributes to the differences. The impact of a community's race and class background may be mixed with that of the policy factor in analyzing development differences. A case in which race and class differences among communities are limited will enable us to focus on the policy issue, and to reveal the dynamic mechanism of how policy works with other factors in a given context.

Shanghai, China's largest city and economic center, provides such a case. Shanghai's population is predominantly Han nationality, and less than 0.1 percent of the city's 1.3 million residents have other ethnic backgrounds. Also, under the control of a socialist government for fifty years, class difference, measured by official social status and especially by their residence, an important indicator of social status in a socialist society, was reduced. However, class difference is reappearing in the post-reform era, as I will discuss below. Compared with American urban communities, Shanghai consists of relatively "homogeneous" communities in terms of race and class composition. On the other hand, even visitors can easily identify differences in urban development among the city's ten administrative districts: more high-rise buildings and new housing in certain districts but fewer in others. The differences and uneven development thus may be attributed to factors other than race and class of the residents.

In exploring the factors contributing to development differences among urban districts in Shanghai, this chapter will focus on how and to what extent public policy, together with other factors such as community features, affects development outcomes. Shanghai is used for the study not only because development differences among the city's districts are apparent, but also because the city's economic and physical restructuring has been so dramatic and impressive that it has been called "the Shanghai miracle." So the case is significant.

I start with a profile of the city and its urban districts, followed by a comparison of development differences among three urban districts: Huangpu, Luwan and Yangpu. Huangpu is Shanghai's traditional CBD. Luwan is the city's main retail and residential area that has been planned to be a "first-class" commercial district since the reforms. And Yangpu is a traditional industrial zone with many factories. The three districts had similar status in the municipal government before the reforms, but their development patterns have shown significant differences since the 1990s. In the conclusion, the main findings of the chapter will be discussed. I argue that the pro-growth policy, which is considered the main contributor to "the Shanghai miracle," has its local costs. One cost is the increasing gap between not only the "advantaged" and "disadvantaged" residents but also the "advantaged" and "disadvantaged" districts. The real challenge is thus not whether to adopt a pro-growth policy, but how to evenly distribute the costs and benefits brought about by the policy and to achieve a balance between social justice and economic growth.

Interviews with six directors of the three district planning bureaus in the summer of 2001 provide data used in the study. Secondary data come from district statistical yearbooks, district history books, and district fact books from 1985 to 2003. In the last two decades, Shanghai's administrative system has been changed several times. The administrative setup, the land and population classification systems (urban versus rural), and district boundaries all have been adjusted. These adjustments make it hard to conduct a consistent analysis on the changes of economic and physical developments at the district level, especially in comparing the changes of district shares of some indicators in the city's total in the last decade. To provide certain consistency, data from China's 2000 census is used.

## A profile of Shanghai and its urban districts

As China's largest city and economic center, Shanghai has a population of 13.2 million (11.4 million in the urban area) and a land area of 6,341 square km (3,924 in the 16 urban districts) as of 2000. The city became China's economic center in the 1900s after Shanghai had been opened as a "trading port" under the Treaty of Nanjing of 1842. Located at the mouth of the Yangzi River in the richest area of China, Shanghai attracted both foreign entrepreneurs and domestic investors. Manufacturing, retailing, financing and trading formed the base of the city's economy. After the revolution in 1949, the city's financial function was removed with the nationalization of the economy. But Shanghai remained China's leading manufacturing center, especially for light industry.

From 1950 to 1976, Shanghai, with only 1 percent of the nation's population, contributed one-sixth of the nation's revenue. The city also functioned as China's gateway to the outside world especially via its harbor. While Shanghai enjoyed the honor of being "China's economic generator" and the major contributor to the central government's revenue, there were high costs to the city including the old and overused infrastructure system. Housing shortage, traffic congestion and poor environmental quality were listed by Shanghai residents as the three top challenges facing the city by the end of 1980s (Kong 2001).

Shanghai's economy has grown significantly in the reform years. The GDP in 2000 (455.1 billion yuan) was 750 percent of that of 1978 (27.3 billion yuan), an annual growth rate of over 10 percent for the period. The growth is partly a result of the change of the city's economic base. Taking advantage of the development opportunities presented by the city's historic and geographic advantages, Shanghai's economy has shifted from a manufacturing center to a service-oriented one. In 1978, the manufacturing industry contributed to 77.4 percent of the city's GDP, but the share dropped to 47.6 percent in 2000, a loss of 38.5 percent. On the other hand, the service industry increased from 18.6 percent of the economy in 1978 to 50.6 percent in 2000, an increase of 172 percent in twenty-two years. Moreover, the composition of the tertiary sector has changed, as shown in Table 8.1. In 1978, almost half (45.6 percent) of the service industry was from retail, wholesale and catering. In 2000, the largest share in the tertiary sector was in financial services (29.7 percent), and the second one was "others" including consulting, tourist, educational and culture-related businesses (24.6 percent). The real-estate industry alone accounted for 10.9 percent in 2000.

Shanghai is not only leading the nation in financial services and trading, it is also becoming a key regional financial and trading center in East Asia. The city's development goal is to become "an international port and trading and financial center of the twenty-first century" (Kong 2001). This goal was initiated by the municipal government in the mid-1990s and supported by the national government. To reach the goal, policies at both the national and local levels have been formulated and implemented. The city has shifted its development direction, giving priority to the service sector while reducing

*Table 8.1* Composition of the tertiary industry in Shanghai: 1978–2000 (%)

| *Sector* | *1978* | *2000* |
|---|---|---|
| Transportation, post and telecommunication | 23.7 | 13.7 |
| Wholesale, retail and catering | 45.6 | 21.1 |
| Banking and insurance | 13.8 | 29.7 |
| Real estate | 0.5 | 10.9 |
| Others | 16.4 | 24.6 |
| Total | 100.0 | 100.0 |

Source: SMSB 2001.

the importance of the industrial sector. This policy change has had major impacts on the development of Shanghai's urban districts, as will be discussed below. (Since the change of China's national leadership in the spring of 2003, Shanghai's development policy has shown a slight "shift back" to emphasize the manufacturing sector. This change may provide an interesting case to be discussed in another study.)

According to the 2000 administration system, Shanghai had sixteen urban districts and three rural counties. (In 2001, all counties under Shanghai's administrative purview were converted to "districts" except for Chongming County.) The average population size of the districts is about 500,000 persons, but some districts are as large as a big city in the US. Yangpu district, for instance, had a population of 1.1 million in 2000. Nine urban districts, including the three studied here, are "central" or "old" urban districts located within the traditional urbanized core. One district, Huangpu, was established after the merge of "the old" Huangpu District with the then Nanshi District in 2000. These central urban districts form the "city proper" and they share some common characteristics such as older infrastructure and housing stock built before the 1990s.

## Differences in district developments after the reforms

This study concentrates on three central urban districts: Huangpu, Luwan and Yangpu. Table 8.2 summarizes their basic conditions. While the three

*Table 8.2* Profile of the three urban districts

| *District* | *Land area (sq. km)* | | *Population* | | *Administrative units**** | |
|---|---|---|---|---|---|---|
| | *1985** | *2000*** | *1985** | *2000*** | *1985* | *2000* |
| Huangpu | 18.7 (5.3) | 12.4 (4.3) | 683,700 (9.8) | 661,800 (10.5) | NA | 9 |
| Luwan | 7.5 (2.2) | 8.05 (2.8) | 480,000 (6.9) | 355,900 (5.7) | NA | 4 |
| Yangpu | 53.0 (15.1) | 60.7 (21.0) | 975,000 (14.0) | 1,070,000 (17.2) | NA | 10 |

Source: SMSB 1986, 2001.

Notes

* Numbers in parentheses indicate the shares of a district in the city's total urban districts in 1985. There were 12 urban districts in 1985, which was increased to 16 in 2000. Most rural areas have been developed and added to the urban area since 1992, and the total urban area has increased to 3,924 square km with a population of 1,136,820 in 2000. "Urban area" of 2000 refers to the "older" central urban area consisting of 9 of the 16 urban districts.

** The 2000 percentage is the share of a district in the city's central urban area.

*** Number of urban sub-district offices. A "Street Office," as called by the Chinese, is the lowest-level government body in China's urban administrative system.

"brothers" share some similarities, they also differ considerably in history, geographic location and economic base, as shown in Table 8.3.

Huangpu is Shanghai's traditional CBD and its administration center because the municipal government is located in the district. The district was the core of the British Concession before the Second World War. In the riverfront area known as "The Bund," there were headquarters of banks and international corporations before the 1949 revolution. The district's main commercial street, Nanjing Road, was and still is the number one shopping street in China. The economy of the district consists of trading and retailing businesses, even after the 1949 revolution. As a result, the district took a large share of the city's office and commercial buildings. For instance, out of a total of 10.65 million square meters of building areas in 1985, 4.84 million (45.4 percent) were for "non-residential uses," including offices and commercial and retail uses.

Located at the central-south of the city, Luwan used to be the French Concession until the 1930s. Fashion shops featured the district's main shopping street, Huaihai Road, that attracted many upper-class customers. Today, the retail business is still the foundation of the district's economy. In 1998, the retail and catering generated 57.5 percent of the district's tax revenue (LSYB 1999). Luwan was also famous for its beautiful villas and luxurious apartment buildings. In 1985, housing stock of the villas (120,000 sq. m.) and apartment buildings (190,000 sq. m.) in Luwan exceeded the total of the two other districts.

Partly due to its location near the Huangpu River, the main waterway of the city, Yangpu became the city's main industrial area in the 1920s. Most industries were established and owned by domestic businessmen and nationalized during the early socialist period in the 1950s. A large amount of public funding was invested in these state-owned industries, especially in heavy industry whose share in the district was 93.9 percent of the total industrial investment from 1949 to 1959. In the 1950s and the 1960s the district contributed one-fifth of the city's total revenue, which peaked in 1964 with a share of 29.3 percent of the city's total revenue (ESYY 1995). Even in the early years of the reforms, the district still had 14 percent of the city's total number of factories that contributed to 26 percent of the city's total industrial output in 1985 (SMSB 1986). The district's economic status was challenged after the city undertook service industry-oriented restructuring in the 1990s.

With Shanghai pursuing a new role as an international trading, port and financial center rather than a manufacturing center, the city has experienced rapid economic growth, especially in the tertiary sector, and apparent differences among districts in development patterns have emerged, as reflected in the their total building areas (Table 8.4).

*Table 8.3* Comparison of district features

| *District* | *Huangpu* | *Luwan* | *Yangpu* |
|---|---|---|---|
| Location | In the CBD | Adjacent to the CBD | Fringe area of the CBD |
| History | Traditional office and retail center, founded by the British authority under the concession treaty in 1845. Today, most residents work in businesses and government entities. Retaining its importance with the shift of the city's economy from manufacturing to services. | Retail/residential center for the "high class" after the district was founded by the French in the 1890s. The traditional economic function has been reborn after the reforms. Many residents have direct or indirect experiences with the marketplace due to history and local culture. Regaining importance with the shift of the city's economy from manufacturing to services. | It was the largest and main industry district in the city for over eighty years before the development of Pudong New District. The port area, founded by the British in the 1900s, played a key role in the city's industrial development. Losing importance with the shift of the city's economy from manufacturing to services. |
| Economy (measured by land use) | A concentration of headquarters of various businesses, and the retail district was recognized as the "number one shopping street in China." | Various retail establishments, especially famous for its fashion shops. A traditional resident district of good quality. The smallest district in the city in terms of land area. | A traditional industry district, a concentration of textile, steel and machinery factories. The manufacturing sector was especially strengthened after the 1949 revolution, but has weakened after the reforms. |
| Leadership | Many leaders are from district government bodies, well connected with the municipal government which is located in the district. Leaders are familiar with business issues as well as the city's bureaucratic system. | Many leaders come from local companies and district government agencies. They are entrepreneurs especially sensitive to development opportunities. | Many leaders have worked in the manufacturing sector. They are familiar with manufacturing rather than the service sector. They tend to have a lower education level as some rose from the rank of workers. |

*Table 8.4* Building construction in Shanghai, 1985–2000 (million sq. m.)

| *District* | *Total building area* | | *Residential use* | | *Non-residential uses* | |
|---|---|---|---|---|---|---|
| | *1985* | *2000* | *1985* | *2000* | *1985* | *2000* |
| Huangpu | 10.65 (9.5) | 16.54 (8.1) | 5.81 (9.0) | 7.55 (6.2) | 4.84 (7.7) | 8.99 (10.9) |
| Luwan | 6.72 (5.3) | 10.62 (5.2) | 3.84 (6.0) | 5.88 (4.9) | 2.88 (4.6) | 4.74 (5.7) |
| Yangpu | 19.43 (15.0) | 36.12 (17.7) | 8.62 (17.1) | 21.35 (17.7) | 10.80 (17.1) | 14.78 (17.9) |
| City total | 127.47 (100.0) | 203.49 (100.0) | 64.44 (100.0) | 120.83 (100.0) | 63.04 (100.0) | 82.66 (100.0) |

Source: SMSB 1986, 2001.

Note
Figures in parentheses are in percentages.

## *Huangpu District*

Shanghai adjusted its district boundary several times. From 1985 to 2000, the land of Huangpu District was reduced by 33.7 percent and the population dropped by 3.3 percent, but the total building area increased by 55.3 percent which resulted in greater population and building density. In 2000, the district's population density is the highest in the three districts.

In Huangpu, areas lost to boundary adjustment were mainly lower-quality residential quarters and older factory zones. The remaining area has been developed more intensively into a "modern CBD" through adding high-quality non-residential buildings. Buildings for non-residential uses increased by 85.7 percent from 1985 to 2000. The amount of office space alone increased from 127 to 423 million square meters, or by 233.1 percent (Table 8.5). Shops increased by 141.1 percent in the same period. The district has become a conventional CBD similar to that of a developed nation with a heavy concentration of office and retail establishments.

The significant increase of new buildings (55.3 percent) and the decrease of slum areas (−73.3 percent) has contributed to the improvement of the overall quality of the district in the fifteen years (Table 8.6). However, the residential area of the district is still lower than that of the two other districts, due largely to the high housing price in the CBD. The average housing price in Huangpu was 5,800 yuan ($707) per square meter in 2000 and 8,000 yuan ($1,000) in 2003, which is the highest among the three districts.

## *Luwan District*

Luwan gained land but lost population (−25.9 percent) in boundary adjustment from 1985 to 2000. More importantly, the district's total building area

*Table 8.5* Non-residential building construction, 1985–2000 (10,000 sq. m.)

| *District* | *Plant* | | *Office* | | *Shop* | |
|---|---|---|---|---|---|---|
| | *1985* | *2000* | *1985* | *2000* | *1985* | *2000* |
| Huangpu | 106.0 (3.0) | 110.0 (3.5) | 127.0 (29.9) | 423.0 (25.5) | 56.0 (18.8) | 135.0 (18.6) |
| Luwan | 160.0 (4.6) | 116.0 (3.7) | 20.0 (4.7) | 154.0 (9.3) | 18.0 (6.0) | 71.0 (9.8) |
| Yangpu | 745.0 (21.4) | 935.0 (29.8) | 25.0 (5.9) | 62.0 (3.7) | 34.0 (11.4) | 62.0 (8.5) |
| City total | 3,485.0 (100.0) | 3,038.0 (100.0) | 425.0 (100.0) | 1,657.0 (100.0) | 298.0 (100.0) | 726.0 (100.0) |

Source: SMSB 1986, 2001.

Note
Figures in parentheses are in percentages.

*Table 8.6* Residential building construction, 1985–2000 (10,000 sq. m.)

| *District* | *Villa* | | *Apartment* | | *Staff housing* | | *Slum* | |
|---|---|---|---|---|---|---|---|---|
| | *1985* | *2000* | *1985* | *2000* | *1985* | *2000* | *1985* | *2000* |
| Huangpu | 2.0 (1.5) | 1.0 (0.6) | 3.0 (3.2) | 3.0 (1.5) | 194.0 (7.1) | 367.0 (3.8) | 15.0 (5.3) | 4.0 (10.8) |
| Luwan | 12.0 (8.8) | 16.0 (9.5) | 19.0 (20.0) | 30.0 (15.3) | 78.0 (2.9) | 342.0 (3.5) | 5.0 (1.8) | 1.0 (2.7) |
| Yangpu | 4.0 (2.9) | 3.0 (1.8) | 1.0 (1.1) | 1.0 (0.5) | 494.0 (18.1) | 1,921.0 (19.7) | 44.0 (15.7) | 12.0 (32.4) |
| City total | 136.0 (100.0) | 168.0 (100.0) | 95.0 (100.0) | 196.0 (100.0) | 2,730.0 (100.0) | 9,743.0 (100.0) | 281.0 (100.0) | 37.0 (100.0) |

Source: SMSB 1986, 2001.

Note
Figures in parentheses are in percentages.

increased by 58.0 percent. With fewer people and more space, the district's overall physical quality has been improved. Increase is especially significant for office buildings: total office space in 2000 is 670 percent of that in 1985, and shop area increased by 294.4 percent in the fifteen years. These changes indicate that the district is becoming an attractive modern retail center and experiencing a "rebirth" of its splendid past of the French Concession period.

Good-quality housing consisting of new villas and apartment buildings has also increased, while the slum area has been reduced by 80 percent in the fifteen years (Table 8.6). The district's average housing price was 5,700 yuan ($695) in 2000 and 7,000 yuan ($855) in 2003, which was more expensive than that of other urban districts which averaged less than 5,000 yuan in 2000, except for the CBD. So Luwan is becoming a better and more expensive place to live. The shift of the land use, such as the decrease of industrial buildings (Table 8.5), also reflects the policy shift from manufacturing to services, which conforms to the city's development strategy.

### *Yangpu District*

In boundary adjustment from 1985 to 2000, Yangpu gained both land (14.5 percent) and population (9.7 percent). A large number of residents from Huangpu, Luwan and other downtown districts moved to Yangpu, due to housing price increase following the improvement of housing quality in the downtown area. As a result, Yangpu is becoming a residential area for the average people: total residential building area increased from 8.6 to 21.4 million square meters or by 148.0 percent, but the share of villa housing decreased by 25 percent (Table 8.6). The average housing price in Yangpu was 3,000 yuan ($366) per square meter in 2,000 and 4,000 yuan ($500) in 2003, not only lower than that in both Huangpu and Luwan, but also one of the two lowest districts of the central city (see also Wu 2002).

Yangpu has a larger share of staff housing (19.7 percent in 2000), the housing for the majority of Shanghai's ordinary residents. The amount of staff housing in the district alone is more than twice as large as the sum of that in the two other districts (Table 8.6). The overall living condition has been improved, evidenced by the decrease in slum area and the increase in average residential space. But the amount of slum space remaining in the district is still more than the sum of the two other districts, which indicates the district's status in the city from the physical quality perspective.

The district has made serious efforts to diversify its economic base. While industrial buildings have increased slightly (25.5 percent), office buildings have increased much faster (148.0 percent). However, the amount of office space is still much less than that in either Huangpu or Luwan. Shops have increased as well, but the amount is less than that in the two other districts. At the same time, factory space in Yangpu is more than the sum of that in the two other districts (Table 8.5). This reflects the district's traditional manufacturing economy. Since no detailed income data is available in Chinese cities, the best measurement for people's economic status is their living conditions, including housing type and housing price. With less good-quality housing and lower housing price, Yangpu is the home to the average Shanghai residents.

## External factors: new policies at the national and local levels

Since China started economic reforms in 1978, a series of new policies have been initiated and implemented, significantly impacting urban development at the city as well as district levels. In the decision-making structure, decentralization has given the municipal government more authority in local development decisions. In Shanghai, both the city and district governments can now sign investment contracts with different foreign corporations or developers. Since "selling a place" depends largely on the "selling point" (especially a place's geographic location) and the "sales person" (the officials), decentralization of decision making thus works together with other internal factors.

More importantly, after the separation of land ownership (urban land is owned by the state) from land use rights which can be transferred with a fee as specified in the 1988 constitution, land use rights have become a critical revenue resource, and land use decision has become a major source of local power (Zhang 2000). The trading (leasing or transferring) of land use rights generates considerable local revenue, which in turn stimulates local government's desire of leasing more land (Yeh and Wu 1996; Zhang 2002). With less vacant land available in the central area, converting land with "lower value" (manufacturing in particular) to "high-value" uses (commercial in particular) through land leasing is a common practice followed by all local governments. This phenomenon is known locally as "increasing land value through [sectoral shift] in land use" (*tongguo tudi zhihuan shi tudi zengzhi*). In the globalizing world, local factories are challenged either to produce for export or face the fate of being shut down. In promoting export, new development strategies give priority to the development of modern factories in industrial parks, especially those established by FDI with high productivity, over the renewal of old plants in Yangpu.

Decentralization of economic decision power, separation of land use rights from land ownership, and emphasizing export are three national policies that have brought significant impacts to district urban development. Urban spatial structure has been changed in all Chinese cities. CBDs have been emerging or re-emerging. The inner core of the city that used to have a mixed residential and manufacturing land uses, has become a place to house high-income residents. Investment in downtown office and commercial buildings has increased rapidly. Factories in the central area have been closed or relocated to peripheral areas to make room for retail establishments, offices and high-quality apartments. In the suburbs, new industrial parks have been built for joint venture enterprises or high-tech factories. Between the central area and the industrial zones are housing projects for average residents. However, the rapid expansion of the urbanized area to the suburbs shows a lack of order leading to the Chinese version of urban sprawl.

At the municipal level, Shanghai has adopted new policies including investing more in downtown rather than in the old industry districts; encouraging service establishments to locate in the most attractive areas to replace existing factories; offering favorite policies to FDI enterprises rather than renovating old manufacturing plants; transforming central city industrial land to housing and commercial development; revising the Master Plan to promote new service centers; and the removal of the equalitarian revenue policy (see below).

In general, Shanghai has experienced three stages in the post-reform development process. From the 1980s to the early 1990s, housing was the focus of the government's development efforts. Starting from 1990, infrastructure, especially transportation projects, became the main development activities. Since 1998, the city shifted to environmental improvement and beautification projects, such as creating more open space and public parks. The outcome of the efforts constitutes part of "the Shanghai miracle" – 120 million square meters of housing, 4,000 high-rise buildings, four bridges and two tunnels across the Huangpu River, two subway lines crossing the city, and 62 km of elevated inner-ring road have been built in the last ten years.

The byproduct, however, is uneven developments among districts. While many local policies have contributed to the unevenness, as discussed above, two policies are especially crucial in restructuring Shanghai's urban space: land leasing and the removal of the equalitarian revenue policy practiced in the pre-reform era.

### *Land leasing*

Among the $45 billion foreign investment injected into the city from 1990 to 2000, a considerable portion is used to pay for land leasing. Zhao (1998) found that land-leasing revenue in the city reached 81 billion yuan (about $10 billion) from 1988 to 1996. Since urban land use rights can be traded, land in downtown and other strategic locations has attracted both domestic and foreign investors. But different district governments have adopted different strategies in response to the high demand. Some ambitious leaders quickly caught the opportunity and designed policies in favor of developers. For instance, a district government can help development companies to relocate residents for land acquisition, sometimes with the use of administration power. The result is a building boom in many districts. Luwan is an example. A total of 340,000 square meters of slum space has been removed and the land has been used for high-quality residential and commercial developments with investment from Hong Kong and Taiwan. Per capita building area in the district has increased by 112.9 percent in 15 years, the highest rate among the three districts. Luwan has been seen as the most successful model of urban renewal in Chinese cities. The district's revenue relies heavily on land leasing. The director of the district planning bureau was proud of the policy: "We grasped the opportunity brought on by the 1990 land management reform. It was a milestone for the development of our district" (interview with Director Mei in July 2001). In fact, the first land-leasing case in Shanghai took place in Luwan. On January 12, 1992, just a week after the municipal government approved the land-leasing regulation, Luwan leased the first piece of land to a Hong Kong developer, followed by a case in Huangpu a month later (Kong 2001).

In other districts, either because land did not attract much attention from foreign developers due to location or because leaders were too conservative to adopt innovative policies, land-leasing revenue was less and the scale of development, especially for high-quality projects, was smaller. Yangpu, in part because of its location, is such a case. In Yangpu, most new buildings have been built with domestic investment to provide affordable housing to the average residents rather than expensive commercial and retail structures invested by foreigners. The amount of staff housing for average residents has increased from 4.9 million square meters in 1985 to 19.2 million in 2000, an increase of 292 percent. The share of staff housing in Yangpu also increased from 18.1 percent in all urban districts in 1985 to 19.7 percent in 2000. In the same period, the share of staff housing in Huangpu dropped from 7.1 percent to 3.8 percent (Table 8.6).

The main reason of building more staff housing in Yangpu can be attributed to its low land price. Shanghai's land price has six grades. Most of the area in Huangpu is in the first grade zone, a large part of Luwan is in the second grade zone, and the entire Yangpu district is in the third grade zone. While building affordable housing in Yangpu is mainly because of its low land price there, the concentration of affordable housing in this district in turn contributes to the lower land price. This situation is somewhat similar to public housing in American cities. The "disadvantaged" district has less land revenue due to low land price which generates less public investment; this in turn makes it even less attractive to future investors.

In addition, with more moderate and lower-income residents moving out of districts where a Chinese version of gentrification has forced population relocation, Yangpu has to bear the load of population redistribution. Both Huangpu and Luwan lost population (most with moderate income), but Yangpu experienced a 9.7 percent population gain from 1985 to 2000, largely because of the construction of affordable housing projects. The demand for schools and clinics is hard for the government to fulfill, and other facilities caused by housing projects to fulfill higher goals of development. It is clear that the land-leasing policy has had strong impacts on land revenue in different locations, which in turn brings about differences in intra-city developments.

### *The removal of equalitarian revenue policy*

Finally, the removal of the equalitarian revenue policy practiced in the pre-reform era has hurt some districts, including Yangpu. Shanghai adopted an equalitarian revenue policy from the 1950s to 1992. Under the planned economy, municipal government collected all revenues from districts then redistributed them based on needs. Yangpu, a district that contributed over one-quarter of the city's revenue, was the main "donor" then. Districts with limited manufacturing industry and less revenue were the main beneficiaries.

After the reforms, a new revenue arrangement called "two levels of government and two levels of financial management (*liangji zhengfu, liangji caizheng*)" has been put in place. Under the new policy, the municipal government shares both tax revenue and development responsibility with the district governments. A fixed proportion of district tax revenue is retained by each district, and the district government is responsible for urban development, environmental protection, price control, labor force development, personnel and business administration in the district. With the shift of development strategy and the challenge imposed by imported goods, districts like Yangpu have become weaker economically today. Should the equalitarian policy still be in place, they would have become new beneficiaries. The removal of the equalitarian tax revenue arrangement thus has hurt these districts severely.

## Internal factors: community features

### *Geographic location*

In market economies, downtown areas in large cities typically receive more development attention and investment than other urban areas (Rast 1999; Brenner and Theodore 2002). In Shanghai, the Huangpu District, the CBD and the municipal administration center is a favorite place for investors and residents. Yangpu, located on the north side of the city with some distance to downtown, is less attractive to commercial developments. Due to its convenient access to both the Huangpu and Yangtze Rivers and with extensive waterfront and port facilities, Yangpu functioned as a manufacturing district for eighty years since the 1920s. But as traditional manufacturing is no longer favored in Shanghai's economic transition, the district's locational advantage has become a disadvantage.

Disadvantaged geographic location itself does not necessarily mean less development opportunity. The Pudong New District is on the east bank of the Huangpu River, and the 500-meter wide river was considered a "transportation threshold" between Pudong and Shanghai's downtown for forty years. Once the city government decided to develop Pudong as Shanghai's new international financial center, four bridges and two tunnels across the river have been built, and Pudong is now the most favorite place for residents and investors alike.

### *History*

A district's past may have considerable impacts on its development pattern. Huangpu has been Shanghai's CBD for over seventy years, and no other district has been able to challenge its central position. Luwan was the French Concession and a high-end retail center before the 1949 revolution. Its history has fostered a local culture rooted in the district's relation with

Western culture and business. The rebirth of the district as a "world-class" commercial center has thus received support not only from the government but also from investors and ordinary residents.

Yangpu was a manufacturing district after the 1920s and manufacturing was Shanghai's largest employment sector till the 1990s. With Shanghai shifting its development goal from manufacturing to services, Yangpu has been weakened economically and administratively. As the number one contributor to Shanghai's revenue, the district contributed to 29.3 percent of the city's revenue in 1964, and remained the second contributor (after Huangpu) in 1989 and 1990. But its financial position has dropped to a much lower rank than Pudong New District, Huangpu, Jingshan and Baoshan in recent years, although no official ranking is available (Interview with Director Wang in July 2001).

Similar to locational disadvantage, the lack of a splendid history does not necessarily mean less development hope. Although Pudong was an agricultural area for over one hundred years, it is now the hottest spot to investors not only in Shanghai but in the whole nation. Government decisions and development policies seem to be more important than history in shaping urban development because whereas policy can make history, history cannot create policy.

### *Economic base*

Huangpu's economic base has always been financial services and retail businesses since it was founded in the 1850s. But its economy was severely weakened after the 1949 revolution. Not until the reforms did Huangpu begin to regain its position as Shanghai's economic center. Although facing challenges from other districts, especially the Pudong New District, it still holds 25.5 percent of the office space and one-fifth (18.6 percent) of shop space in the city's central urban districts (Table 8.5).

Luwan, with a smaller population and limited land, was a "little brother" among the urban districts from 1949 to 1990. But reforms provided the district with a redevelopment opportunity. With less manufacturing and less pollution in the district, it is relatively easy to convert land to service and residential uses here than in the industrial districts. Luwan's share in the city's office buildings increased from 4.7 percent in 1985 to 9.3 percent in 2000. The share of shop space increased from 6 percent to 9.8 percent in the same period. These are strong indications that the district's economic base has been strengthened.

Manufacturing was, and still is, Yangpu's economic base. Its textile, steel, shipbuilding, machinery and other factories contributed to 26.5 percent of Shanghai's industry output in 1965, but it dropped to 14.6 percent in 1990 (ESYH 1995). The importance of the district's manufacturing sector to Shanghai's economy was reflected in Yangpu's land use: its share in the number of factories was 21.4 percent of the city's total in 1985. The share increased to 29.8 percent in 2000 as a result of other districts converting more industrial land to service and residential uses. However, with the shift of the city's development strategy, many factories are facing the fate of being moved to planned industrial zones in periphery areas. The removal of most of the city's textile industry had resulted in 400,000 workers being laid off, the majority of them in the Yangpu district. With the direction of development shifted to services and housing, the district government has been trying to strike a balance of development between manufacturing and services. Thus it is obvious that policy change can have very strong impacts on the economic base of a district.

### *Local leadership*

Scholars have found that leadership makes a difference in urban development outcomes (Fitzgerald and Leigh 2002). In China, local government officials' jobs and career advancement largely depend on the officials of the superior government. Connection to the municipal government is critical to district officials for their careers. Because of the location of the district that hosts the municipal government, Huangpu district leaders have more chances to contact and build connections with officials in the municipal government. Chen Liangyu, the current Party Secretary of Shanghai, was the director of the Huangpu District in the 1980s. Many ex-leaders of the district now work in the municipal government, and they are the channels through which current district officials get access to the municipal government.

In Luwan district, many leaders started their careers from local government branches, but they are also well connected with local businesses. Officials of the district have a reputation of being "official-entrepreneurs" because they are sensitive to business opportunities. Officials in Luwan often react quickly to development opportunities presented by policy changes. Hang Zheng, the current Shanghai Mayor, was once the district director in charge of urban development. Two deputy mayors, Zhou Yupeng and Zhou Muyao, also served in Luwan.

Most leaders of Yangpu came from the manufacturing industry. They are more familiar with manufacturing than with urban development. In the manufacturing sector, experience is more valuable than education. Many leaders from manufacturing backgrounds tend to have a modest level of education, which makes them less competitive in seeking promotion in the new economy. No major leaders in the municipal government have come from Yangpu, although Yangpu is the second largest district in the city. In turn, this situation makes it hard for the district's leaders to bring the district's needs to the municipal agenda.

## Conclusion: interaction of policy and community features

If applying the concept of the Regime Theory to Shanghai, we may find a strong pro-growth coalition in the city not only at the city level but also at the district level (Zhang 2002). Urban development often reflects the wishes of a city's power holders and resource controllers. With China's national emphasis shifted to the growth of the market economy, local policy makers have in turn pushed for high rates of GDP growth while paying less attention to social justice. Policies such as the removal of the equalitarian tax revenue arrangement are the results of the attitude of the nation's top leaders. Such policies have contributed to the differences in urban development in the three districts.

Policy matters a great deal to local development, as shown in this study. But policies are uniformly applied to all districts in a city. Why are there significant differences among these districts? The study has found that urban development differences are largely the outcomes of the interaction between policies and community features. Without the opening of the policy window, such as the urban land use reform, development opportunities would not emerge. Without entrepreneurship, however, district leaders may miss these opportunities. Leaders of Luwan District, for instance, are more ambitious and willing to take political risks in initiating land leasing early in the reforms. On the other hand, the politically more conservative government in Yangpu cost the district some development opportunities.

The extent to which a policy may bring about differences in urban development relies heavily on community features: the geographic location, the history, the quality of residents and leaders, and the economic base of the community. People in Luwan have traditionally been sensitive to business opportunities, and this local culture and the location of the district make it easier for the district leaders to build consensus over innovative policies and get support from the constituency. Various community features themselves interactively react to opportunities brought on by new policies. Thus national and local factors interwoven with global forces have affected the three districts' development outcomes.

Since 1998, Yangpu has started to adopt aggressive policies and become more active in urban development under a new district leadership. Meanwhile, a new opportunity for the district has also emerged. Given that most land in Huangpu and Luwan has been acquired and developed in the last decade and under the central government's new policy to control sprawl in rural areas, developers are seeking any available land in the central city. With many factories moved out or shut down, land in Yangpu has become attractive again and it is available. With such changing conditions in the district, a new policy window has appeared that may enable Yangpu to reduce the development gap with other districts in the future. But again, the future of the district will depend on the interaction of internal and external factors that have proven to be the main force determining a district's development future.

## References

Brenner, N. and Theodore, N. (2002) *Space of Neoliberalism: Urban Restructuring in North America and West Europe*, Oxford: Blackwell.

ESYH (1995) *Huangpu jingji shehui nianjian (Economic and Social Yearbook of Huangpu)*, Shanghai: Shanghai Statistics Press.

ESYY (2000) *Yangpu jingji shehui nianjian (Economic and Social Yearbook of Yangpu)*, Shanghai: Shanghai Statistics Press.

Fitzgerald, J. and Leigh, N. (2002) *Economic Revitalization: Cases and Strategies for City and Suburb*, Thousand Oaks, CA: Sage Publications.

Kong, Y. (2001) *Jiedu Shanghai (Understanding Shanghai)*, Shanghai: Shanghai People's Publishing.

LSYB (1999) *Yangpu nianjian (Yangpu Statistical Yearbook)*, Shanghai: Shanghai Statistics Press.

Rast, J. (1999) *Remaking Chicago: The Political Origins of Urban Industrial Change*, De Kalb, IL: Northern Illinois University Press.

SMSB (Shanghai Municipal Statistical Bureau) (1986, 2001 and 2002) *Shanghai Statistical Yearbook*, Shanghai: China Statistics Press.

Wu, F. L. (2002) "Sociospatial differentiation in urban China: evidence from Shanghai's real estate markets," *Environment and Planning A*, 34: 1591–1615.

Yeh, A. G. O. and Wu, F. L. (1996) "The new land development process and urban development in Chinese cities," *International Journal of Urban and Regional Research*, 20: 330–353.

Zhang, T. W. (2000) "Urban sprawl in China: land market force and government's role," *Cities*, 17(1): 123–135.

—— (2002) "A socialist pro-growth coalition in Shanghai," *Urban Affairs Review*, 4: 475–499.

Zhao, M. (1998) *The Reform of Land Use Regulation and Urban Development*, Shanghai: Tongji University Press.

# 9 Industrial restructuring and urban spatial transformation in Xi'an

*Huaiting Yin, Xiaoping Shen and Zhe Zhao*

## Introduction

An ancient Chinese capital, Xi'an became a major industrial city only after 1949. During the periods of the First Five-Year Plan (1953–1957) and the Third Front Construction (*sanxian jianshe*) (1965–1971), Xi'an received a tremendous amount of investment from the state and built seven major industrial districts – Electric Machinery City (*Diangong cheng*; EMC hereafter) in the west, Textile City in the east, Sanqiao Industrial District in the northwest, West Suburb Industrial District, Northeast Suburb Industrial District, Hujiamiao Industrial District in the northeast, and Hansenzhai Machinery Industrial District in the east (Figure 9.1). After the 1978 economic reforms, Xi'an's state-owned industries, such as textile, machinery and other state-owned enterprises (SOEs), experienced a serious decline. In 2001, the share of SOEs in the city's total industrial output was only 48 percent, compared to more than 80 percent from 1957 to 1980. The decline of SOEs and the growth of high-tech industries and the tertiary sector after reforms have significantly changed the city's industrial structure and urban space.

Although it is well documented that urban spatial changes in Western countries are related to stages of economic development (Hall 1984; Berry 1976), China's urbanization experience is very different from that of the United States (Zhou and Ma 2000; Hu 2000). Whereas there is considerable literature on industrialization and urbanization in China, this literature tends to focus on the cities in the eastern region, such as Shanghai, Beijing, Shenyang, Guangzhou and Suzhou (Zhou and Ma 2000; Hu *et al*. 2000; Zhou and Meng 1997; Yeung and Hu 1992; Lo 1994; Ning and Yan 1995; Wu 1998). Although some Chinese scholars have studied industrial restructuring in Xi'an (Ge and Zhang 1998; Li 1995; Wan *et al*. 2001) and the development of specific industrial sectors such as the sewing machine industry (*Shaanxi ribao*, Xi'an, April 22, 1998, 5) and electric equipment industry (Wang 1999) and the city's industrial future (Zhang 2001; Zhang 1998), relatively little attention has been paid to the effects of industrial restructuring on urban land use change. Notable exceptions are Yin and Liu (2002)

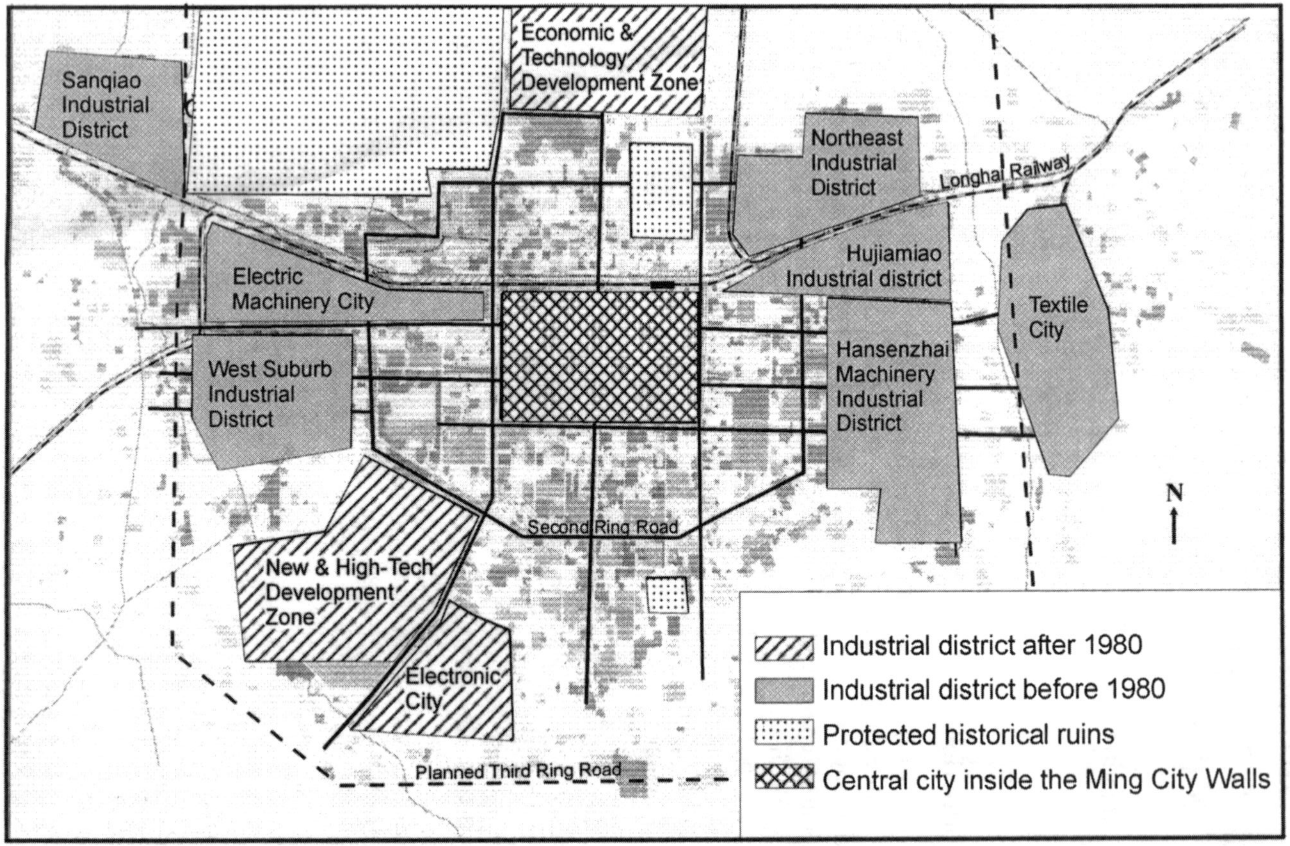


*Figure 9.1* Industrial districts in Xi'an.

who have studied urban land use problems, Yang and Yin (2002) who have examined the central city's land prices, and Liu *et al.* (2002) who have analyzed the redevelopment of the city's old industrial districts focusing on predicting its future land use. As the largest city in northwest China and a leading industrial center of the nation, Xi'an's industrial restructuring and the changing land use patterns demand scholarly attention. Based on field surveys by Shen in September 2002 and by Yin and Zhao in May 2003, this chapter examines the impact of industrial restructuring on urban land use patterns in Xi'an and analyzes the factors affecting the changes.

## Industrial development in Xi'an

Despite the fact that Xi'an had been the capital of the Chinese empire for over a thousand years in history, it was not officially designated as a city until 1928. This new identity, however, did not help raise the level of the city's industrial development. In 1949, there were only a few medium-sized plants and about a thousand small privately owned factories in the city. Due to low production capacity and poor technology, most consumer goods had to be brought in from other regions and foreign countries.

### *The First Five-Year Plan*

During the First Five-Year Plan when the development of inland regions was emphasized, Xi'an received a very large amount of state investment for industrial development. Among the state's 156 key development projects, 17 were located in Xi'an which accounted for 10.8 percent of the total national investment, topping the list of all cities in China. Designed as a "light and precision machinery and textile industrial city," all six of the nation's key electric machinery factories were built in Xi'an, and the Electric Machinery City was created from scratch in five years on the western side of the central city (Figure 9.1). For the textile industry, four large-scale factories, Number 3, 4, 5 and 6 State Textile Plants, were built in the eastern part of Xi'an, creating another industrial district, the Textile City. The overall state investment in Xi'an reached 1.4 billion yuan during the First Five-Year Plan which quickly made the city a leading industrial center in the nation with five functional zones: EMC in the west, defense industrial district and the Textile City in the east, cultural and educational district in the south, preserved historical sites in the north, and central business district within the city walls of the Ming dynasty.

The total GDP of Xi'an increased from 337 million yuan in 1952 to 802 million yuan in 1957 with an annual growth rate of 15.7 percent (Xi'an Statistical Bureau 1999). Industrial output increased even faster from 235 million yuan in 1952 to 715 million yuan in 1957 at an annual growth rate of 25 percent (Xi'an Statistical Bureau 2002). Although investment poured into the city during the same time period, the ratio of industrial output between light and heavy industries changed more slowly due to the greater time lag between investment and production for the heavy industries. The ratio was 74.5:25.5 in 1957, but in 1965 it was changed to 48.3:51.7.

### *The Cultural Revolution and the Third Front Construction*

From 1958 to 1965, Xi'an, as one of the strategic rear bases in China, continued to receive state investment and established more large-scale factories in various industries. In addition, a number of universities and research institutes were relocated from the coastal provinces, the strategic first front region, to Xi'an. By 1965, Xi'an had become a major space, aviation, electronic, textile and machinery industrial base in China. In less than twenty years, Xi'an's industrial output increased twenty-fold, from less than 100 million yuan in 1949 to 2,004 million yuan in 1965. In terms of industrial output, Xi'an was the eleventh largest industrial city in China (CSSB 1999).

During the Cultural Revolution (1966–1976), political movements and class struggles adversely affected the nation's industrial production and economy, and Xi'an was no exception. Many factories halted production and almost all of the functioning factories operated under capacity. Xi'an's industrial output experienced three big plunges in the ten-year period, and in just one year from 1966 to 1967 the loss in output was as much as 30 percent. From 1969 to 1974, as a consequence of perceived international threat to national security, the state launched the Third Front Construction (*sanxian jianshe*) campaign that involved the relocation of a large number of defense-oriented industries from the coastal region to the inland provinces (Naughton 1988) and the construction of many new industries in the western region including Xi'an. Following the locational principles prevalent at the time known as "mountains (*shan*), dispersal (*san*) and tunnels (*dong*)," newly built industries were scattered in isolated tunnels in the mountainous areas (Ma and Wei 1997).

As a major city in the Third Front region, Xi'an again received a vast amount of state investment, and many large and medium-sized factories in the coastal region were relocated therein. Heavy industry, particularly the electric machinery industry, received the largest state support, and a large number of engineers and workers were brought in. Light industry also benefitted from factory relocation, affecting the emergence of such factories as the Shaanxi Sewing Machine Plant, Hongqi Watch Factory, Clock Parts Factory, Fenglei Instrument Factory and Light Machine Factory. Among these, the Sewing Machine Plant has become one of the largest in the nation in recent years. Several new industrial districts were established to accommodate the relocated and new enterprises: the Sanqiao Industrial District in the northwest for transportation equipment industry, the Western Suburb Industrial District for chemical industry, and the Northeastern Suburb Industrial District, the Hujiamiao Industrial District and the Hansenzhai Machinery Industrial District for heavy machinery, automobile and military industries (Figure 9.1). As a consequence of these developments, Xi'an's industrial output increased by 240 percent from 1965 to 1978, which was higher than many coastal cities such as Shanghai, Tianjin, Shenyang, Anshan and Fushun (CSSB 1999). Meanwhile, Xi'an's population increased by 273 percent from 770,900 in 1949 to 2.1 million in 1978 (not including the counties administered by the city). This percentage increase was much greater than those of the coastal cities during the same time period, such as Shanghai (133 percent), Guangzhou (199 percent), Tianjin (194 percent) and Qingdao (165 percent) (CSSB 1999).

Without state investment, China's inland region would never have been developed to the extent it did during the socialist period. However, there were some inherent problems in socialist industrial development. First, heavy industries, particularly traditional and military industries, took more time than light industries to reach production capacity and a much longer time to recover investment and generate revenue due to large initial investment. In Xi'an, a large number of heavy industrial enterprises depended on state subsidies and investment for survival, with little money available for technological innovation and equipment upgrade. Second, most of these enterprises were built in Xi'an not because local industrial resources were available nor because of any local demand. Raw materials had to be brought in to many of the transplanted factories. In addition, these enterprises reported to and were managed directly by various ministries of the central government in Beijing that had no idea about and no interest in Xi'an's economic development. Third, as the enterprises were state-owned, they suffered from the same problems as the SOEs in other places, such as excessive administrative intervention, heavy bureaucracy, permanent employment, "eating from the same big pot," pollution and low efficiency and low productivity (Shen 2001). By 1978, the SOEs produced 84 percent of the city's total industrial output, and 60 percent of the total came from large and medium-sized enterprises (CSSB 1985).

### *Industrial restructuring after economic reforms*

After economic reforms were launched in 1978, industries in Xi'an started their long and difficult journey of restructuring. In the early 1980s, Xi'an was one of the most inefficient large industrial cities in the nation. In 1983, the tax and profit ratio of the original value of fixed assets in Xi'an was only 15 percent, which was the second lowest after Haerbin among the seventeen largest industrial cities in China (CSSB 1985). Because significant reforms of large and medium-sized SOEs did not begin until the 1990s, Xi'an's economic development was slow.

Xi'an's industry has been significantly restructured in the reform era. One major aspect of restructuring took place in the 1980s when a number of military and related industries that had been constructed during the Third Front period in the nearby Qinling Mountains were moved to the city. The relocation of such industries was necessary as they were no longer able to survive in isolated locations in the mountains where they had long suffered from economic inefficiency and operational problems. However, such industries were not allowed to move back to the coastal provinces. Thus, Xi'an, the largest industrial city closest to them, became their new home. Some of these relocated enterprises were sent to the Electronic City in the southwest suburb of the city (Figure 9.1).

Industrial reforms in Xi'an are also reflected in the changes in both ownership form and sectoral structure. Table 9.1 shows the changes in industrial ownership structure from 1957 to 2001. It is clear that the SOEs that had once dominated the industry in Xi'an with 91 percent of its total industrial output, declined rapidly after 1980, producing only 48 percent of the output in 2001. Collectively owned enterprises, mostly rural industries, had grown since 1980 (Shen 1998). In 1995, their share in the city's total industrial output reached 40 percent and after that some of them were privatized. This change coincided with the nationwide trend of ownership restructuring of the SOEs due to their poor industrial productivity and low efficiency (Shen 2001). Although ownership restructuring in Xi'an has moved at a greater speed than other northwestern cities and provinces, the share of Xi'an's SOEs in the city's total industrial output in 2001, at 48 percent, was still higher than the national average (44 percent) and much higher than that of some of the coastal provinces and cities, such as Zhejiang (15 percent), Jiangsu (26 percent), Fujian (30 percent), Guangdong (26 percent) and Suzhou (8 percent) (CSSB 2002; Suzhou Statistical Bureau 2002). Nonetheless, ownership change, the decline of the SOEs and the emergence of an urban land market prompted a significant number of the city's factories to change the ways their land properties were used (discussed below).

In addition to ownership restructuring, changes in the sectoral structure of the city's industries are even more significant and have led to the creation of new leading industries in Xi'an. Table 9.2 shows the percentage share of

*Table 9.1* Percentage share of industrial output by ownership in Xi'an

| *Year* | *State* | *Collective* | *Other* |
|---|---|---|---|
| 1957 | 87.5 | 10.8 | 1.7 |
| 1965 | 91.4 | 8.6 | 0.0 |
| 1978 | 83.9 | 16.1 | 0.0 |
| 1980 | 82.8 | 17.2 | 0.0 |
| 1985 | 74.2 | 25.7 | 0.2 |
| 1990 | 67.6 | 31.0 | 1.5 |
| 1995 | 51.0 | 41.0 | 8.0 |
| 2001 | 48.2 | 28.9 | 22.9 |

Source: Xi'an Statistical Bureau (2002).

*Table 9.2* Percentage of industrial output by selected sectors in Xi'an

| | *1980* | *1985* | *1990* | *1995* | *2001* |
|---|---|---|---|---|---|
| Textile | 43 | 20 | 15 | 8 | 3 |
| Machinery | 28 | 18 | 19 | 11 | 10 |
| Chemical | 6 | 5 | 7 | 6 | 4 |
| Electronic | 2 | 5 | 15 | 5 | 13 |
| Medicine | n.a. | n.a. | 1 | 6 | 12 |
| Electric Equipment | n.a. | n.a. | n.a. | 11 | 9 |
| Transportation Equipment | n.a. | n.a. | n.a. | 14 | 20 |

Sources: Xi'an Statistical Bureau (2002); Xi'an Statistical Bureau (1996); CSSB (1986, 2002).

Xi'an's major industrial sectors between 1980 and 2001. The dramatic decline of the textile industry is most noticeable. Its percentage share in the total industrial output plunged from 43 percent in 1980 to 3 percent in 2001. Other major changes include the decline of the machinery industry by almost two-thirds, and the weakening of electric equipment industry. New industries, such as electronics, biomedical and automobile industries have emerged since 1990. Industrial restructuring has significantly affected urban land use in Xi'an which we will discuss in the next section.

## Effects of industrial restructuring on urban space

In the last half century, urban built-up areas in Xi'an increased from less than 40 square kilometers in 1949 that were found mostly inside the Ming dynasty city walls to 187 square kilometers in 2000 spreading out from the old city center in all directions (CSSB 2001). Industrial development and restructuring has been the major driving force for urban spatial change. During the first thirty years in the Maoist era, urban land use was controlled by state planning that guided the development of urban built-up areas that were found mainly in the eastern and western suburbs where the seven industrial districts identified early in the chapter were established. Among the leading industries developed in the socialist era, machinery and electric equipment industries were located closer to the urban center than the textile industry which was located farther away from the central city between the Ba and Chan Rivers. All of the industrial districts and some of the large enterprises have had feeder railways connected to the Longhai trunk railway, one of the nation's major railroads running in an east–west direction that passes through Xi'an along the northern side of the city walls. Areas on the northern side of the Longhai railway were used mostly for warehouses before 1978 due to the inconvenience of railway crossings and the presence of some large protected historical sites in the region, such as Han dynasty Changan city ruins and Tang dynasty Daming Palace ruins. The southern part of the city outside of the city walls was designed as the city's cultural and educational district where a number of universities are located. However, some factories were built in this area during the Cultural Revolution when urban planning was abandoned. In 1978, Xi'an's urban built-up area had a rectangular shape with a long east–west axis that almost doubled the length of the north–south axis. This spatial pattern reflected the influences of the Longhai railway, the main means of transportation linking the city to other parts of the nation. During the Cultural Revolution, urban construction in Xi'an was chaotic. Residential housing took over spaces wherever possible in the industrial districts while factories infiltrated in the neighborhoods of universities and occupied museums in the southern suburban areas. Pollution problems became serious and the city was getting more and more crowded.

During the reform era, Xi'an has experienced tremendous urban spatial transformation as a consequence of industrial restructuring and marketization of urban land. Two major changes are particularly significant. First, various new and high-tech development zones have been built, mainly in the northern and southern suburbs. These zones are rapidly converting farmland for industrial and urban uses. Given that Xi'an's urban built-up area before reforms was extended mainly along the Longhai railway from east to west, the tendency to locate businesses closest to the urban center plays an important role in the development of new industrial districts in the northern and southern suburbs. In addition, this pattern of urban expansion is also affected by the construction of new highways; by the expansion of the water supply system and other urban infrastructure in the northern and southern suburbs after the reforms; and by the available farmland which is cheaper and easier for development compared to the built-up areas. Almost all newly built development zones are located in the areas between the Second Ring Road and the planned Third Ring Road in Xi'an. Many of today's leading industrial enterprises are located in such development zones.

The second major change is the retreat of the secondary sector and the growth of the tertiary sector or tertiarization in the built-up areas, a trend that Zhou and Ma (2000) have observed in the coastal cities. Given that Xi'an is a major industrial city built by central planning in the first thirty years after 1949, the transition of the land use pattern in the old industrial districts is more dramatic than many other Chinese cities where the impact of socialist industrial development was less apparent. Based on official statistics, maps and field surveys, we present below the most salient cases of the transformation of urban industrial land in Xi'an.

### *Transforming urban industrial land for commercial use*

As a result of continuous expansion of urban residential and commercial areas in the past decades, old industrial districts are no longer peripheral in location. The land value in these old industrial districts has climbed so high that inefficient enterprises make less profit from industrial production than the revenue they could generate if they sold their land use rights to developers for other purposes. Under the process of marketization of urban land use rights, enterprises located closest to the central city were among the first to change their land use. Also, enterprises that were run at a loss or were insolvent were under pressure to generate some income, and land was their only marketable asset. Generally speaking, there are two types of land use transformation: total and partial.

Two types of industrial enterprises have changed all of their land to commercial use. First, most of the enterprises that occupied prime retail locations in or near the city center have torn down their factories and replaced them with newly built commercial buildings. The new commercial establishments are owned either solely or jointly by the land-owning enterprises and are staffed with some of the former factory workers. (The rest of the former employees are retired or laid off with limited living allowances.) For example, a former Xi'an enamel factory, next to the Chaoyang Gate on the east side of the city where it was surrounded by wholesale markets on Fukang Road and the city retail center, converted the entire factory premises into a clothing market and a residential area in the mid-1990s. Another example is the Xi'an Sewing Machine Plant that also converted all of its land into a wholesale shoe market on Daqing Road (Figure 9.2). Second, complete land use change can occur when a factory is relocated to a distant suburban location. This type of land use change involves mainly factories that are still profitable and wish to continue production but the difference of land prices between their urban location and suburban areas lures them to relocate. For such factories, the money made from selling or leasing their land use rights is used to cover relocation costs and to upgrade their equipment and technology. For example, the Xi'an Steel Rod Plant was moved to the Xi'an New and High-tech Development Zone, and its former location at Taibai Bei Road is now occupied by a bank, an investment firm and an auction company.

Instead of relocating or dissolving their industrial facilities, more factories have opted for partial transformation of their land to commercial use. One major morphological feature of Chinese cities is the presence of large numbers of state-owned units, including government agencies and SOEs that are typically walled in to form individual spatial cells in the cities. The stuccoed walls are often located along the streets. Given that the most valuable retail space is the narrow area (about 10 to 20 meters) facing the main urban streets (such area is known as *men mian fang* area), most factories that have such spaces would either tear down the walls and use the space for stores or office buildings to generate rent income, or lease the space to the highest bidder for commercial use. Development of such space is very popular on Daqing Road in the EMC (Figure 9.2B) and in the area of Xingfu Road and Wanshou Road in the Hansenzhai Machinery Industrial District and the Hujiamiao Industrial District (Figure 9.3B).

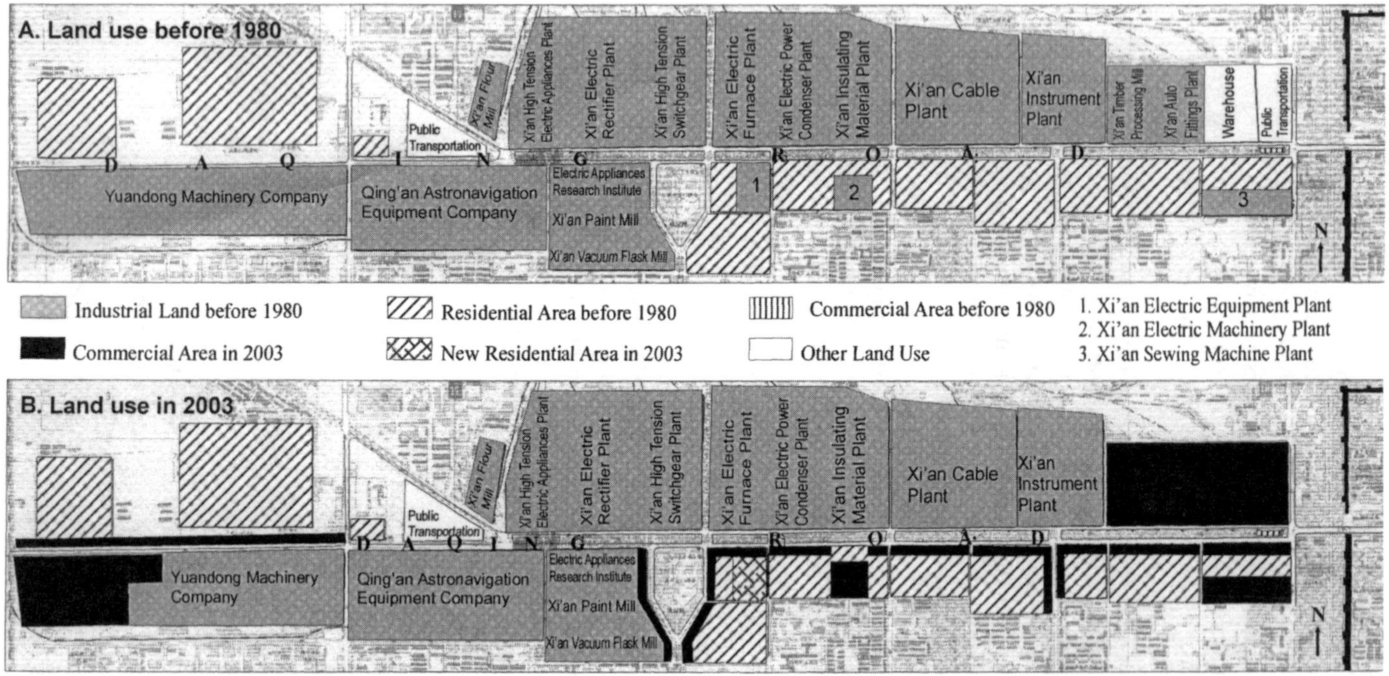


*Figure 9.2* Daqing Road Land Use Change (source: Xi'an ditu chubanshe (2002) *Xi'an jiaotong luyou tu (Xi'an transportation and tourism map*), Xi'an ditu chubanshe and field surveys by the authors).

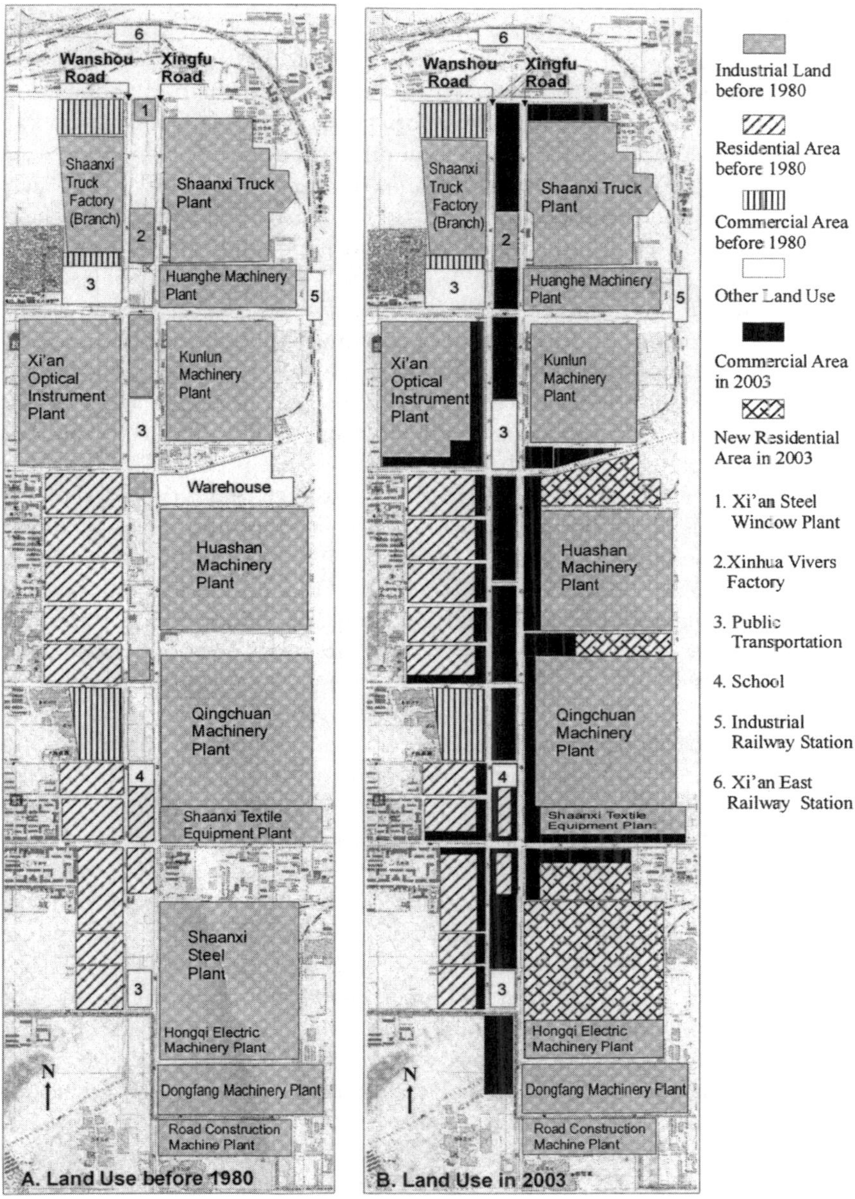


*Figure 9.3* Xingfu Road and Wanshou Road land use change (source: Xi'an ditu chubanshe (2002) *Xi'an jiaotong luyou tu (Xi'an transportation and tourism map)*, Xi'an ditu chubanshe and field surveys by the authors).

### *Transforming industrial land to commodity housing space*

Xi'an's population grew from 2.27 million in 1949 to 4.98 million in 1978 and reached 6.95 million in 2001 (Xi'an Statistical Bureau 2002; CSSB 1999). Planned permanent migration in the pre-reform period and spontaneous rural–urban migration in the reform era are the main reasons for the increase in the city's population. Large numbers of permanent migrants were sent in by central plan from coastal regions during the first thirty years after 1949 and, after the reforms, migrants from suburban and rural areas have poured into the city in large numbers. Rapid population growth has led to severe housing shortage in the city. In addition, under the policy of "production first, living conditions second (*xian shengchan, hou shenghuo*)," residential housing construction in the first thirty years of the socialist rule, particularly during the Cultural Revolution, lagged so far behind industrial development that per capita living space was only 3.4 square meters in 1978. It was not uncommon for three generations to share one room. After the economic reforms, numerous residential buildings have been erected throughout Xi'an. However, unlike the cities in the US, the central city and the near suburbs are the most favored locations. Per capita living space increased rapidly in Xi'an from 6.19 square meters in 1990 to 13.72 square meters in 2000 (CSSB 2001).

Marketization of urban land use rights and the opening up of the urban housing market have made real-estate development one of the most profitable businesses in urban China. Many industrial enterprises have transformed their land for residential use either partially or completely. Most factories that have some space available on their premises are motivated to build apartment buildings for the improvement of their employees' living conditions and for generating extra income for the factories. Many economically inefficient factories have torn down some of their factory buildings to make space available for lease to real-estate companies. Total land use transformation occurs in the factories that are either bankrupt or forced to move out of the urban area because of pollution problems. Two typical examples are the Shaanxi Steel Plant on the south end of Xingfu Road (Figure 9.3) and the Xi'an Iron and Steel Plant on Hongguang Road. The latter was built in the 1950s along with other factories in the Electric Machinery City. Shaanxi Steel Plant was relocated from Dalian to Xi'an in 1965. Although both factories had transported raw materials from and shipped most of their products to other regions since the first day of operation, they were profitable during the pre-reform period and managed to survive the competition of the market economy until the early 1990s. When the mandate to reform reached the point where inefficiency was no longer acceptable to the state, medium- and large-scale SOEs were no longer protected from bankruptcy. With the economy in China changed from a producers' market to a buyers' market in the mid-1990s, both factories quickly became unprofitable after 1995, and they were closed down permanently a

*Figure 9.4* New commodity housing under construction on the land of former Xi'an Iron and Steel Plant.

few years later (Xi'an Statistical Bureau 1996). Besides their financial losses, these two factories were big polluters located near the central city within the Second Ring Road, the prime locations for real-estate development. Figure 9.4 shows that commodity apartment buildings were nearly finished that were mostly pre-sold on the location of the former Xi'an Iron and Steel Plant. Figure 9.3B indicates that the entire land area of the former Shaanxi Steel Plant has also been changed to housing development which is currently under construction by the former factory and a real-estate company.

## Land use transformation in old industrial districts: two case studies

Among the seven industrial districts built before 1978, EMC in the west and the Hansenzhai Machinery Industrial District and the Hujiamiao Industrial District in the east provide considerable details on how industrial land use changes have actually occurred. Their land use shifts have come about not only because of their prime location for tertiary activities but also because the dominating industry in the two districts, machinery, has been declining very rapidly in the last decade. Land use changes along two main streets, Daqing Road in EMC and the parallel Xingfu Road and Wanshou Road in the Hensenzhai Machinery Industrial District and the Hujiamiao Industrial District, are discussed in detail below.

### *Daqing Road in EMC*

Built during the First Five-Year Plan in the 1950s, EMC is one of the oldest industrial districts in Xi'an. Due to its early establishment, it occupies a prime location that covers the area from west of the Ming Dynasty city walls to the planned Third Ring Road. It is the only industrial district in Xi'an that is adjacent to the city walls and has a large area inside the Second Ring Road.

Daqing Road, a portion of the east–west axis in Xi'an, is a 6.75 kilometers long arterial road that runs through the entire industrial district from Yuxiang Gate on the east to Afang Road on the west (Figure 9.2). Therefore, to the people of Xi'an, Daqing Road means EMC and vice versa. Being a planned industrial area, this road typifies the design style of the 1950s – one side of the road is reserved for factories and the other side has residential areas for each factory directly across the street except for a few factories at the west end of the street. There are more than ten large and medium factories in the area producing mainly electric equipment and machines, such as the Xi'an High-Tension Electric Appliances Plant, Xi'an Cable Plant, and Xi'an Electric Furnace Plant (*bianya dianlu chang*). Before 1980, there were only a few retail stores and service establishments scattered in the residential areas on the south side of the road.

After 1980, the location of the area has attracted many businesses. High land price of the location and low revenue from industrial production have provided incentives for the land owners – the factories – to develop, sell or lease their land for commercial uses. Figure 9.2B shows that the transformation of urban space along Daqing Road has the following characteristics. First, the areas closest to the city walls have the most significant changes. Almost all the land areas, including those for public transportation, warehouses and industries, have been changed to retail and commercial uses. Several large retail and wholesale markets are located in the area next to each other. Second, the narrow areas facing the road (the *men mian fang* area) are occupied more and more by stores and other businesses. The development of commercial properties along the major street has taken about 10 percent of the total land area in the industrial district and most of the new development has occurred on the south side of the road but some stores are also found on the north side of the road near the west end. There are more than ten large hotels, supermarkets, wholesale and retail markets and markets for agricultural products. More than three hundred stores and businesses are here that provide a variety of goods and services, including electrical equipment. There are also machinery sales offices, electronics stores, hardware stores, restaurants, bars, entertainment centers and health clubs and medical offices (Field survey). The total *men mian fang* area is about 150,000 square meters and retail takes up almost half of it (Liu *et al.* 2002). In addition, sales offices for the factories take a significant portion of the land area. Such offices never existed during the pre-reform era. Third, besides the factories near the city walls, the factories that gave up their entire or a large portion of their land for real-estate development or commercial use are mostly on the south side of the road such as the Xi'an Electric Machinery Plant, Xi'an Electric Equipment Plant and Xi'an Sewing Machine Plant. Fourth, almost all intersections along the south side of Daqing Road have become local commercial centers, such as the intersections at West Ring Road, Laodong Road, Xinyuan Road, Taoyuan Road, Fengdeng Road, West Second Ring Road, Tumen Road and Tuanjie Road.

Several factors affect the transformation of the land use pattern on Daqing Road. First, although almost half of the factories on Daqing Road had operated at a loss sometime during the past two decades, they have not reached the point of insolvency and can still manage to survive. In addition, because the factories along the road are mostly electrical equipment and machinery industries, they are not major polluters as the chemical or iron and steel industries are. Therefore, they are not under pressure either from the government or local residents to relocate. Second, the north side of Daqing Road is a one-way street going out from the central city and only a few intersections allow U-turns. People going out from the old city are not likely to stop for shopping and turn back right away. That makes the land on the north side less interesting to businesses. Third, given that Daqing Road is about 100 meters wide with a 40-meter central green island dividing the road, local people who live in the residential areas on the south side are not interested in walking over 100 meters to cross the busy road for shopping. Consequently, while the south side of the road is crowded with businesses and people, most *men mian fang* areas on the north side of the road remain undeveloped to this day.

However, the *men mian fang* areas on the south side of the road will be fully developed soon. When prime locations on the south side of Daqing Road are totally occupied in the near future, the north side of the road will attract businesses. In addition, given the low-density development and availability of land in the premises of factories, many factories, particularly those not making a profit, are interested in developing or leasing their land for commercial use. In fact, 73 percent of the factories have already expressed such an interest, and some of them, such as the Xi'an Cable Plant and the companies of Yuandong and Qing'an, already know which specific pieces of land they are willing to lease out (Liu *et al.* 2002).

### *Xingfu Road and Wanshou Road*

Built during late 1950s and early 1960s, the Hansenzhai Machinery Industrial District and the Hujiamiao Industrial District are located east of the central city between the Second Ring Road and the planned Third Ring Road (Figure 9.1). The districts are designed mainly for large-scale heavy machinery plants. Therefore, the Xi'an East Railway Station and Industrial Railway Station are located north and east of the districts and most plants in the districts have feeder railways linked to the Longhai trunk railway. Some factories in the districts even have in-plant railways connecting workshops. Xingfu Road and Wanshou Road are two main streets that are four kilometers long, parallel to each other and running through the two industrial districts from north to south. Again, factories and residential areas along the two roads form the cores of the two districts. Similar to the land use pattern on Daqing Road, areas east of Xingfu Road are for factories and areas west of Wanshou Road are mainly residential areas for employees (Figure 9.3A). Different from Daqing Road, the areas between Xingfu Road and Wanshou Road are wider. Although the areas between the two roads were originally designed as green islands as those on Daqing Road, the areas have been occupied gradually by small factories, schools and other units over the years. Before economic reforms, there were no stores and services on Xingfu Road and only a few retail stores located on Wanshou Road.

Since 1980, the changes of land use pattern on both Xingfu Road and Wanshou Road have been remarkable (Figure 9.3). On the east side of Xingfu Road, 60 percent of the *men mian fang* areas have been occupied by businesses, particularly automobile sales and services. There are seventeen car sales offices, forty-five auto-parts stores, twenty auto repair garages, and about thirty other stores and services (Field survey). On the west side of Xingfu Road, the green space has been changed to a specialized auto repair and service area. From East Changle Road intersection on north to Weishi Street intersection on south, some 230 auto repair and service shops take up 90 percent of the developed *men mian fang* (Field survey). There are only a few small stores, restaurants and three markets for flowers, motorcycles and food.

Although Wanshou Road is less than one hundred meters away from Xingfu Road, it is on the residential area's side, and consequently has developed different sets of businesses. On both sides of Wanshou Road, over 80 percent of the *men mian fang* areas have been taken by businesses serving local residents. There are about four hundred businesses on both sides of Wanshou Road including convenience stores, restaurants, department stores, supermarkets, retail and wholesale markets, health clubs, entertainment and other services (Field survey).

Besides partial transformation of industrial land to commercial and residential uses, the most significant change in these two industrial districts is the closing of Shaanxi Steel Plant and the development of real estate on the premises of former industrial establishments. Together with three smaller residential areas on the east side of Xingfu Road (Figure 9.3B), real-estate development has significantly reconfigured the land use pattern of the industrial district and this is likely to put pressure on the other factories to follow suit.

In short, tertiarization and commodification of housing in the built-up areas in Xi'an have greatly altered the spatial patterns in the old industrial districts. Several features of land use change can be identified. First, the development of the tertiary sector is clearly driven by market forces, with the prime retail locations commanding the highest land use price. Such areas tend to be close to the city walls or suitable for the development of *men mian fang*. Second, similar types of businesses tend to locate together, as shown along Xingfu Road and Wanshou Road. Third, the profitability of individual factories influences land use transformation. Besides bankrupt factories that have been forced to transform their entire landed property, other land use changes are based on comparative profitability between industrial production and land development for non-industrial uses. The higher the land price, the more land factories are willing to redevelop or to lease out.

## Conclusion

As a major industrial city built by socialist planning, earlier industrial development and subsequent restructuring in Xi'an have significantly affected its spatial patterns in the last half century. During the Maoist era, government policies on inland development and the Third Front Construction campaign resulted in the allocation of vast amounts of state investment that made Xi'an the largest industrial center in northwest China. As a large socialist industrial city, Xi'an's urban land use structure differs from that of the coastal cities in at least three respects. First, relative to other Chinese cities, Xi'an has more large-size industrial districts (seven in total) built during the Maoist era with standardized land use patterns. Second, SOEs in Xi'an are mostly located in industrial districts, while a large number of factories in coastal cities, such as Shanghai and Tianjin, are scattered in urban areas, often mixed with residential and commercial areas due to their long history of industrial development. Third, the vast industrial districts in Xi'an such as EMC are located next or close to the urban center which, over time, have become part of the urban built-up area, whereas in coastal cities the planned industrial districts are either far away from the urban centers (such as, in Beijing, the eastern suburb industrial district and the Shijingshan iron and steel industrial district are both located outside of the third ring road) or found in satellite cities (such as Baoshan Iron and Steel in Shanghai's and Beijing's petroleum industry in Fangshan).

Xi'an's industrial restructuring after the reforms has been prompted by the decline of the socialist industrial sector and by the simultaneous tertiarization of the urban economy. These two related urban processes have occurred largely in the old industrial areas in Xi'an. Relative to the coastal cities such as Shanghai and Guangzhou (Wu 1998), foreign investment has played a less significant role in reshaping the urban landscape. But as in other Chinese cities, the trend toward land use conversion from industrial to commercial and residential uses in Xi'an is clear. Through detailed field surveys, this chapter has revealed the locational patterns of the newly established retail and residential areas along Xingfu Road, Wanshou Road and Daqing Road. The new land use patterns have been significantly affected by pre-existing street patterns earlier land uses along a street (i.e., the residential or industrial side of a street), the types of factories nearby and distance from the central city. We also found that while a number of industrial land parcels have been partially or completely changed to commercial or residential areas, old residential areas remain largely unchanged except for the *men mian fang* areas that are mostly occupied by retail and businesses serving local residents.

The results of this study suggest that both market forces and government policies have played important roles in Xi'an's industrial restructuring and urban spatial reconfiguration. With the deepening of economic reforms emphasizing efficiency and competitiveness at the firm level and with the maturity of the rapidly expanding real-estate and housing markets in the city, more land use shifts can be expected in Xi'an in the immediate future, which will no doubt further reshape the city's economy and its built environment.

## Acknowledgments

Xiaoping Shen is grateful for the financial support of the Connecticut State University Research Grant and the hospitality of Northwest University in Xi'an during her fieldwork. The authors would like to thank the editors Professor Laurence J. C. Ma and Dr Fulong Wu and two anonymous reviewers for their constructive comments and detailed suggestions for revision.

## References

Berry, B. J. L. (1976) *Urbanization and Counterurbanization*, Beverly Hills: Sage.

CSSB (China State Statistical Bureau) (1985) *Zhongguo gongye jingji tongji ziliao, 1949–1984* (*China Industrial Economic Statistics, 1949–1984*), Beijing: China Statistics Press.

—— (1986, 2001, 2002) *Zhongguo chengshi tongji nianjian 1986, 2001, 2002* (*China Urban Statistical Yearbook 1986, 2001, 2002*), Beijing: China Statistics Press.

—— (1999) *Xin Zhongguo chengshi 50 nian* (*50 Years of New China's Cities*), Beijing: China Statistics Press.

Ge, Q. and Zhang, S. P. (1998) "Xi'an gongye jiegou zhuanhuan yu zhudao chanye xuanze (Industrial restructuring and the choice of key industries)," *Shaanxi jingmao xueyuan xuebao* (*Journal of the Shaanxi College of Economics and Business*), 2: 51–56.

Hall, P. (1984) *The World Cities*, New York: St Martin's Press.

Hu, B. (2000) "Cong gongyehua yu chengshihua de guanxi tantao woguo chengshihua wenti (Exploring China's urbanization through the relationship between industrialization and urbanization)," *Caijing yanjiu* (*Finance and Economics Research*), 8: 46–51.

Hu, X. W., Zhou, Y. X. and Gu, C. L. (2000) *Zhongguo yanhai chengzhen miji diqu kongjian jiju yu kuosan yanjiu* (*Studies on the Spatial Agglomeration and Dispersion in*

*China's Coastal Areas with Dense Concentrations of Cities and Towns*), Beijing: Kexue chubanshe.

Li, Y. H. (1995) "Yunyong xitong guandian yanjiu quyu jingji (Studying regional economy using systems perspective)," *Renwen zazhi* (*Humanities Journal*), 6: 57–59.

Liu, Y. M., Zhang, J. and Cao, M. M. (2002) "Shilun chengshi lao gongyequ tudi de zaikaifa (On the redevelopment of industrial land in old urban industrial districts)," in Cao, M. M. and Li, T. S. (eds), *Chengshi yu quyu yanjiu de lilun he shijian* (*Urban and Regional Studies – Theories and Applications*), Xi'an: Shaanxi renming chubanshe.

Lo, C. P. (1994) "Economic reforms and socialist city structure: a case study of Guangzhou, China," *Urban Geography*, 15: 128–149.

Ma, L. J. C. and Wei, Y. H. (1997) "Determinants of state investment in China: 1953–1990," *Tijdschrift voor economische en sociale geografie*, 88(3): 211–225.

Naughton, B. (1988) "The Third-Front: defence industrialization in the Chinese interior," *China Quarterly*, 115: 351–386.

Ning, Y. M. and Yan, Z. M. (1995) "Changing industrial and spatial structure in Shanghai," *Urban Geography*, 16: 577–594.

*Shaanxi ribao* (*Shaanxi Daily*, Xi'an), April 22, 1998, 5. "Biaozhun jituan you wang chengwei Zhongguo zuida fengzhi shebei gongye jidi (It is possible that the Biaozhun Group may become the largest production base for sewing machines in China)."

Shen, X. P. (1998) "Spatial inequality of rural industrial development in China, 1989–1994," *Journal of Rural Studies*, 15(2): 179–199.

—— (2001) "Regional variations and changes of industrial productivity in China, 1980–1995," *Asian Geographer*, 20(1–2): 53–78.

Suzhou Statistical Bureau (2002) *Suzhou Statistical Yearbook 2002*, Beijing: China Statistics Press.

Wan, D. F., Wang, Y. G. and Zhu, W. M. (2001) "Xibu chanye chuangxing fazhan ji chongzu de chubu yanjiu (A preliminary study on innovation and restructuring of industries in western China)," *Xi'an Jiaotong daxue xuebao: shekeban* (*Journal of the Xi'an Jiaotong University: Social Sciences Series*), 1(1): 6–29.

Wang, L. M. (1999) "Woguo xin yaou dalu qiao yanxian diqu jidian gongye fazhan yanjiu (On the development of engineering and electronic industry in the region along the new Euro-Asian continental bridge of China)," *Diyu yanjiu yu kaifa* (*Regional Research and Development*), 3: 21–24.

Wu, F. L. (1998) "Polycentric urban development and land-use change in a transitional economy: the case of Guangzhou," *Environment and Planning A*, 30: 1077–1100.

Xi'an Statistical Bureau (1996) *Xi'an Statistical Yearbook 1996*, Beijing: China Statistics Press.

—— (1999) "Xi'an," in *Xin Zhongguo chengshi 50 nian* (*Fifty Years of New China's Cities*), Beijing: China Statistics Press.

—— (2002) *Xi'an Statistical Yearbook 2002*, Beijing: China Statistics Press.

Yang, H. J. and Yin, H. T. (2002) "Xi'anshi zhongxin chengqu jizhun dijia pinggu yanjiu (A study on the evaluation of land base price of Xi'an city)," in Cao, M. M. and Li, T. S. (eds), *Cheng shi yu qu yu yan jiu de li lun he shi jian* (*Urban and Regional studies – Theories and Applications*), Xi'an: Shaanxi renming chubanshe.

Yeung, Y. M. and Hu, X. W. (eds) (1992) *China's Coastal Cities*, Honolulu, HI: University of Hawaii Press.

Yin, H. T. and Liu, K. W. (2002) "Xi'an chengshi wenti jiqi xinshiji chengshi kongjian fazhan gouxiang (Thoughts on the urban problems of Xi'an and its spatial development in the 21st century)," *Renwen dili* (*Human Geography*), 17(4): 85–88.

Zhang, H. N. (ed.) (2001) *Maixiang 21 shiji de Shaanxi gongye* (*Shaanxi industry Entering the 21st Century*), Xi'an, China: Shaanxi renmin chubanshe.

Zhang, J. Q. (1998) "Neilu diqu gongye waixianghua de daidong zhanlue – yi Xi'an diqu weili (The strategy of externalizing industries to drive the economy of the inland region – A case study of the Xi'an region)," *Shaanxi jingmao xueyuan xuebao* (*Journal of Shaanxi College of Economics and Business*), 6: 5–10.

Zhou, Y. X. and Ma, L. J. C. (2000) "Economic restructuring and suburbanization in China," *Urban Geography*, 21(3): 205–236.

—— and Meng, Y. C. (1997) "Shenyang de jiaoquhua – jianlun zhongxifang jiaoquhua de bijiao (Shenyang's suburbanization: with a comparison of suburbanization between China and the West)," *Dili xuebao* (*Acta Geographica Sinica*), 52: 289–299.

# 10 Residential mobility and urban change in China

## What have we learned so far?

*Si-ming Li*

### Significance of residential mobility for urban studies

The study of residential mobility is intimately tied to that of urban housing and residential differentiation. Housing as a commodity is characterized by durability, spatial fixity, lack of malleability, multi-dimensionality and expensiveness (Maclennan 1982). While *in situ* adjustments in housing consumption are not uncommon, their scope is very limited. By and large a change in housing consumption means a change in residence. Ever since the publication of Rossi (1955) classic piece, residential mobility has been conceptualized as a spatial adjustment process: to adjust housing consumption to changing needs. The latter may be caused by demographic changes within the family; equally it may be caused by changes that take place in the workplace, such as advancement along the job ladder. In the words of economists, a move is an attempt to restore consumption equilibrium from a state of disequilibrium (Hanushek and Quigley 1978). Impediments to residential mobility amount to impediments to attainment of Pareto optimality. Institutional arrangements such as providing adequate and accurate information about housing vacancies and enacting laws governing behaviors of real-estate agents should therefore be established to facilitate housing exchange.

While this largely demand-oriented spatial adjustment model has dominated mobility research for almost a half-century, its applicability depends to a significant extent on how the housing market is organized. In Britain, prior to the "right-to-buy" policy under Thatcher, council housing comprised a substantial proportion of the stock (Burrows 1999). Access to housing was significantly defined by eligibility criteria. In continental Europe the social housing sector is even larger, and so is the role played by the relevant housing authority (Bourne 1981). Incidentally it may be noted in both Britain and continental Europe the mobility rate tends to be lower than that in the United States and other countries where the market prevails. The relatively high rate of mobility in the United States, which used to be about 20 percent per annum but has declined somewhat in recent years (Long 1992), has been seen as evidence of market efficiency (Strassmann 2001).

But even in the United States, there are nonetheless gatekeepers who have substantial influence on housing accessibility. These include the financial institutions and real-estate agents (Bourne 1981). The state also plays an important part. Measures such as deduction of mortgage interest payment from taxable income and non-taxation of implicit rental income have been adopted to promote homeownership. They have been instrumental in fostering purchase of suburban homes by the upper and upper-middle income groups. With the dwindling of the rental market, the choice facing the lower income households, who have to rely on renting, is very much circumscribed (Moore and Rosenberg 1993). Particularly vulnerable are the single-parent and minority households (van Kempen and Ozuekren 1998). It is true that high mobility rates tend to characterize low-income renters (Clark and Dieleman 1996). Yet this does not mean that they can move out of "socially excluding" zones (Strassmann 2001). In a similar vein, "white flight" caused by redlining by the financial institutions and real-estate agents can be seen as manifestations of market failure (Bourne 1981).

From the above it is clear that an understanding of residential mobility is crucial to understanding how a city is composed and recomposed and how individuals and socioeconomic groups fare within the city. While the literature on residential mobility is voluminous, to date the bulk of studies are on cities in the West. Only in more recent years have scholars begun to undertake serious examination of mobility behaviors and the resulting patterns of move in Chinese cities. Some of the works published are Zhou's (1996) earlier work on Guangzhou, Chai *et al.*'s (2002) study of Tianjin, Dalian and Shenzhen, Li and Siu's (2001a and b) study of moves to newly constructed "commodity housing" in Guangzhou and Beijing, Li's (2003) analysis of residential mobility and tenure change, Li's (2004) and Li and Wang's (2003) analysis of the changing mobility rates in Guangzhou and Beijing, and Wu's (2004) study of Shanghai. *Environment and Planning A* recently published a special issue on residential mobility and housing choice in China (Li and Wu 2004). To these one may add Zhou and Ma's (2000) study of suburbanization in Beijing. Perhaps it is about time to review our stock of knowledge, assessing to what degree residential mobility in China differs from that in the West, and how residential mobility underlies residential structuring in Chinese cities.

Below I first examine evidence on the frequency of move. Next I review findings on the spatial patterns of residential moves. I then look at how the mobility incidence varies between demographic and socioeconomic groups. New housing construction almost by definition will result in moves. Efforts are made to examine how differential mobility inclinations and the phenomenal housing construction programs over the past two decades have together brought about new dimensions of residential differentiation in urban China. The twenty-plus years of urban housing reform is about to come to conclusion with the cessation of welfare allocation of housing in 1998. Based on what we have learned in the past, I make conjectures on the socio-spatial implications of this latest round of housing reform.

## How frequent do urban dwellers in China move?

Earlier studies provided only anecdotal evidence. Wu's (1990) study must be among the earliest on this topic. Wu surveyed four neighborhoods in Beijing, Nanjing, Tianjin and Wuxi. In his sample, 2–8 percent of the respondents had stayed in their residence for less than five years, 37–52 percent for 5–10 years, 26–40 percent for 10–20 years and 14–21 percent for more than 20 years. The implied mean duration of stay would be slightly more than ten years, and the implied mobility rate would be in the mid to high single digits. Zhou's (1996) survey of inner-city districts of Guangzhou in 1990 gives somewhat longer durations of stay: 17 percent of the respondents had lived in the current residence for less than five years, 8 percent for 5–10 years, 31 percent for 10–20 years and 43 percent for more than twenty years. But there were also quite a number of moves that had taken place within five years prior to the survey. With the deepening of the economic reform in the 1990s, a number of authors (Li and Siu 2001a and b; Zhou 1996) have hypothesized an increase in mobility in more recent times.

To my knowledge, the first serious attempts at estimating the residential mobility rate in urban China are Li (2004) for Beijing, and Li and Wang (2003) for Guangzhou, both based on retrospective residential histories. The Beijing residential mobility rate obtained averaged 4.29 percent per annum over the period 1980–2001. This rate should be adjusted upward by about 0.6 percent per annum after taking into account age cohort effects arising from use of retrospective life history data. The Guangzhou rate for the same period averaged 5.3 percent per annum. Again, this rate should be raised by 0.6 percent after accounting for age cohort effects.

We noted the relatively high mobility rate of the United States. But in most West European countries the residential mobility rate generally is in the region of 7–9 percent per annum. In this light the rates obtained for Beijing and Guangzhou were on the low side, but they were not particularly low. In fact, in comparison with cities in other economies in transition, these rates might even be considered high. In Moscow in 1992, the residential mobility rate was only 2.5 percent per annum (Daniell and Struyk 1997). In Slovenia where housing reform was implemented as far back as the mid-1960s, the mobility rate varied between 2.1–3.1 percent per annum over the period 1990–1994 (Mandic 2001).

In both Beijing and Guangzhou, the mobility rate has oscillated with broad movements in the macro-economy (Figure 10.1). However, in Beijing the general trend is a slightly downward one after peaking at about 5.5 percent per annum in 1985; in Guangzhou the general trend is a rapidly rising one, with the rate staying below 1 percent in the early 1980s but topping 10 percent per annum in 1998. Further evidence is given by Chai *et al.* (2002) for Shenzhen, again based on retrospective data. Similar to the case of Guangzhou, in Shenzhen the implied mobility rate was less than 1

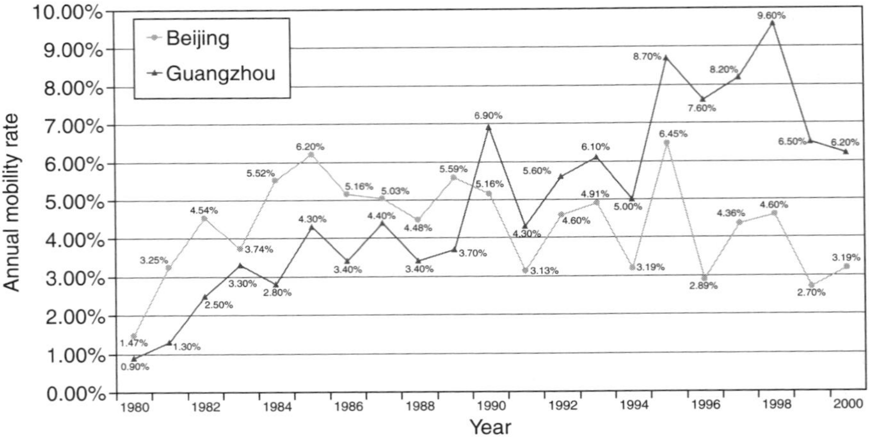


*Figure 10.1* Annual mobility rate in Beijing and Guangzhou, 1980–2000 (source: Li 2004; Li and Wang 2003).

percent per annum in the early 1980s; it increased rapidly to more than 10 percent per annum in the late 1990s.

Thus, in cities with a more open economy such as Guangzhou and Shenzhen the mobility rate has increased sharply in recent years. This may suggest increased applicability of the spatial adjustment model. A corollary of this is the presence of increasing similarities between dimensions of residential differentiation in these Chinese cities and cities in the West. For instance, lifecycle stages are likely to manifest spatially as series of concentric zones; and income and socioeconomic status in general are likely to become an important differentiator of urban neighborhoods. However, the same may not be said for cities in which state sectors still dominate, such as Beijing. In the latter cities, growing homeownership with only partial property rights in conjunction with the selling of housing formerly owned by state work-units to their employees might have caused the slight decline in the mobility rate observed. This would lead to a freeze of the residential structure over a rather prolonged period of time. I shall come back to this later.

## Spatial patterns of residential moves in Chinese cities

### *Spatial variations in the mobility incidence*

In North American cities residential mobility varies systematically over space. In particular mobility is generally much higher in the central city than in the suburbs. Mobility in run-down neighborhoods in the "zone of transition" (Burgess 1925) is especially high. To the Chicago School, this is an indicator of social pathology. In contemporary China this social pathology thesis may find some relevance in the so-called "urban villages" located on the former urban–rural fringe where the new migrants to the city congregate (Wang and Murie 2000). Despite their high mobility, however, these urban transients are very much trapped in such "villages," as they have largely been denied access to housing opportunities elsewhere. In the main parts of the urban area, the mobility rates are generally much lower. Between the central core and the newly developed districts there is evidence that mobility rates in the latter are somewhat higher than in the former (Li and Wang 2003).

### *Distance and direction of move*

Previous studies have established that moves tend to take place within the same neighborhood (Cadwallader 1996). However, longer moves, especially those that can be described as "flight" to the suburb, are not uncommon (South and Crowder 1997). The latter, in particular, have produced contrasting social and physical landscapes between central city and suburban communities in America and elsewhere. In urban China, evidence so far shows that the majority of residential moves are also of short distance. Li and Siu's (2001a) survey of recent movers in Guangzhou in 1996 shows that 40 percent of the moves in their sample involved a distance of less than 2 km. The great majority of moves took place within the same urban district or were to an adjacent district. Long-distance moves and leapfrogging were rare. But outward moves (35.9 percent of the sample) far outnumbered inward moves (3.0 percent of the sample), suggesting a general trend towards suburbanization.

Data from a survey of Beijing movers in 1996 also show that large proportions of the moves were within the same urban district; indeed 30 percent of moves were within a distance of 2 km (Li and Siu 2001b). But in the Beijing sample longer distance moves were also important. Again, there was a clear process of suburbanization, with out-migration far exceeding in-migration. The end outcome was a steady decline in central city population and rapid increase in the population of the inner suburban districts (Zhou and Ma 2000). Similar movement patterns are found in other cities. For example, Chai *et al.*'s (2002) survey shows that in Dalian, 48 percent of moves were within 2.5 km; in Shenzhen the figure was 64 percent. But moves to newly developed districts tended to be of longer distance, and leapfrogging was not uncommon.

## New housing construction and residential mobility

A major impetus underlying residential moves is new housing construction. The past two decades witnessed unprecedented building activities in

Chinese cities. Total housing floor space increased from 1,130 million $m^2$ in 1985 to 6,650 million $m^2$ in 2001, or close to 600 percent in a period of sixteen years (SSB 2002a: 364). Correspondingly, per capita gross housing space consumed increased from 10.0 $m^2$ in 1985 to 20.8 $m^2$ (SSB 2002a: 319). Data from the 2000 Population Census show that for the entire nation, an astonishing 34.01 percent of the housing stock was constructed in the 1980s and a further 47.57 percent was constructed in the 1990s. In Shanghai, 55.5 percent of the 2000 stock was completed in the 1990s alone (SSB 2002b: 1852). Commodity housing constitutes the bulk of this new stock.

Most of the urban housing built since the mid-1980s has been in the form of large housing estates variously labeled as *xiaoqu* (small [residential] districts). A large proportion of these are new developments in the inner suburban ring. But there have also been large-scale redevelopments of both the pre-1949 urban core and the work-unit compounds that surround it (Sit 1999; Wang and Murie 1999). Under housing reform, work-units were encouraged not to build their own housing but to buy housing built by development companies at supposedly market price (Li 2000a). Such housing was termed commodity housing (*shangpinfang*), although throughout much of the late 1980s and 1990s work-units were the main buyers. Once bought, the housing units would be taken out of the sphere of circulation, i.e., decommodified. Within the work-units, established rules of housing allocation still applied. Nevertheless, this production side commodification has added a great deal of variety to the housing stock, both in terms of building layout and internal fixtures, and of neighborhood composition and geographical location. Thus, there are now housing estates constructed for the masses; there are also gated communities with twenty-four-hour security and clubhouse facilities constructed for expatriates and other high-income groups. Even American-style suburban developments are now evident in Beijing, Shanghai and other major cities of the country.

Because the great majority of buyers of commodity housing were state work-units and the sellers were also state enterprises, problems of soft-budget constraint prevailed. In particular, there were tendencies to over-speculate, resulting in high vacancies and ever-escalating prices. In Beijing, for example, the average selling price of residential buildings rose from RMB 1,613 per $m^2$ in 1992 to RMB 5,337 per $m^2$ in 1997 (Lau and Li 2004). Housing in the open market was beyond the reach of all but the few proprietors of private enterprises, prominent actors in the entertainment industry, and expatriates and overseas investors (Wu 2001). Excessive speculation was also fueled by dual-track land allocation under the system of paid transfer of land use rights, which amounted to land recommodification. This caused re-emergence of the urban land rent gradient and hence possibilities of rent gap exploitation, i.e., exploiting the difference in land rent under the existing use and under alternative uses upon redevelopment (Wu 1999). Attempts to exploit the rent gap by the work-units and also by the municipal government have instigated extensive redevelopment activities. Resettlement of households affected can be seen as involuntary migration. In the 1996 sample of Beijing movers, 21.2 percent belonged to this group; in the Guangzhou sample, 14.9 percent (Li 2000b). Resettled households also account for similar percentages of movers in Chai *et al.*'s (2002) study of Shenzhen, Tianjin and Dalian. Wu's (2004) study of Shanghai reports an even larger percentage of resettled households: 34.5 percent. Wu further shows while a move generally brings about improvement in housing consumption, the improvement associated with involuntary moves is smaller than those with more voluntary ones.

## Socio-demographic differentials of residential mobility

### *Demographic variations*

One of the most important variables that underlie residential mobility is a person's age, which largely determines a person's position in the family lifecycle, another major differentiator of residential mobility (Clark 1982). A number of authors have argued that it is change in lifecycle stage rather than lifecycle stage *per se* that is of importance in triggering a move. There are other triggers in one's life course, for example, promotion and job change (Clark and Dieleman 1996). The frequency that these triggers occur tends also to vary systematically with age. Li's (2004) analysis of retrospective life histories in Beijing shows that the relation between age and mobility rate exhibits the usual curvilinear shape. However, mobility peaks at around thirty-five years of age, as compared with less than twenty years of age for the United States (Long 1992). In the Guangzhou study (Li and Wang 2003), the mobility rate peaks at an even later age, around forty-five years of age. The relatively high mobility peaking age for Chinese cities is perhaps indicative of availability of disproportionately more housing opportunities for people more advanced in their career in transitional economies.

As for the relation between age and direction of move, Li and Siu (2001b) reveal that in Guangzhou, as of the mid-1990s, younger households were more likely to be associated with suburban residence than were households headed by middle-age persons. However, in Beijing, the effect of age on suburban residence was not significant. Apparently, the relatively high mobility rates of the middle-age group are not a major force behind China's suburbanization.

### *Socioeconomic status*

#### *Occupational rank and educational attainment*

In redistributive economies, occupational rank is perhaps the single most important factor affecting access to resource of all kinds, housing included

(Szelenyi 1978). In the 1996 survey of recent movers in Guangzhou and Beijing, a disproportionately large percentage of the respondents belonged to the professional and managerial class (48 percent of the Guangzhou sample and 37 percent for the Beijing sample) (Li 2000b). Similarly, there were also disproportionate shares of white-collar workers in Chai *et al.*'s (2002) samples of recent movers in Dalian, Tianjin and Shenzhen. Education correlates closely with occupational rank. Li's (2004) study on Beijing and Li and Wang's (2003) study on Guangzhou show that level of education attainment was significantly and positively related to probability of move. In addition, advancement along the job ladder was a highly significant variable triggering a move.

### *Income*

In redistributive economies income levels tend to be kept artificially low. Workers are as much rewarded in kind as in cash. With implementation of the reform, the effect of income on housing consumption and on residential location decisions would have increased over time. Daniell and Struyk (1997) observed that in Moscow the move probability increased significantly with income after the wholesale disposal of public-sector housing in 1992. Because income data based on retrospective data were quite dubious, neither Li (2004) nor Li and Wang (2003) included this in their list of regressors. More indirect evidence, however, suggests that there is a positive relation between income and mobility in Chinese cities. In the 1996 survey of movers in Guangzhou and Beijing, the mean household incomes in both samples were substantially higher than the mean family incomes given by the statistical yearbooks of these two cities (Li 2000b).

### *Party membership*

Reforms in China have largely been confined to the economic sphere. To date, China remains a one-party state. Membership in the Chinese Communist Party (CCP) enhances access to housing and other resources (Li 2000b; Logan *et al.* 1999). Li's (2004) and Li and Wang's (2003) analysis of retrospective life histories in Beijing and Guangzhou, respectively, show that CCP membership also enhances mobility probability.

It may be concluded from the above that in urban China higher socioeconomic status is generally associated with higher move probability. How these socioeconomic differentials in the mobility rate are translated to the composition and recomposition of urban social space is another question. It depends, *inter alia*, on where the housing opportunities are located. Most new housing built since the late 1980s has been in the form of massive housing estates located in more outlying areas, outside the ring of work-unit compounds. The positive correlation between socioeconomic status and mobility probability implies that the high-status groups would be the first to move out of work-unit compounds. Given the distribution of housing opportunities, it is likely that a large proportion of the high-status groups would move to the outer ring of new housing estates. This would be the case because work-units until quite recently still dictated who got where and how much. Indeed, analyses based on the 1996 surveys of Beijing and Guangzhou have shown that in both instances membership in the CCP, which influences not only the amount of housing one consumes but also one's position in the job ladder, is positively related to suburban residence. Bivariate analysis also shows high correlation between occupational rank and suburban residence.

### *Work-unit type*

In China a major source of housing inequality is the nature of the organization in which an individual works (Logan *et al.* 1999). Li and Wang (2003) reveals that in Guangzhou workers in collective-owned and private enterprises are less likely to move than are workers in state and party organizations and in state-owned enterprises, but workers in share-holding companies are the most likely to move. It may be noted that the majority of share-holding companies are in fact state-owned enterprises having undergone corporate transformation. Insofar as a move generally brings about improvement in housing consumption (Wu 2004), the higher mobility rates for people working in state-owned enterprises and state and party organizations, i.e., people working in the traditional work-units, are indicative of their better position to benefit from improvements in housing provision. And insofar as most new housing is located in the suburban ring, the traditional work-units have been a major mover behind China's suburbanization (Zhou and Ma 2000). This point is substantiated by Li and Siu (2001a and b) in their analysis of the 1996 Beijing and Guangzhou samples of recent movers.

## Housing tenure and residential mobility in urban China

Cadwallader (1996) and others have pointed out that housing tenure is among the most important factors affecting residential mobility. In particular, homeownership results in substantially lower mobility. The quarter century of housing reform in China has produced a complex mix of housing tenures (Li 2000a, 2003). Traditionally, work-units monopolized housing provision. Rental occupancy was the norm, with rents set at nominal levels (Li 1995). Wu (1996) observes that work-units' dominating position had been strengthened even further in the 1980s and 1990s with devolution of decision-making powers to the enterprises. This is supported by Li's (2003) analysis of recent movers in Beijing, although the same study shows that in Guangzhou many former occupants of work-unit housing have become owner occupiers in the open market sector.

In China, housing provided by work-units is regarded as public housing. The housing reform until quite recently involved largely the raising of public housing rents and the selling of work-unit housing to workers at highly discounted prices (Wang and Murie 2000). Buyers of housing sold by the work-units, variously known as reform housing, do not enjoy full property rights. More specifically, ownership of reform housing is subject to strict resale restrictions and cannot be rented out. This may be contrasted with ownership of homes purchased directly in the market, which enjoys full property rights. As of the mid- and late-1990s, owner-occupation in the open market sector only comprised a very small percentage of the total stock, although ownership in general already comprised 46.6 percent of a national sample surveyed in 1996 (Huang and Clark 2002). According to the 2000 Population Census, as of 2000, in Beijing ownership of reform housing outnumbered full owner-occupation by a ratio of 7:1; in Shanghai the ratio was 2:1 (SSB 2002b: 1865). Until quite recently, then, homeownership in urban China essentially means ownership of reform housing.

Li's (2004) analysis of retrospective life histories in Beijing shows that homeownership in China, like homeownership elsewhere, results in a significant drop in the mobility rate: from around 4 percent per annum to less than 1 percent per annum. Li and Wang's (2003) study on Guangzhou reveals further that owning reform housing is associated with the lowest mobility. Recall that a reform housing unit, once sold, will be removed from the sphere of circulation. Therefore, unless the resale restrictions are relaxed, housing privatization would have curtailed rather than enhanced residential mobility. Work units might be the most important agent behind suburbanization in urban China in the 1980s and 1990s. Yet by selling reform housing to their workers, work-units also helped freeze the pattern of socio-spatial mix of given neighborhoods at the very time they were formed. The freezing of the housing stock also means that the vacancy chains generated by new housing construction would be short. This helps explain why despite the massiveness of housing construction activities in 1990s, in Beijing where work-units still dominate housing provision residential mobility remained quite low.

## Intra-urban migration and changing patterns of residential differentiation

High levels of construction activities together with differential intra-urban migration have helped engender a rather unique pattern of residential differentiation in Chinese cities. The new suburban ring of commodity housing estates built since 1985 now dominates the urban landscape. Above, it was hypothesized that the high-status groups are the first to move out of work-unit compounds to these housing estates. In addition to the evidence presented above, this hypothesis is also supported by two factorial ecological studies of Guangzhou. The first is Yeh, Xu and Hu (1995) using data collected in 1984, and the second is Zheng, Xu and Chen (1995) based on the 1990 Population Census. While the major dimensions of residential differentiation extracted and the major classes of social areas identified differed very little, the zones labelled as "cadres areas" and "intellectual areas" extended outward to the east by quite a substantial margin between 1984 and 1990.

Nonetheless, this correlation between socioeconomic status and distance to the city center would not be strong. Wu (1996) points out that in an increasingly uncertain environment, in order to solicit support from their subordinates cadres of work-units resorted to providing better housing to the ordinary workers. As a result, large numbers of ordinary workers also moved out of the work-unit compounds to the suburban commodity housing estates. At the same time, substantial numbers of people affected by redevelopment projects were resettled in suburban resettlement housing blocks. Thus, unlike suburbs in America, suburbs in Chinese cities are far from uniform. There are housing estates accommodating the cadres and professionals. There are also residential blocks for lower status groups. Moreover, there are the urban villages with housing of dubious legal status catering for the millions of migrants. The suburban rings too contain highly exclusive gated communities of luxurious condominiums and detached housing targeted at the very rich. As such, the suburban residential space in China is highly differentiated; residential segregation takes place at a refined geographic scale.

Extensive redevelopment of the inner-city core and work-unit compounds has also drastically changed the landscape of the pre-1985 city. Central business district (CBD) functions have been revived (Gaubatz 1999). There is evidence that urban Chinese have strong preference for central locations (see below). The revival of the CBD has further increased their attractiveness. Responding to the structure of housing preference, developers have built luxurious residential towers in centrally located sites to target at proprietors of private business and other high-income groups. Socially, like the post-1985 suburban ring, the areas developed before 1985 are also quite heterogeneous. There are low-status groups including workers laid off by the work-units and retirees who still occupy run-down housing in work-unit compounds and in central city neighborhoods that have not yet been redeveloped. There are also the urban new rich residing in upmarket apartments in gentrified neighborhoods. Wang and Murie (2000) construct an idealized map depicting this pattern of residential distribution that characterized Chinese cities at the turn of this century.

## Recent housing reform

In 1998 the State Council decided to end welfare allocation of housing. This means that work-units will no longer provide housing to their workers, although many work-units continue to provide cash subsidies to support home purchase (Li 2003). In order to "catch the last train," many bought homes in 1998 and 1999, the final years in which reform housing was sold at prices 80–90 percent below prevailing market levels. Another major breakthrough concerns the conferment of full property rights to owners of reform housing. Owners of reform housing are now allowed to sell or rent their homes in the market and retain the bulk, if not the whole, of the resale proceeds or rental incomes. This helps develop a thriving second-hand housing market and rental market. But conferment of full property rights amounts to conferment of windfall profits to owners of reform housing. This will set in train a series of moves, the consequences of which are destined to be profound.

Given that the current residence has been frozen for quite some time, it is likely that the housing needs and circumstances external to the household would have experienced quite drastic changes since the time the household bought the reform housing dwelling. In addition, the introduction of new housing types such as luxurious condominium complexes, and developers' aggressive advertising campaigns could result in substantial changes in housing preference. For example, estate management would now become important in housing decisions. Also, the production arm of the work-unit might have been relocated to a more outlying location, necessitating a change in the commuting pattern. Furthermore, construction of metro systems and rising car ownership has offered a much wider choice of residential location possibilities. With the resale proceeds, the individual household may think about trading its old reform housing dwelling for a place that better suits its current needs and financial means. Alternatively, the household may opt to mortgage its existing home and make use of the money borrowed to finance the new home purchase. In other words, it may decide to simultaneously own two homes. The household can also choose to rent his old property out.

The rapidly rising mobility in Guangzhou and Shenzhen in recent years suggests that in cities which have been pioneers in market transformation quite a larger number of households are ready to enter the market again. With conferment of full property rights, it is likely that the mobility rate in other cities will also increase. Most likely, it would be those who have benefitted most from the housing reform, in particular party and government cadres and managerial and professional workers, who are also the first to take advantage of the new opportunities. These higher-status workers were the ones that had the means and opportunities to buy reform housing. Also the homes that they bought tend to be of better quality and larger size and hence command higher resale values (Li 2000b). As such, they stand to benefit most from the windfall gains arising from the conferment of full property rights to their reform housing units.

These high-status workers are also better informed of market opportunities. Their position in the workplace allows them much better access to information. *Guanxi* or personal relations would enable them to get favorable treatments such as price discounts from the developers. In addition, housing and other in-kind subsidies are now incorporated in the monthly salary. With the phenomenal rise in the income spread between managerial and professional staff and ordinary workers under enterprise reform, these high-status workers now also possess the financial means to purchase homes directly in the market. Ready availability of mortgage loans will facilitate purchase. Some seventy years ago, Hoyt (1939) hypothesized that movement of the high-status groups largely defines how a city is organized spatially. Perhaps there is some truth in this hypothesis in Chinese cities today.

Both Zhou and Ma (2000) and Li and Siu (2001b) note the strong preference for central locations in China. Wang and Li's (2004) preference experiments demonstrate that it is only those inner-city districts with good reputation that are preferred. Given the structure of location preference, it is likely that sizeable numbers of high-status workers, who had previously bought reform housing and moved to more outlying locations, will now consider moving back to central city districts. Developers will undertake inner-city redevelopment projects accordingly. As was noted above, there are immense incentives for the municipal government to facilitate land resumption so as to generate incomes from rent gap extraction. Resistance by local communities, though not entirely absent, is generally weak, given the continual dominance of the state. Under such circumstances, gentrification of inner-city neighborhoods is likely to take place at a massive scale. Of course, the "back to city" move may not be the choice for everyone. With rising incomes and availability of the private car, a sizeable proportion of the high-status groups may opt for more suburban living. There are signs that this is already happening in Shanghai, Beijing, Guangzhou and other major cities in the country.

Residential moves generate vacancy chains. Yet, to my knowledge, as far as research on urban China is concerned, no work has been done on this topic. There is a lack of systematic data on the series of "sequence occupancies" or the "filtering" of a residential premise from one occupant to another. Much of the reform housing, especially the ones occupied by the high-status groups, is quite new and of quite good quality; hence they should command reasonably high prices. It is highly unlikely that the lower income strata, such as the middle-age workers laid off by the work-units and the recent migrants from remote villages, can afford to move to such premises. Instead, there is a good chance that the premises vacated by owners of reform housing will be taken up by young professionals and administrative workers who have just started their career. These young entrants to the workforce are no longer eligible for work-unit housing. The reform housing units vacated would be a good place to start their quest for housing. As for the urban poor, the situation does not look very promising. There are of course those displaced by redevelopment projects who are lucky enough to be provided with compensation housing. But there are many more that do not have this privilege. With the withdrawal of the state from housing provision, the recent migrants and other low-income groups will continue to have to seek residence in the urban villages, if the local government still tolerates their presence, or to find housing in the increasingly dwindling stock of old inner city tenements and run-down work-unit housing blocks.

## Concluding remarks

The past few years have seen an increase in interest in micro-level studies on urban housing and residential change in Chinese cities. This chapter highlighted some recent research findings. We now have a much better idea about how frequent and under what conditions residential move takes place. For instance, we know that under gradualist reform the housing market in different cities is organized differently. As a consequence, in some cities mobility has increased while in others it has not. More specifically, while the selling of reform housing by work-units is a major impetus underlying China's suburbanization, ownership with strict resale restrictions has resulted in the freezing of the residence subsequently. High-status groups such as cadres and professionals were the first to buy reform housing and hence the first to move to suburban locations. But there are strong preferences in China for central locations. With the cessation of welfare allocation of housing, the Chinese authority now permits reform housing to enter the market. This will likely set off a series of moves, with the high-status groups again leading the way. Some will move back to central city locations, causing extensive gentrification. Others, with availability of the private car and much improved urban highway networks, may seek residence in luxurious condominiums and single family housing in suburban gated communities. The premises the high-status groups vacate will be filtered downward to others down the social ladder. But very likely the urban poor will have to be contented with the dwindling stock of rundown work-unit housing and old tenements and the dubious housing in the urban villages. It is as yet too early to assess the full impacts of the latest round of housing reform. Concrete evidence is lacking. The above is meant to be speculative and represents no more than an educated guess. Nevertheless, such conjectures on how different socioeconomic groups may respond to the policy change are useful as they point to areas that demand future research.

## Acknowledgments

This research is supported in part by Hong Kong Research Grant Council Grant No. HKBU 2080/99H and Hong Kong Baptist University Faculty Research Grant No. FRG FGR 30-02-0254.

## References

Bourne, L. S. (1981) *The Geography of Housing*, London: Arnold.

Burgess, E. W. (1925) "The growth of the city: an introduction to a research project," in Park, R. E., Burgess, E. W. and McKenzie, R. D. (eds), *The City*, Chicago: University of Chicago Press, 47–62.

Burrows, R. (1999) "Residential mobility and residualization of social housing in England," *Journal of Social Policy*, 28(1): 27–52.

Cadwallader, M. (1996) *Urban Geography: An Analytical Approach*, Upper Saddle River, New Jersey: Prentice Hall.

Chai, Y., Lin, Z., Li, Z., Gong, H., Shi, H. and Wu, Z. (2002) *Zhongguo Chengshi de Shikongjian Jiekou (Spatial-Temporal Structure of Chinese Cities)*, Beijing: Peking University Press.

Clark, W. A. V. (1982) "Recent research on migration and mobility: a review and interpretation," *Progress in Planning*, 18(1): 1–56.

—— and Dieleman, F. M. (1996) *Households and Housing: Choice and Outcomes in the Housing Market*, New Brunswick, New Jersey: Center for Urban Policy Analysis, Rutgers University.

Daniell, J. and Struyk, R. (1997) "The evolving housing market in Moscow: indicators of housing reform," *Urban Studies*, 34(2): 235–254.

Gaubatz, P. (1999) "China's urban transformation: patterns and processes of morphological change in Beijing, Shanghai and Guangzhou," *Urban Studies*, 36(9): 1495–1521.

Hanushek, J. and Quigley, J. M. (1978) "An explicit model of intra-metropolitan mobility," *Land Economics*, 54: 411–429.

Hoyt, H. (1939) *The Structure and Growth of Residential Neighborhoods in American Cities*, Washington, DC: Federal Housing Administration.

Huang, Y. Q. and Clark, W. A. V. (2002) "Housing tenure choice in transitional urban China: a multilevel analysis," *Urban Studies*, 39(1): 7–32.

Lau, K. M. and Li, S.-M. (2004) "Housing affordability in Beijing, 1992–2001," a paper presented at the Asia-Pacific Network for Housing Research Conference on Housing and Social Development, 5–6 February 2004, University of Hong Kong.

Li, S.-M. (1995) "*Zhongguo chengzhen de zhufang tizhi gaige* (Institutional reforms in China's urban housing)," in Li, S. M., Siu, Y. M. and Mok, T. K. (eds), *Zhongguo shehui fazhan (Social Development in China)*, Hong Kong: Commercial Press, 324–344.

—— (2000a) "Housing market and tenure decision in Chinese cities: a multivariate analysis of the case of Guangzhou," *Housing Studies*, 15(2): 213–236.

—— (2000b) "Housing consumption in urban China: a comparative study of Beijing and Guangzhou," *Environment and Planning A*, 32: 1115–1134.

—— (2003) "Housing tenure and residential mobility in urban China: a study of commodity housing development in Beijing and Guangzhou," *Urban Affairs Review*, 38(4): 510–534.

—— (2004) "Life course and residential mobility in Beijing, China," *Environment and Planning A*, 36(1): 27–44.

—— and Siu, Y-M (2001a) "Residential mobility and urban restructuring under market transition: a study of Guangzhou China," *Professional Geographer*, 53(2): 219–229.

—— (2001b) "Commodity housing construction and intra-urban migration in Beijing: an analysis of survey data," *Third World Planning Review*, 23(1): 39–60.

—— and Wang, D. G. (2003) "Life course and residential mobility in Guangzhou," Hong Kong Baptist University, Centre for China Urban and Regional Studies, Occasional Paper No. 39.

—— and Wu, F. L. (2004) "Contextualizing residential mobility and housing choice: evidence from urban China," *Environment and Planning A*, 36(1): 1–7.

Logan, J. R., Bian, Y. and Bian, F. (1999) "Income inequality in urban China in the 1990s," *International Journal of Urban and Regional Research*, 23: 7–25.

Long, L. (1992) "Changing residence: comparative perspectives on its relationship to age, sex, and marital status," *Population Studies*, 46: 141–158.

Maclennan, D. (1982) *Housing Economics: An Applied Approach*, London and New York: Longman.

Mandic, S. (2001) "Residential mobility versus 'in-place' adjustments in Slovenia: viewpoint from a society 'in transition,'" *Housing Studies*, 16(1): 53–73.

Moore, E. G. and Rosenberg, M. W. (1993) "Migration, mobility and population redistribution," in Bourne, L. and Ley, D. (eds), *The Changing Social Geography of Canadian Cities*, Montreal and Kingston: McGill-Queen's University Press, 121–137.

Rossi, P. H. (1955) *Why Families Move*, New York: Macmillan.

Sit, V. F. S. (1999) "Social areas in Beijing," *Geografiska Annaler*, 81B(4): 203–221.

South, S. J. and Crowder, K. D. (1997) "Residential mobility between cities and suburbs: race, suburbanization, and back-to-city moves," *Demography*, 34(4): 525–538.

State Statistical Bureau of China (SSB) (2002a) *China Statistical Yearbook 2002*, Beijing: China Statistical Press.

State Statistical Bureau of China (SSB) (2002b) *Tabulations on the 2000 Population Census of China*, Beijing: China Statistical Press.

Strassmann, W. P. (2001) "Residential mobility: contrasting approaches in Europe and the United States," *Housing Studies*, 16(1): 7–20.

Szelenyi, I. (1978) "Social inequalities in state socialist redistributive economies," *International Journal of Comparative Sociology*, 19: 63–87.

Van Kempen, R. and Ozuekren, A. S. (1998) "Ethnic segregation in cities: new forms and explanations in a dynamic world," *Urban Studies*, 35(10): 1631–1656.

Wang, D. G. and Li, S.-M. (2004) "Housing preferences in a transitional housing system: the case of Beijing, China," *Environment and Planning A* 36(1): 69–88.

Wang, Y. P. and Murie, A. (2000) "Social and spatial implications of housing reform in China," *International Journal of Urban and Regional Research*, 24(2): 397–417.

Wu, F. L. (1996) "Changes in the structure of public housing provision in urban China," *Urban Studies*, 33(9): 1601–1627.

—— (1999) "The 'game' of landed-property production and capital circulation in China's transitional economy, with reference to Shanghai," *Environment and Planning A*, 31: 1757–1771.

—— (2001) "Housing provision under globalization: a case study of Shanghai," *Environment and Planning A*, 33: 1741–1764.

—— (2004) "Intraurban residential relocation in Shanghai: modes and stratification," *Environment and Planning A*, 36(1): 8–26.

Wu, J. (1990) *Zhongguo Chengshi Xingtai, Jiekou, Tezheng Ji Yanbian (The Form, Structure and Characteristics of Chinese Cities and Their Evolutions)*, Nanjing: Jiangsu Keshue Jishu Chubanshe.

Yeh, A. G. O., Xu, X. Q. and Hu, H. Y. (1995) "The social space of Guangzhou city, China," *Urban Geography*, 16(7): 595–621.

Zheng, Z., Xu, X. and Chen, H. (1995) "Guangzhou shi shehui kongjian de yinzhi shengtai zai fengxi (A re-analysis of the factorial ecology of Guangzhou City)," *Dili Yanjiu (Geographical Research)*, 14(2): 15–26.

Zhou, C. S. (1996) A *Study on Population Distribution and Mobility in Large Cities Since Reform: The Case of Guangzhou City (Gaige Kaifang Yilai Dadushi Renkou Fenbu Yu Qiangyi Yanjiu: Yi Guangzhoushi Weili)*, Guangzhou: Guangdongsheng gaodeng jianyu chubanshe.

Zhou, Y. X. and Ma, L. J. C. (2000) "Economic restructuring and suburbanization in China," *Urban Geography*, 21(3): 205–236.

# 11 From work-unit compounds to gated communities

## Housing inequality and residential segregation in transitional Beijing

*Youqin Huang*

### Introduction

Housing provision and consumption in Chinese cities have undergone profound changes since the nationwide launch of housing reform in 1988. While previously subsidized public housing is being privatized through massive sell-off, new private housing provided by developers ("commodity housing" or *shang ping fang*) has mushroomed. In the past, urban households had no choice but to wait for public rental housing, but now they can choose between rental and homeownership, and between public and private housing. With the newly granted freedom of housing choice and residential mobility, urban households in China began to be sifted and sorted socially, economically and spatially. A relatively homogeneous society organized around work-unit compounds in socialist urban China is evolving into one with significant stratification and segregation. Wealthy "gated communities" and dilapidated "migrant enclaves," neither of which existed in the socialist era, are now emerging side by side in many Chinese cities.

This chapter examines the dynamics of increasing housing inequality and emerging residential segregation in Chinese cities conceptually and empirically. I hypothesize that both market forces and socialist institutions such as the household registration system and work-units contribute to housing inequality and neighborhood sorting. Using the latest 2000 Census and fieldwork data, the case study of Beijing shows significant housing inequality across education and occupation, and unprecedented residential segregation at the neighborhood level. In addition to providing an understanding of housing inequality, this research sheds some light on the overall inequality emerging in transitional China.

### Housing inequality and residential segregation: a conceptual understanding

Housing inequality and residential segregation is one of the central topics in social sciences. Despite the longstanding debate on what is the main driving force, scholars generally agree that a set of factors including socioeconomic status, racial discrimination, personal preference, and urban economic structure contributes to housing inequality and residential segregation in the USA (e.g. Burgess 1967; Clark 1986; Galster 1988; Massey and Denton 1993). Yet, much of the existing literature focuses on housing systems where private housing dominates and market mechanism prevails, which makes it problematic to apply existing theories directly to a different housing system. Even in Western European housing systems that have a relatively large share of public housing, a different conceptualization taking government behavior into account is needed, as demonstrated by the recent "residualization" in public housing in Britain as a result of the "Right to Buy" program initiated in 1980 (Forrest and Murie 1999; Burrow 1999).

A different conceptualization is important to understand housing inequality and residential segregation in Chinese cities, where the housing system has been welfare-oriented and the residential space has been organized with very different rules. Till the end of the 1990s, housing in Chinese cities has been mainly provided by employers (work-units) and local governments (municipal housing bureaus), whose allocation is based on a set of non-monetary factors such as job rank, seniority and marital status (Bian *et al.* 1997). While there were evidences of housing inequality (Logan *et al.* 1999), the overall level of housing consumption was low (Zhang 1998; Huang 2003a) and the magnitude of housing inequality was relatively small compared to Western standards. Furthermore, the prevailing philosophy of combining work and residential space in socialist urban planning has resulted in a cellular residential landscape dominated by work-unit compounds built next to employment centers (Ma 1981; Gaubatz 1995). With uniform apartment buildings accommodating employees from high-ranking officials to ordinary workers, the residential pattern in socialist Chinese cities has been characterized by relatively homogeneous work-unit compounds, often surrounding a historical old housing district in inner cities.

The housing reform in Chinese cities aims to introduce market mechanisms into the welfare-oriented housing system. While private housing is allowed to be constructed and public housing is being privatized, households are given freedom of housing choice for the first time in decades, and a spatial and social sorting of households and neighborhoods is in progress. Consequently, "gated communities" with luxury villas are developed to meet the desire of the new rich, in sharp contrast to the emergence of dilapidated "migrant enclaves" that provide shelters to millions of rural migrants (Hu and Kaplan 2001; Ma and Xiang 1998; Zhang 2001). However, due to the gradualism adopted in housing reform, socialist institutions continue to play a significant role in housing consumption, although somewhat differently from the previous era (Li 2000a, b; Huang and Clark 2002; Huang 2003b). Thus, the housing context in Chinese cities and the dynamics of housing inequality and residential segregation are more complex than those in both market and socialist economies.

I argue that both socialist institutions and market mechanisms contribute to increasing housing inequality and residential segregation in transitional Chinese cities. While they lead to different forms of housing inequality and residential patterns, they also work together often to aggravate existing housing inequality (Figure 11.1). First of all, despite that the housing reform was launched more than a decade ago, socialist institutions such as the household registration (*hukou*) system, work-units and municipal housing bureaus are still functioning in the urban housing system. The *hukou* system is one of the most important institutions in China that defines a person's socioeconomic status and access to welfare benefits (Chan 1994; Cheng and Selden 1994). It divides the population into one with urban *hukou* and one with rural *hukou* mainly based on birthplace, and one with permanent and temporary *hukou* based on the place of registration. Only residents with urban and permanent *hukou* in cities are qualified for public housing provided by work-units and municipal housing bureaus, while the others, including rural residents in suburbs and migrants with rural or temporary *hukou*, are not qualified for public housing except for temporary shelters such as dormitories provided by work-units.

Although the *hukou* system is currently under reform, migrants still face institutionalized discrimination in accessing public housing. Furthermore, migrants are either not allowed to (especially at the early stage of the reform) (BMG 1992) or cannot afford to purchase newly built commodity housing, because most of them are peasants coming to cities for low-wage jobs. Thus they have no choice but to live in construction sites, trading

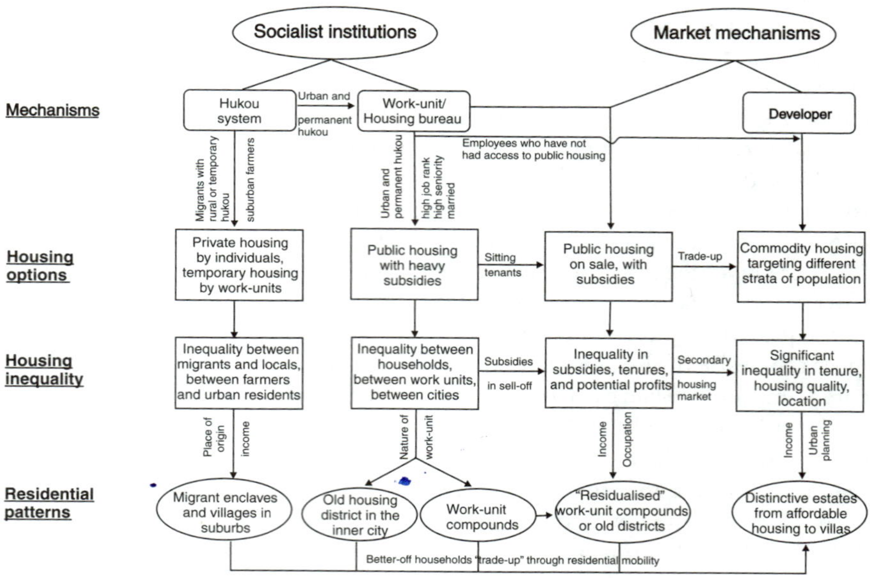


*Figure 11.1* Housing inequality and residential segregation in transitional urban China.

markets, hotels and individual homes, especially peasants' homes in suburbs due to availability of extra rooms (Solinger 1999; Ma and Xiang 1998; W. Wu 2002). Without housing subsidies, migrants often live in extremely poor and crowded housing to save rent, and suburban farmers themselves sometimes live in crowded housing in order to sublease part of their dwellings (Zhang 2001). Thus the *hukou* system has generated significant housing inequality between migrants and local urban residents, and between suburban farmers and urban residents. The spatial manifestation of housing inequality resulting from the *hukou* system is islands of rural villages with dilapidated and crowded bungalows, illegal shacks in narrow alleys, unpaved and often filthy streets amidst typical urban landscape in the suburbs. Because of the availability of cheap housing, many villages have experienced an influx of migrants, forming so-called "migrant enclaves," which further encourages illegal constructions and exacerbates residential crowding (cf. Zhang Chapter 8, this volume). In addition, in Beijing at least, migrants from the same province tend to live in the same village and engage in similar occupations, forming different types of "migrant enclaves" such as "Zhejiang Village" and "Henan Village" (Ma and Xiang 1998).

Among the households with urban and permanent *hukou*, there is also housing inequality because of persisting socialist housing allocation system through work-units and housing bureaus. Employees working in resourceful work-units, mostly large, state-owned and high-ranking work-units, can access heavily subsidized housing from their work-units; and those with high-level jobs and job seniority are more likely to access better and larger housing units (Huang and Clark 2002; Bian *et al.* 1997). In contrast, employees in small, low-ranking, or non-state-owned work-units who have no capability of providing housing, often have to rely on municipal housing bureaus for shelters. Compared to recently built apartment buildings by work-units, housing bureaus usually provide old housing inherited from the pre-1949 period, often dilapidated bungalows in inner cities. Thus the socialist housing allocation system has generated housing inequality between households with different statuses and work-units (Logan *et al.* 1999). Despite the central government's determination to end public housing provision in 1998 (State Council 1998), work-units continue to provide subsidized housing to their employees. In 2000, more than one-third of housing completed (36.97 percent of floor space) was built by the state, mainly through state-owned enterprises and government agencies (SSB 1999, 2001). Thus, housing inequality based on individual political status and the nature of work-units continues in the reform era. Old housing districts with dilapidated bungalows and work-unit compounds with uniform apartment buildings remain an important part of the residential landscape in the transitional era.

Second, the recent housing reform has introduced new types of housing inequality and residential pattern. One main component of the housing reform is to privatize the housing system through both the construction of private housing and the massive sale of public housing (Tolley 1991). Households are encouraged to purchase their occupied public dwellings at subsidized prices or to buy commodity housing at market prices. As a result, more than 70 percent of urban households were homeowners in 2000, compared to less than 20 percent in the 1980s (SSB 2002). With the emergence of a new class of homeowners, stratification along housing tenure (own vs. rent) is in the making. Compared to previous housing inequality mainly in the form of different flat sizes, tenure inequality is more significant, and has profound socioeconomic ramifications. Furthermore, compared to uniform utilitarian apartment buildings in work-unit compounds, commodity housing in general has better but varied quality. There are different types of private housing, from low-end affordable housing[1] to regular commodity housing, and to upscale townhouse and villa, sheltering different strata of urban population. With the demand of the new rich and foreigners, detached houses similar to the upscale housing in American suburbs are developed in Chinese cities. In other words, with the availability of private housing, the spectrum of housing consumption and thus inequality has been significantly widened. Thus, the privatization process has resulted in not only unprecedented stratification along housing tenure but also significant inequality between commodity housing and public housing, and inequality within commodity housing.

Housing inequality generated by market mechanisms also has its footprints on the residential landscape. With the development of a housing market, income is becoming a main factor determining where households can live. Only the wealthiest strata can afford single-family houses in beautiful suburbs and upscale apartments in well-serviced downtown, while low–medium income households have to live in affordable housing complexes that are often constructed in less convenient suburbs for cheap land. Often walled and gated, these private housing estates form distinctive and exclusive neighborhoods. Despite some similarities in physical forms between private housing estates and work-unit compounds (e.g. surrounding walls), they are fundamentally different neighborhoods. In contrast to occupational homogeneity within work-unit compounds and housing homogeneity between work-unit compounds, private housing estates are mainly stratified by income and housing quality.

Third, while socialist institutions and market forces result in different types of housing inequality and residential patterns, they often work together to aggravate existing housing inequality. During the privatization of public housing, only those who are currently living in public housing can purchase their occupied dwellings at subsidized prices, while those still on the waiting list are now expected to purchase housing from the private sector at much higher prices. Thus, the previously temporary difference in the timing of housing access has now become permanent with those who have not had access to public housing at a disadvantage. Clearly it is much easier for those living in public housing to become homeowners, and to "trade-up" into larger and better commodity housing, because of heavy subsidies in the sale of public housing. In addition, people with higher statuses receive more subsidies during the privatization process,[2] and they are likely to gain more economically from the secondary housing markets. Thus, different access to public housing in the socialist era now may lead to a significant differentiation in homeownership, the size of owned homes and potential economic gains. In other words, the newly introduced market forces aggravate inequality resulted from the socialist housing allocation system. Of course, it is undeniable that market forces may reduce the importance of socialist institutions in housing inequality to a certain degree. For example, households who could not access public housing in the past such as those working in private sectors now can purchase private housing if they have the financial means.

Spatially, the interaction between market mechanisms and socialist institutions are creating "residualized" work-unit compounds. While most work-unit compounds are still physically in place, their tenant profiles are rapidly changing. Despite heavy subsidies in the privatization of public housing, some households still cannot afford to purchase their dwellings. Thus differentiation in tenure is taking place within work-unit compounds. At the same time, the new homeowners, especially those with means, are moving out of work-unit compounds into private housing estates, while their dwellings are filtered down to people who may not be associated with the work-unit. With better-off households moving out of work-unit compounds, and more desirable properties shifted into the private domain, a process similar to the "residualization" in British social housing (Burrows 1999) is taking place in public housing in Chinese cities. While the physical features of work-unit compounds such as walls and uniform apartment buildings remain, the original concept of work-unit compounds and associated connotation of homogeneity are disappearing.

Thus, with ongoing spatial sorting, different types of neighborhoods have been formed, including migrant enclaves/rural villages, old housing districts and work-unit compounds, "residualized" work-unit compounds, and different private housing estates ranging from affordable housing complexes, commodity housing to luxury townhouse and villa estates. The socialist residential landscape characterized by the dominance of homogeneous work-unit compounds surrounding old housing districts is undergoing profound transformation, and residential segregation is emerging in Chinese cities. Segregation represents voluntary or involuntary physical separation of members of one racial or socioeconomic group from members of other groups (Lieberson and Carter 1982). Not surprisingly, residential mobility and a maturing housing market have contributed to residential separation between different strata of population – class segregation, demonstrated by the emergence of wealthy "gated communities." In addition, as discussed earlier, only households with urban and permanent *hukou* can access public housing in old housing districts and work-unit compounds, and private but subsidized affordable housing. In contrast, migrants with rural or temporary *hukou* can only access certain private housing. Thus, there is segregation between migrants and local urban residents due to institutionalized housing discrimination against migrants – *hukou* segregation, demonstrated by the emergence of "migrant enclaves." In addition to physical separation between migrants and urban residents, psychologically there is a certain degree of fear and avoidance between them as migrants are afraid of mistreatment by urban residents, and urban residents attribute high crime rates to migrants (Zhang 2001).

While segregated neighborhoods are emerging, the overall residential pattern is different from that in most Western cities. In contrast to a clear separation between decayed inner cities and wealthy suburbs in most North American cities (Burgess 1967), wealthy and poor neighborhoods coexist in both inner cities and suburbs in Chinese cities. Furthermore, compared to clearly demarcated and geographically separated elite districts in either suburban or central locations in Western European and Latin American cities (White 1984; Knox 1993; Griffin and Ford 1993), geographic distance between the rich and poor in Chinese cities sometimes can be negligible. A wealthy "gated community" can be built right next to a dilapidated "migrant enclave" in the suburbs or an old housing neighborhood in the inner city. In other words, despite that differentiated neighborhoods are taking shape and there is an increasing segregation between neighborhoods, residential sorting in Chinese cities has not yet led to decayed inner cities, nor all wealthy suburbs. While there is a significant separation at micro-level (neighborhood), residential segregation at the macro-level (city) has not taken place yet.

In addition to the still strong desire to live in well-serviced central cities and the early stage of spatial sorting, several factors such as the continuing role of the local government in urban planning and housing provision, the dual land systems in suburbs, centralized provision of public goods and ethnic homogeneity contribute to the unique residential pattern in Chinese cities. First, while the local government is becoming less important in housing provision, it continues to shape residential pattern through urban planning and its persistent role in the housing system. For example, the local government's decision of urban renewal often leads to the development of wealthy neighborhoods in dilapidated housing districts in inner cities, and the resettlement of poor households to the suburbs, a different pattern of population movement from suburbanization in most Western cities. Furthermore, to provide adequate shelters to low–medium income households, developers are required by the government to provide affordable housing with government-controlled prices.[3] Because of its low prices, affordable housing is mostly constructed in the suburbs for their cheaper land, which ironically results in its coexistence with some of the wealthiest communities. Second, different ownerships between urban and rural land[4] and different degrees of land reform in these two sectors often lead to inconsistent development in suburbs (Deng and Huang 2004). With collective ownership, rural villages in the suburbs are often underdeveloped with dilapidated bungalows despite the large demand for housing from migrants, while surrounding farmlands are rapidly converted into upscale housing estates and other urban projects due to eminent domain. Thus at the interface between rural and urban areas, the dynamics for residential pattern is rather complex, often resulting in a mix of poor and rich neighborhoods. Third, while the quality of public goods such as schools and security is an important factor in residential choice in the West, the centralized provision of public goods in Chinese cities has led to its relatively even quality across neighborhoods, and thus reduced its importance in residential choice. For example, the majority of schools in Chinese cities are public schools, whose quality depends more on the municipal government's education policy and investment than the socioeconomic profile of local residents and their tax contribution. Thus there is no need for the rich to live apart from the poor for the purpose of public goods. Furthermore, ethnic homogeneity in Chinese society has significantly reduced the degree of residential segregation. According to the 2000 Census, about 96 percent of urban population are ethnic Han Chinese, and the rest include fifty-five other ethnicities (SSB 2002). While place of origin, or "Chinese ethnicity" (Ma and Xiang 1998), often leads to social grouping and identification, it is generally not a basis for social discrimination and residential avoidance, as race is the case in American cities (Massey and Denton 1993). The most important social and psychological division in Chinese cities is between urban residents and rural migrants, which leads to residential segregation based on household registration with migrants congregated in "migrant enclaves". Overall, residential separation is mostly between classes, and there is tremendous tolerance for geographic proximity.

In summary, because of the transitional nature of the housing system, housing inequality and residential segregation in Chinese cities have more complex dynamics than those in both socialist and market economies. While housing reform has not completely changed the roles of socialist institutions in the housing system and consequent housing inequality, it has not only generated new dimensions of housing inequality but also aggravated existing housing inequality. At the same time, the increasing housing inequality and freedom of residential mobility have resulted in increasing residential segregation that is based on not only traditional factors in market economies such as income but also institutional factors unique to China such as household registration status. Despite the emergence of segregated neighborhoods, the overall residential pattern in Chinese cities is currently characterized by the coexistence of and geographic proximity between the poor and wealthy neighborhoods in both inner cities and suburbs, a pattern different from the central city–suburb division in the West.

## Beijing: an empirical analysis

Beijing provides an ideal context to study increasing housing inequality because it had a classic socialist housing system that is being transformed by a burgeoning private housing market. Before the housing reform, most households in Beijing lived in public housing. As the national capital, Beijing hosts many high-ranking work-units such as those at the central or ministry level such that work-units often play more important roles in housing provision and consumption than elsewhere. Beijing has experienced two housing booms, one in the 1980s due to massive construction of public housing by work-units and the other in the late 1990s as a result of both public and private housing construction (Figure 11.2). Although the State Council declared the end of public housing provision in 1998, work-units in Beijing continue to provide housing, and the investment in public housing even increased from 5.3 billion yuan in 1998 to 8.4 billion in 1999 and 10.9 billion in 2001 (BSB 2002b). Despite the rapid development of private housing in Beijing in the last two decades, work-units continue to be the main consumers of private housing in the late 1990s while elsewhere individuals have become the predominant consumers.[5] At the same time, Beijing attracts both the most elite in China and millions of poor migrants. With the introduction of market forces, housing inequality in Beijing has increased significantly with the rich living in luxury housing estates such as Purple Jade Villas (*Ziyu shanzhuang*) ($410,000 to $2 million/house) and poor migrants living in crowded shacks (Hu and Kaplan 2001; Zhang 2001; W. Wu 2002). Thus it is important to understand how socialist legacy and market forces interact, and contribute to housing inequality and residential segregation in a socialist capital city.

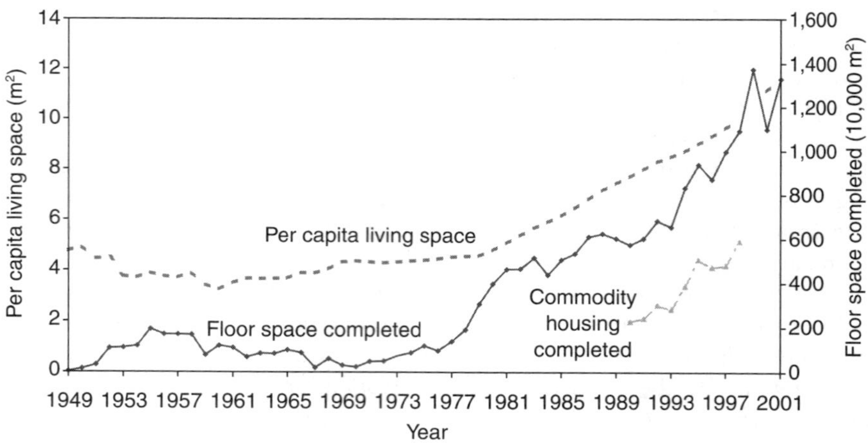


*Figure 11.2* Housing construction and per capita living space in Beijing (source: BSB 1999: 111, 119, 353).

The 2000 Census is the first census in China that collected housing information. Basic housing questions such as number of rooms and total amount of floor space were asked to every household in the short form, and additional questions on housing source, cost and facilities were asked to 10 percent of the population in the long form. Since the micro-level census data has not been made public yet, the aggregated data (city and district) will be used in this study. The study area includes all city districts, including four urban districts (Dongcheng, Xicheng, Xuanwu and Chongweng), four inner-suburban districts (Haidian, Shijingshan, Fengtai and Chaoyang), and five outer-suburban districts (Mengtougou, Fangshan, Tongzhou, Shunyi and Changping).[6] In addition, a database on private housing on the market since 1998 has been compiled based on my fieldwork in Beijing.[7] In this study, only the most luxury housing estates (villas and townhouses), low-end private housing (affordable housing) from the database, and migrant enclaves are geocoded to illustrate the formation of different neighborhoods and the spatial pattern of residential segregation.

## *Housing inequality in Beijing*

The overall level of housing consumption in Beijing has increased significantly over time, from less than 4 $m^2$ of living areas (*juzhu mianji*) per capita in the 1950s to 11.64 $m^2$ in 2001 (Figure 11.2). However, housing inequality is on the rise as well. It is clear that there are significant differences in per capita floor space across education and occupation (Table 11.1). Households with educated heads, especially those with above high-school education, clearly occupy larger housing units. In housing consumption by occupation, households seem to fall into three groups. The first group includes households headed by cadres/officials,[8] professionals and clerks/staff who mostly work in high-ranking state work-units and have high job ranks. In other words, they have a close institutional relationship with their work-units and the state (Huang and Clark 2002), and they consume the largest amount of floor space (>22 $m^2$). The second category includes commerce/service workers and machine operators, who work in either private sectors or in the declining state sector (e.g. SOEs). They consume the least amount of floor space (15.68 $m^2$, 17.96 $m^2$). The third group is agricultural workers, mostly suburban farmers, who own large houses. Yet, it is a small group with only about 2 percent of households. It is clear that those who occupy larger dwellings in the socialist era continue to do so in the reform era, which to a certain degree indicates the persistence of the influence of socialist institutions in housing allocation.

One main feature of the housing reform is to promote homeownership. As a result, the rate of homeownership has increased significantly in Beijing, reaching 55 percent in 2000. Similar to inequality in the amount of floor space, people with above high-school education are more likely to own than those who do not, and people who work in public, high-ranking work-units

*Table 11.1* Housing consumption in Beijing by household head's education and occupation (2000)

| | *Per capita floor space ($m^2$/person)*[a] | *Percent of households* |
|---|---|---|
| *Education* | | |
| University+ | 25.63 | 13.60 |
| Professional | 22.03 | 18.93 |
| High school | 18.13 | 17.85 |
| Middle school | 17.90 | 32.40 |
| Elementary school | 17.31 | 12.43 |
| Illiterate/some literacy | 17.16 | 4.78 |
| All | 19.57 | 100.00 |
| *Occupation*[b] | | |
| Cadres/officials | 24.97 | 9.70 |
| Professionals | 23.30 | 20.89 |
| Clerks/staff | 22.06 | 15.24 |
| Commerce/services | 15.68 | 24.20 |
| Machine operators | 17.96 | 27.70 |
| Agricultural workers | 22.53 | 2.27 |
| All | 19.97 | 100.00 |

Source: BSB 2002.

Notes

a Per capita floor space measures "construction area" (*jianzhu mianji*), which includes all areas within a housing unit and some shared common areas such as stairs and common corridors.

b The "other" category for occupation is not included as it has a very small number of households.

and with high job-ranks (cadres/officials, professionals, clerks/staff) are more likely to own than workers in commerce/service and industry (Table 11.2). In particular, as the commerce and service sectors have been mostly privatized during the economic reform, their workers are only about half as likely to own as cadres/officials (33.70 vs. 63.43 percent). It is clear that those who benefit more from the socialist housing system are also more likely to own because they are allowed to purchase public housing and they usually receive more subsidies during the privatization process. This shows that the housing reform reinforces housing inequality from the socialist era. While agricultural workers are the most likely to own, they in fact experience housing discrimination because of their rural *hukou* status and they receive the least amount of housing subsidies, as will be shown later.

While the rate of homeownership shows current difference in tenure, housing source can reveal how households gain their dwellings and whether they receive subsidies. First of all, most households have obtained their dwellings from the public sector, indicating most households have received

*Table 11.2* Housing choice in Beijing by household head's education and occupation (2000)

| (%) | Rate of homeownership | Share of public housing | Self-build housing | Purchase commodity housing | Purchase affordable housing | Purchase public housing | Rent public housing | Rent commodity housing | Others | Total |
|---|---|---|---|---|---|---|---|---|---|---|
| University+ | **69.26** | **88.08** | 1.27 | 3.77 | 1.99 | 62.23 | 25.86 | 2.12 | 2.77 | 100.00 |
| Professional | **60.95** | **83.13** | 4.75 | 3.41 | 2.81 | 49.98 | 33.15 | 3.25 | 2.66 | 100.00 |
| High school | 46.87 | 74.34 | 10.48 | 2.74 | 2.19 | 31.46 | 42.87 | 6.44 | 3.82 | 100.00 |
| Middle school | 51.24 | 57.69 | 20.72 | 2.48 | 1.86 | 26.17 | 31.52 | **12.57** | 4.67 | 100.00 |
| Elementary school | 53.62 | 61.64 | 21.99 | 1.72 | 1.65 | 28.26 | 33.38 | 9.30 | 3.70 | 100.00 |
| Illiterate/some literacy | 50.68 | 65.85 | 23.03 | 1.51 | 1.74 | 24.40 | 41.45 | 4.34 | 3.52 | 100.00 |
| All | 55.18 | 70.77 | 13.34 | 2.75 | 2.09 | 36.99 | 33.78 | 7.37 | 3.68 | 100.00 |
| Cadres/officials | **63.43** | **73.92** | 7.72 | 6.73 | 2.57 | 46.41 | 27.51 | 5.56 | 3.51 | 100.00 |
| Professionals | **60.72** | **83.29** | 4.40 | 3.51 | 2.75 | 50.05 | 33.24 | 2.87 | 3.17 | 100.00 |
| Clerks/staff | **60.36** | **78.19** | 9.60 | 3.32 | 2.66 | 44.78 | 33.41 | 2.70 | 3.53 | 100.00 |
| Commerce/services | 33.70 | 52.13 | 11.79 | 2.34 | 1.58 | 17.99 | 34.14 | **24.86** | 7.30 | 100.00 |
| Machine operators | 50.13 | 61.73 | 18.60 | 2.55 | 2.23 | 26.75 | 34.98 | 10.49 | 4.40 | 100.00 |
| Agricultural workers | 74.44 | 14.26 | 67.75 | 1.39 | 1.39 | 3.91 | 10.35 | 9.11 | 6.11 | 100.00 |
| All | 52.04 | 66.78 | 12.66 | 3.20 | 2.27 | 33.90 | 32.88 | 10.46 | 4.62 | 100.00 |

Source: BSB 2002.

Note

Rate of homeownership is the sum of self-build housing, purchased commodity housing, affordable housing and public housing. Share of public housing includes those who purchased public housing and those who are renting public housing.

or are continuing to receive heavy housing subsidies (Table 11.2). In particular, people with above high-school education (>83 percent) are even more likely to live in public housing than the rest. While they are more likely to purchase public housing, those with lower education are more likely to rent public housing. Despite the rapid development of private housing in Beijing, a very small proportion of households rent (7.37 percent) or purchase commodity housing (2.75 percent), or purchase subsidized affordable housing (2.09 percent). While better-educated households are more likely to own private housing, those with lower education especially those with middle-school education (12.57 percent) are more likely to rent commodity housing, probably because of the lack of access to public rental housing. Although about 50 percent of the people with lower than high-school education are owners, they are much more likely to live in self-build housing (>20 percent) with virtually no subsidies, in contrast to those with higher education who mostly own public housing with heavy subsidies. In other words, people with lower education are less likely to receive housing subsidies.

Similarly, people with different occupations access their housing through different paths and enjoy different subsidies. Cadres/officials (73.92 percent), professionals (83.29) and clerks/staff (78.19) are much more likely to live in public housing than commerce/service workers (52.13), machine operators (61.73) and agricultural workers (14.26), indicating the former are more likely to receive housing subsidies. The former are also much more likely to purchase while the latter are more likely to rent public housing. While the former may have higher income, it is also true that the former receive more subsidies when they purchase public housing. Ironically, commerce/service workers, many of who are migrants, are the most likely to rent expensive commodity housing (24.86 percent), probably because they are denied access to public housing. Not surprisingly, two-thirds of agricultural workers live in self-build housing with no subsidies.

Another way to look at housing subsidies is to examine actual housing costs (Table 11.3). Despite the extremely high housing prices in Beijing, more than 90 percent of owners paid less than 100,000 yuan ($11,947), mainly because about two-thirds of the owners purchased public housing at heavily subsidized prices.[9] In general, people with higher education and those at the top of occupational hierarchy are more likely to pay 50,000+ yuan for their dwellings, and they pay the highest "average" prices.[10] This is consistent with the findings in market economies. However, they are as much likely to pay medium prices (10,000–50,000 yuan) as those at lower rungs of education and occupational hierarchy, indicating many of them received heavy subsidies. In other words, although those at the top of education and occupational hierarchy consume the largest amount of floor space, most of them did not necessarily pay much more than others.

There are heavy subsidies in the rental sector too, as more than half of all renters pay less than 100 yuan/month, and the "average" rent is less than

*Table 11.3* Housing cost in Beijing by education and occupation (in yuan) (2000)

| *Education* | *Housing prices (%)* | | | | | | | *"Average" prices* | *Monthly rent (%)* | | | | | | *"Average" rents* |
|---|---|---|---|---|---|---|---|---|---|---|---|---|---|---|---|
| | *<10k* | *10–30k* | *30–50k* | *50–100k* | *100–200k* | *200k+* | *Total* | | *<50* | *50–100* | *100–200* | *200–500* | *500+* | *Total* | |
| University+ | 2.79 | **32.16** | **25.95** | 26.89 | 7.45 | 4.76 | 100.00 | 62,574 | 32.53 | 21.61 | 30.17 | 7.28 | 8.41 | 100.00 | 158 |
| Professional | 6.42 | **37.54** | **28.83** | 19.55 | 4.74 | 2.92 | 100.00 | 49,886 | 40.45 | 20.77 | 29.17 | 5.91 | 3.70 | 100.00 | 118 |
| High school | 12.33 | **37.86** | **26.94** | 16.11 | 4.28 | 2.48 | 100.00 | 44,913 | 44.09 | 19.22 | 26.22 | 7.24 | 3.23 | 100.00 | 114 |
| Middle school | 13.93 | 39.73 | 24.78 | 15.16 | 4.49 | 1.92 | 100.00 | 42,397 | **33.67** | **17.76** | 31.35 | **13.44** | **3.78** | 100.00 | **144** |
| Elementary school | 17.41 | 43.03 | 22.92 | 11.52 | 3.48 | 1.64 | 100.00 | 37,438 | 38.50 | 20.24 | 30.05 | 9.13 | 2.09 | 100.00 | 117 |
| Illiterate/some literacy | 22.99 | 38.76 | 22.36 | 11.50 | 3.04 | 1.35 | 100.00 | 35,080 | 48.58 | 21.21 | 25.13 | 4.37 | 0.70 | 100.00 | 86 |
| All | 10.95 | 37.98 | 25.85 | 17.71 | 4.84 | 2.66 | 100.00 | 47,010 | 38.40 | 19.47 | 29.26 | 9.17 | 3.70 | 100.00 | 128 |
| *Occupation* | | | | | | | | | | | | | | | |
| Cadres/officials | 4.70 | **24.88** | **25.35** | 26.11 | 11.01 | 7.95 | 100.00 | 75,294 | 31.10 | 16.79 | 28.12 | 11.71 | 12.27 | 100.00 | 196 |
| Professionals | 4.25 | **29.20** | **28.45** | 27.96 | 6.48 | 3.65 | 100.00 | 59,076 | 39.05 | 21.20 | 28.18 | 5.98 | 5.59 | 100.00 | 131 |
| Clerks/staff | 6.07 | **30.59** | 29.44 | 24.29 | 6.69 | 2.91 | 100.00 | 55,194 | 40.08 | 19.86 | 31.73 | 5.43 | 2.91 | 100.00 | 113 |
| Commerce/services | 13.36 | 33.31 | 26.50 | 18.23 | 5.10 | 3.50 | 100.00 | 49,762 | **24.63** | **14.02** | 31.35 | **22.30** | **7.69** | 100.00 | **199** |
| Machine operators | 13.33 | 35.89 | 26.82 | 17.85 | 4.56 | 1.55 | 100.00 | 43,455 | 39.69 | 19.86 | 31.08 | 7.86 | 1.51 | 100.00 | 110 |
| Agricultural workers | 25.36 | 28.61 | 20.10 | 20.24 | 4.51 | 1.19 | 100.00 | 40,531 | 31.91 | 21.14 | 38.52 | 7.39 | 1.04 | 100.00 | 115 |
| All | 9.14 | 31.34 | 27.26 | 22.65 | 6.26 | 3.36 | 100.00 | 54,072 | 34.18 | 18.05 | 30.59 | 12.03 | 5.15 | 100.00 | 149 |

Source: BSB 2002.

Note
"Average" prices and rents are the sum of the product of the midpoint price and the percentage of households falling in each category. The midpoint of the price range is treated as the "average" price for the range. For the last open-end category, 300,000 is used as the midpoint of the range of 200,000+, and 750 is used for 500+.

150 yuan/month. While people at the top of education and occupation hierarchy are more likely to pay high rents (500+) than those at the lower rungs, the former are as likely to pay lower rents as the latter. In other words, people across education and occupational hierarchy enjoy rental housing subsidies. Yet, people with middle school education and those working in commerce/service sectors are the least likely to pay low rents (<100 yuan) and the most likely to pay high rents (>200) respectively, but they pay similar "average" rents as those with university+ education (144 vs. 158) and those who are cadres/officials (199 vs. 196), despite the fact that they occupy a much smaller dwelling. This shows again that they are much less likely to receive housing subsidies probably because of their migrant status or affiliation with the private sector.

In summary, there are significant differentiations in the amount of floor space, rate of homeownership, housing source and cost across education and occupation. People with above high-school education and people with closer institutional relationships with their work-units and the state such as cadres/officials, professionals and clerk/staff consume larger housing units and are more likely to own. But most of them do not necessarily pay much more than others. One main reason is that they are more likely to access public housing and receive heavy subsidies during the privatization process. In contrast, commerce/service workers and people with middle school education are the most likely to rent commodity housing and pay high rents mainly because they are less likely to enjoy housing subsidies due to possibly their migrant status and association with private work-units. Clearly, socialist institutions continue to lead to inequality in housing consumption. At the same time, market mechanisms begin to shape housing consumption in Beijing as people with higher education and those at the top of occupational hierarchy are not only more likely to own and consume large floor space, but also more likely to own commodity housing and pay higher prices.

### *Residential patterns and emerging residential segregation in Beijing*

The main spatial ramification of housing inequality is residential segregation. While I argue for emerging residential segregation in Beijing due to recent housing reform, it should be acknowledged that residential segregation is not a new phenomenon in Beijing. As the capital of several dynasties, residential separation between the ruling class and the masses was always maintained through physical walls and strict building code[11] (Chang 1977). During the Qing period (1644–1911), ethnic Chinese in Beijing were forced to move out of the Inner City to the Outer City (Wakeman 1985; Sit 1995). Thus both class and ethnic segregation existed in Imperial Beijing. Skinner (1977) and Belsky (2000) argue for two distinct residential and social spaces in late imperial Beijing – one for scholar–official elite and the other for merchants and tradesmen. It is interesting to note that these two "nuclei" were both characterized by occupational homogeneity and personal-wealth heterogeneity (Belsky 2000).

Since the 1949 revolution, massive provision of public housing and socialist transformation of private housing have significantly transformed the residential and social space in Beijing. While traditional courtyard bungalows still dominate the inner city, sheltering mainly people working in low-ranking and non-state work-units, the suburbs have been characterized by work-unit compounds with uniform apartment buildings (Sit 1995, 2000). Often walled and gated, these work-unit compounds created residential and social spaces based on occupations and industries. For example, heavy industries and large factories were concentrated in the west and south of the central city such as Shijingshan and Fengtai districts where work-unit compounds shelter mostly industrial workers. In contrast, work-unit compounds in the northwest of the city are mainly for scholars and professionals as it is an education district with hundreds of academic institutions. Land use and urban planning played important roles in shaping socialist residential pattern. Despite evidence of housing inequality (Logan *et al.* 1999), geographic proximity of different households (living in the same compound) and physical uniformity of housing across work-unit compounds (similar 3–7 storey utilitarian walk-up apartment buildings) have minimized social separation. In other words, residential patterns in the socialist era are characterized by occupational separation but with relative homogeneity in income and housing consumption in the form of work-unit compounds.

Since the launch of housing reform, commodity housing, usually of higher quality than public housing, has been added to the housing stock rapidly. Better-off households are moving to desired dwellings and neighborhoods, usually from the inner city to the suburbs and from work-unit compounds to private housing estates. The ongoing spatial sorting is transforming the socialist residential pattern. Recognizing that there are huge differentiations within a district, it is still helpful to compare housing consumption between districts to show spatial differentiation at the macro-level. It is clear that households in the four old urban districts are more likely to live in crowded, old bungalows with poor facilities than those in the suburban districts (Table 11.4). The former all have less than two bedrooms per household, and more than 40 percent of the households are multi-generation (2+) households with only one bedroom. Residential crowding and poor housing conditions are particularly acute in Chongwen District. In contrast, households in four suburban districts have more than two bedrooms/household, and more than 40 percent of them live in new housing built in the 1990s, and only less than one-third do not have a private bathroom and about 20 percent do not have a private kitchen. In addition, households living in suburban districts are more likely to own than those in urban districts because they are more likely to purchase public housing or build houses themselves. The former are also more likely to pay higher prices (>50,000 yuan) for their dwellings. While households in the suburbs

*Table 11.4* Housing condition and housing consumption by district in Beijing in 2000

| *Housing consumption* | *Urban districts* | | | | *Suburban districts* | | | | *All* |
|---|---|---|---|---|---|---|---|---|---|
| | *Dongcheng* | *Xicheng* | *Chongwen* | *Xuanwu* | *Chaoyang* | *Fengtai* | *Shijingshan* | *Haidian* | |
| Rooms per household | 1.96 | 1.93 | 1.69 | 1.80 | 2.33 | 2.31 | 2.02 | 2.28 | 2.21* |
| Per capita living space ($m^2$) | 14.97 | 16.77 | 12.86 | 15.78 | 21.29 | 19.84 | 17.62 | 21.40 | 19.57* |
| 2+ generation HHs with one room (%) | 46.10 | 44.58 | 64.33 | 52.15 | 27.90 | 26.46 | 23.83 | 23.50 | 32.58 |
| **Housing condition** | | | | | | | | | |
| Bungalows (%) | 55.43 | 39.17 | 57.25 | 41.41 | 32.73 | 34.12 | 31.47 | 27.78 | 35.43 |
| Year built (% of household) 1990–2000 | 15.63 | 24.11 | 13.07 | 26.00 | 44.23 | 48.75 | 41.94 | 44.06 | 38.51 |
| *Facilities (% of households)* | | | | | | | | | |
| Shared or no kitchen | 51.01 | 25.71 | 39.97 | 36.19 | 18.50 | 18.09 | 21.67 | 17.83 | 23.35 |
| Shared or no bathroom | 59.42 | 45.19 | 67.19 | 48.90 | 32.15 | 30.30 | 33.42 | 26.28 | 36.16 |
| **Housing tenure** | | | | | | | | | |
| *Ownership (%)* | **36.37** | **46.70** | **28.38** | **43.56** | **57.71** | **52.98** | **60.91** | **58.78** | **52.76** |
| Self-build housing | 8.71 | 7.71 | 7.57 | 7.45 | 16.26 | 16.77 | 9.79 | 11.76 | 12.65 |
| Purchase commodity/affordable housing | 1.48 | 1.71 | 1.39 | 2.37 | 3.67 | 4.85 | 2.46 | 3.76 | 3.30 |
| Purchase public housing | 26.17 | 37.28 | 19.43 | 33.75 | 37.79 | 31.37 | 48.66 | 43.26 | 36.81 |
| *Rental (%)* | **61.97** | **50.76** | **60.32** | **53.34** | **39.07** | **41.71** | **35.07** | **38.30** | **43.51** |
| Rent public housing | 58.49 | 47.57 | 57.16 | 50.77 | 30.06 | 33.36 | 22.01 | 29.89 | 36.07 |
| Rent commodity housing | 3.49 | 3.18 | 3.16 | 2.57 | 9.01 | 8.35 | 13.06 | 8.41 | 7.43 |
| *Others (%)* | **1.66** | **2.54** | **11.30** | **3.09** | **3.21** | **5.31** | **4.02** | **2.92** | **3.73** |
| Total | 100.00 | 100.00 | 100.00 | 100.00 | 100.00 | 100.00 | 100.00 | 100.00 | 100.00 |

| | | | | | | | | | |
|---|---|---|---|---|---|---|---|---|---|
| **Housing cost** | | | | | | | | | |
| *Housing price (% of households)* | | | | | | | | | |
| <10,000 yuan | 20.93 | 14.51 | 24.29 | 14.82 | 6.21 | 13.96 | 8.02 | 6.68 | 9.93 |
| 10,000–50,000 | 64.30 | 66.71 | 59.55 | 65.29 | 65.01 | 61.60 | 72.42 | 64.35 | 64.78 |
| 50,000+ | 14.77 | 18.78 | 16.15 | 19.89 | 28.78 | 24.44 | 19.56 | 28.97 | 25.29 |
| Total | 100.00 | 100.00 | 100.00 | 100.00 | 100.00 | 100.00 | 100.00 | 100.00 | 100.00 |
| *Rent (% of households)* | | | | | | | | | |
| <100 yuan/month | 83.24 | 76.29 | 85.27 | 78.05 | 41.88 | 44.22 | 42.04 | 50.58 | 57.13 |
| 100–500 | 15.36 | 21.40 | 13.05 | 20.19 | 54.28 | 51.46 | 53.75 | 43.05 | 39.10 |
| 500+ | 1.41 | 2.31 | 1.69 | 1.77 | 3.84 | 4.32 | 4.21 | 6.37 | 3.78 |
| Total | 100.00 | 100.00 | 100.00 | 100.00 | 100.00 | 100.00 | 100.00 | 100.00 | 100.00 |

Source: BSB 2002.

Note
*City areas in the five outer-suburban districts (Mengtougou, Fangshan, Tongzhou, Shunyi and Changping) are also included for calculation.

are much less likely to rent, they are more likely to rent commodity housing (>8 percent) than those in the urban districts, especially in Shijingshan District (13 percent), and consequently more likely to pay high rents.

It is clear that there are significant differentiations between urban and suburban districts, with households in the latter more likely to own, live in newer, larger, better-equipped and more expensive housing. While this spatial differentiation is partly a result of the residential pattern in the socialist era (work-unit compounds in the suburbs vs. old bungalows in the inner city), the recent boom of private housing in the suburbs is another important factor. Since the late 1990s, thousands of private housing estates have been developed, mostly in the suburbs. There are many different types of private housing in Beijing, ranging from the most luxury villas, townhouses, irregular commodity housing to subsidized affordable housing. While villas and townhouses – mostly detached or semi-detached, cater to the need of the elite, affordable housing is price controlled to shelter low–medium income households, especially civil servants, teachers and employees of SOEs and public institutions (BMG 1998). To demonstrate the formation of different neighborhoods and the spatial pattern of residential segregation, a map on the two ends of private housing spectrum – high-end villa and townhouse estates, and low-end affordable housing estates and migrant enclaves – is presented (Figure 11.3).

At least two findings can be noted. First, noticeable wealthy and low–medium income residential areas are emerging. With convenient transportation and attractive amenities, wealthy neighborhoods made of villas and townhouses are forming in the north near the Asian Games Village/Olympic Forest Park and Xiaotangshan Warm Springs, in the northeast along Jingshun Road near Wenyu River and Chaobai River, in the northwest along Jingchang Express, and in the southeast next to Beijing Economic and Technology Development Zone. Housing quality in these estates are superb, as shown in Purple Jade Villas (*Ziyu shanzhuang*) (Figure 11.4) and Near-Forest Garden (*Yilin jiayuan*) (Figure 11.5). Because of extremely high prices, only foreigners, overseas Chinese and the most elite Chinese can afford to live in these estates. A survey of 92 households in Near-Forest Garden[12] shows that most buyers are entrepreneurs (34 percent) and managers of joint ventures (29 percent), and more than 80 percent of them have university or higher education and have more than 10,000 yuan monthly income. As only 28 percent are first-time owners and more than half of them owned commodity housing with more than $100\,m^2$ before, moving into Near-Forest Garden is clearly a trade-up in the housing hierarchy.

At the same time, with concentration of several large-scale affordable housing estates, low–medium income neighborhoods are taking shape in areas such as Huilongguan in the northwest along Jingchang Express, Ping-
and Fengtai in the west, Nanyuan in the south, Guanzhuang in the

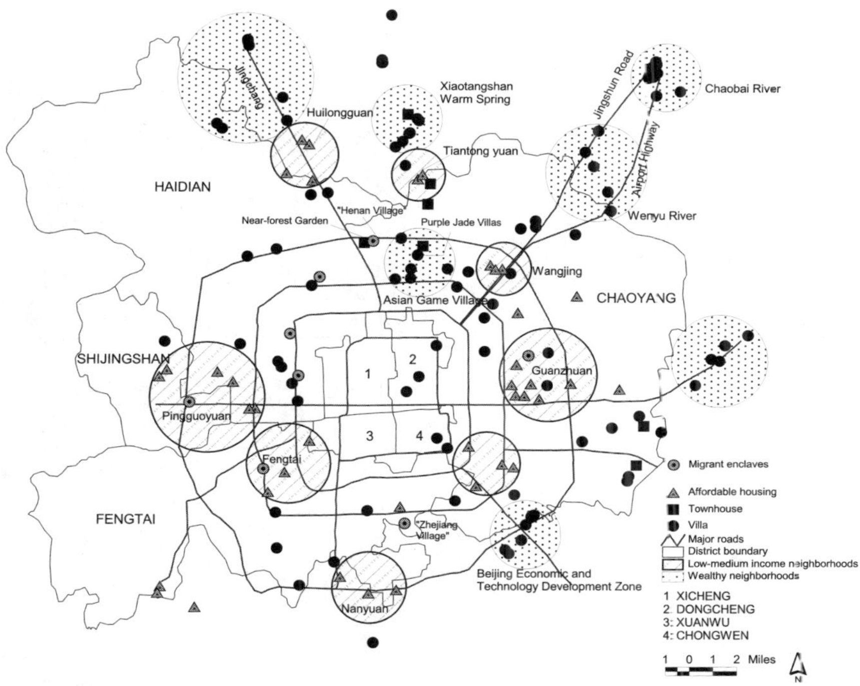


*Figure 11.3* Distribution of high-end and low-end private housing estates (source: locatin for migrant enclaves are based on Ma and Xiang 1998 and fieldwork; other housing estates are based on *Beijing Real Estate Report* 2000–2001, 2001–2002, Beijing Real Estate Information and Data Collection 2001).

east and Wangjing in the northeast along the Airport Highway. While these estates are recently built, amenities and facilities are rather limited compared to other private housing estates (Table 11.5; Figure 11.6). Only households with annual income less than 60,000 yuan can purchase affordable housing (BCB 2000).[13] The majority of residents are first-time owners, and most of them are government employees and workers/staff in the public sector.

Second, different from the inner-city–suburb contrast in the West, the wealthiest and the poorest neighborhoods can be found in the suburbs and the inner city, often next to each other. While most villa/townhouse estates are built in the suburbs, there are several such estates in the poor old housing districts in the inner city. At the same time, most affordable housing estates and migrant enclaves are located in the suburbs, often next to luxury housing estates. While the affordable housing complexes may not be the poorest neighborhoods, migrant enclaves usually host the poorest

*Figure 11.4* Detached houses in a gated community – Purple Jade Villas (Ziyu shanzhuang).

people in the city. For example, with dilapidated bungalows and shacks, Wali Village (or "Henan Village") provides housing to rural migrants from Henan province who mainly engage in garbage collection (Figure 11.7). Yet, it is less than a block away from Near-Forest Garden, and within a short distance of the villa area around Purple Jade Villas. While segregated neighborhoods are taking shape as a result of both market-oriented factors such as income and institutional factors such as *hukou* status, the overall residential pattern in Beijing is still characterized by the mixing of different

*Figure 11.5* Townhouses in Near-Forest Garden (Yilin jiayuan).

*Figure 11.6* Apartment buildings in an affordable housing complex – Dragon-Turn Vista (Huilongguan).

*Table 11.5* A comparison between different private housing estates in Beijing

| *Name of estate* | *Ziyu shanzhuang (Purple Jade Villas)* | *Yilin jiayuan (Near-Forest Garden)* | *Huilongguan (Dragon-Turn Vista)* | *Wali Village ("Henan Village")* |
|---|---|---|---|---|
| *Housing type* | Villa<br>Detached or semi-detached | Townhouse<br>Duplex or triplex | Affordable housing<br>Multi-storey buildings | Migrant enclaves<br>Bungalows/shacks |
| Unit size ($m^2$) | 150–600 | 178–233 | 60–180 | |
| Price (yuan/$m^2$) | 24,800 (\$3,000/$m^2$) | 8,000 (\$1,000/$m^2$) | 2,600 (\$314/$m^2$) | n.a. |
| Price for a flat/house | \$450,000–\$1.8 million | \$178,000–\$233,000 | \$18,840–\$56,520 | n.a. |
| Monthly rent (yuan/room) | n.a. | n.a. | n.a. | 100–800 (\$12–\$100) |
| *Amenity/facilities* | | | | |
| Club/gym | Yes | No | No | No |
| Cable, satellite TV | Yes | No | No | No |
| Broadband Internet access | Yes | Yes | No | No |
| 24-hour security | Yes | Yes | No | No |
| Garage/parking | Yes | Yes | No | No |
| Hot water | Yes | Yes | No | No |
| Central heating/cooling | Yes | Yes | Individual heating | No |
| Private kitchen and bathroom | Yes | Yes | Yes | No |
| Green space | 80% | Next to Olympic Forest Park | 37% | 0 |

Source: *Report on Beijing Real Estate Market 2000–2001, 2001–2002*, fieldwork and http.//www.purplejade.com.cn.

*Figure 11.7* Bungalows and shacks in Wali Village, a village crowded with migrants from Henan Province ("Henan Village").

neighborhoods. As discussed earlier, the geographic proximity between the rich and the poor is a result of many factors such as different land systems in the suburbs, ethnic homogeneity, centralized provision of public goods and urban planning.

## Conclusions and discussions

As part of the overall transition toward a market economy, housing reform in Chinese cities has significantly improved residents' housing consumption. Yet, it has also led to unprecedented housing inequality and residential segregation, demonstrated by the coexistence of wealthy "gated communities" and dilapidated "migrant enclaves." This chapter analyzes the dynamics of increasing housing inequality and emerging residential segregation in Chinese cities both conceptually and empirically through a case study of Beijing. I argue that housing inequality and residential segregation in transitional Chinese cities can be attributed to persistent socialist institutions, newly introduced market forces, and their interactions.

Despite the decade-long reform, socialist institutions such as the *hukou* system and work-units continue to play important roles in housing consumption. While migrants with rural or temporary *hukou* are still denied long-term public housing, urban residents continue to receive housing subsidies from their work-units. Thus housing inequality resulted from the socialist housing allocation system, such as inequality between migrants and urban residents and between urban residents with different political status and work-units, remains in the reform era. In addition, with the availability of private housing and the possibility of homeownership, housing reform has generated new types of housing inequality such as stratification along housing tenure, which has significant socioeconomic ramifications. With the availability of luxury private housing, housing reform has also widened the spectrum of inequality in housing consumption. Furthermore, socialist institutions and market forces often work together to aggravate existing housing inequality. One main component of the housing reform is the privatization of the existing public housing stock through subsidized sale. Yet, only those who have had access to public housing can purchase their dwellings with subsidies, and they are obviously in an advantaged position in the housing market. In other words, those who have benefitted from the socialist housing system can gain more during the process of privatization due to the interaction between socialist institutions and market forces, which results in a larger inequality in housing consumption.

The spatial manifestation of increasing housing inequality is emerging residential segregation, which is transforming the socialist residential landscape dominated by homogeneous work-unit compounds. With increasing mobility, households are sorted into differentiated neighborhoods, ranging from detached/semi-detached villa/townhouse complexes, to commodity housing and affordable housing communities with apartment buildings/

towers, and to migrant enclaves with dilapidated shacks/bungalows. At the same time, previous work-unit compounds are being "residualized" with better-off households moving out and better properties shifted into the private domain. Despite the emergence of differentiated neighborhoods, the overall residential pattern is different from the inner-city–suburb contrast in the West, as both poor and wealthy neighborhoods form in the suburbs and inner cities, and the geographic distance between the two can be very small. The resultant residential pattern is characterized by segregation at the micro-level (neighborhood) but mixing at the macro-level (city). In addition to the persisting role of local government through urban planning and housing subsidies, factors such as dual land systems in the suburbs, centralized provision of public goods, ethnic homogeneity as well as early stage of neighborhood sorting contribute to the unique residential pattern.

The case study of Beijing supports above arguments. The latest census data show significant housing inequality across education and occupation. While people with higher education and in the public sector enjoy larger housing units and are more likely to own, most of them do not necessarily pay much more than others because of housing subsidies. In contrast, those in the private sector are more likely to pay higher rents for private housing, probably because they are denied public housing. It is clear that socialist institutions contribute to housing inequality. At the same time, market forces begin to shape housing consumption, as those at the top of education and occupational hierarchy are more likely to purchase expensive private housing. Spatially, households in suburban districts in general have more spacious housing with newer and better facilities and are more likely to own than those in the urban districts, due partly to the socialist residential pattern and partly to the recent development of private housing in the suburbs. While low- to medium-income neighborhoods based on affordable housing estates and wealthy neighborhoods based on villas and townhouses are emerging, they are located in the inner city as well as the suburbs, and often next to each other. However, the ongoing neighborhood sorting and residential segregation in Beijing has not yet led to a decayed inner city, nor all wealthy suburbs.

Increasing housing inequality and residential segregation have led to the observation that China has moved away from being one of the most egalitarian to one of the most unequal societies in the world (World Bank 1997). This seems to be inevitable as the housing market matures, especially when the government has not played any significant role to mitigate it. Instead, its housing programs appear to have aggravated housing inequality. During the transition era, the government tends to reward those who have already benefitted from the previous socialist housing system, while those who have not benefitted (e.g. migrants and workers in private sectors) are pushed into the private market without any government assistance. This has already resulted in severe housing problems for the latter as is demonstrated by the emergence of "migrant enclaves." The government's recent promotion of affordable housing and cheap rental housing (*lian zu fang*) is a welcome effort to provide decent housing to low- and medium-income households, as private housing is becoming increasingly unaffordable especially in large cities like Beijing. Yet, the allocation of affordable housing and cheap rental housing is still discriminative as in the previous socialist housing system, because only households with both urban and permanent *hukou* can have access, while migrants, regardless of their income level, are not qualified (BCB 2000; BLHMB 2001). Government assistance is needed for the lowest strata of the society if urban China is to have a more equitable housing system.

## Acknowledgments

I am grateful to the editors of the volume and two referees for their helpful comments. I also thank David Chunyu, Zhou Yu and Jingjing Li for their help in data collection and analysis.

## Notes

1 Affordable housing is a special type of commodity housing, with its prices set by the local government so that developers can recover their costs and earn some profit (<3 percent profit rate) (*bao ben wei li*) (State Council 1998).

2 For example, people with higher job ranks are allowed to purchase larger dwellings with subsidies. In Beijing, staff and workers can only purchase 60 $m^2$ at subsidized prices, technicians and cadres at the urban district (*qu*) level are qualified for 80 $m^2$, and cadres at the bureau (*ju*) level are qualified for 105 $m^2$ (BMG 1999). People can own more than they are qualified to buy but they have to pay market prices for the extra amount of floor space.

3 Each developer is required to construct affordable housing that accounts for at least 20 percent of their total housing development (State Council 1994).

4 Rural land is officially owned collectively by the villagers while urban land is owned by the state.

5 Work-units were the largest consumer of private housing in the early stage of housing reform (F. Wu 1996; Huang and Clark 2002). In 1991, less than 3 percent of commodity housing sold in Beijing was purchased by individuals, in contrast to 33.7 percent nationwide. Since then, the share sold to individuals has increased significantly. Yet, in Beijing, only 40.2 percent of commodity housing sold in 1998 was purchased by individuals, compared to 83.2 percent in Shanghai and 73.7 percent nationwide (ECCRESY 1999).

6 Ideally the five outer-suburban districts should be excluded from the analysis. But it is impossible to do so due to the aggregation of available census data. However, the areas in the five districts are relatively small, with only 10.5 percent of the total city population (BSB 2002a).

7 The main sources for the database is *Beijing Real Estate Yearbook 1998–99*, *Report on Real Estate Market in Beijing 2000–2001, 2001–2002*, and *A Data Collection on Real Estate Market in Beijing 2001.*

8 Cadres/officials refer to leaders of government and Party institutions, enterprises and organizations.

9 The average price for commodity housing was 4,815 yuan/$m^2$ in 1998 in

Beijing, and the average price for a 100 m² unit would be 481,500 yuan. In comparison, the price for public housing was 1,485 yuan/m² (BHRO 1998), and the price for a 100 m² unit would be 148,500 yuan.

10 See the note under Table 11.3 for the calculation of the "average" prices.

11 Traditional housing in Beijing is courtyard bungalows (*siheyuan*). Some building designs (e.g. golden roof, red column) were reserved for the royal families and officials. Ordinary city dwellers had to use other designs (e.g. gray roof).

12 The survey was conducted by Goldnet Real Estate Investment Corporation.

13 Households with both spouses as teachers or civil servants, and those resettled due to urban renewal or the construction of major government projects can purchase affordable housing without meeting the low income requirement (BCB 2000).

## References


Beijing Construction Bureau (BCB) (2000) "Beijing shi chenzhen jumin goumai jingji shiyong fang youguan wenti de zanxing guiding (A temporary regulation on urban residents' purchase of affordable housing in Beijing)," No. 31. Beijing: BCB.

Beijing Housing Reform Office (BHRO) (1998) "Guanyu 1999 nian xiang zhigong chushou gongyou zhuzai loufang de jiage ji youguan zhence de tongzhi (A notice on the price of public housing sold to employees in 1999 and related policies)," No. 265. Beijing: BHRO.

Beijing Land and Housing Management Bureau (BLHMB) (2001) "Beijing shi chengzhen lianzu zhufang guanli shixing banfa shishi yijian (An implementation plan on the management of cheap rental housing in Beijing)," No. 1005. Beijing: BLHMB.

Beijing Municipal Government (BMG) (1992) "Beijing shi chengzhen jumin goumai shangping zhuzhai loufang guangli banfa (A regulation on residents' purchase of commodity housing in Beijing)," No. 35. Beijing: BMG.

—— (1998) "Beijing shi guanyu jiakuai jingji shiyong zhufang jianshe de ruogan guiding (A regulation on promoting the construction of affordable housing in Beijing)," No. 54. Beijing: BMG.

—— (1999) "Beijing shi jinyibu shenhua chenzhen zhufang zhidu gaige jiakuai zhufang jianshe shishi fangan (An implementation plan to further deepen housing reform and speed up housing construction in Beijing)," No. 21. Beijing: BMG.

Beijing Statistical Bureau (BSB) (2002) Beijing shi 2000 nian renkou pucha ziliao (Tabulation on 2000 Population Census in Beijing), Beijing: China Statistics Press.

—— (2002) *Beijing shi 2002 tongji nianjian (Beijing Statistical Yearbook 2002)*, Beijing: China Statistics Press.

Belsky, R. (2000) "The urban ecology of late imperial Beijing reconsidered: the transformation of social space in China's late Imperial Capital City," *Journal of Urban History*, 27(1): 54–74.

Bian, Y. J., Logan, J. R., Lu, H. L., Pan, Y. K. and Guan, Y. (1997) "'Working units' and the commodification of housing: observations on the transition to a market economy with Chinese characteristics," *Social Sciences in China*, 18(4): 28–35.

Burgess, E. W. (1967) "The growth of the city," in Park, R. E., Burgess, E. W. and McKenzie, R. D. (eds), *The City*, Chicago and London: University of Chicago Press.

Burrows, R. (1999) "Residential mobility and residualisation in social housing in England," *Journal of Social Policy*, 28: 27–52.

Chan, K. W. (1994) *Cities with Invisible Walls: Reinterpreting Urbanization in Post-1949 China*, Hong Kong: Oxford University Press.

Chang, S. D. (1977) "The morphology of walled capitals," in Skinner, G. W. (ed.), *The City in Late Imperial China*, Stanford, CA: Stanford University Press.

Cheng, T. J. and Selden, M. (1994) "The origins and social consequences of China's hukou system," *China Quarterly*, 139: 644–668.

Clark, W. A. V. (1986) "Residential segregation in American cities: a review and interpretation," *Population Research and Policy Review*, 5: 95–127.

Deng, F. F. and Huang, Y. Q. (2004) "Urban land reform and urban sprawl in China," *Progress in Planning*, 61: 211–236.

Editorial Committee of China Real Estate Statistical Yearbook (ECCRESY) (1999) *Zhongguo fangdichan tongji nianjie 1999 (China Real Estate Statistical Yearbook 1999)*, Beijing: China City Press.

Forrest, R. and Murie, A. (1999) "Residualisation and council housing: a statistical update," SAUS Working Paper 91, School for Advanced Urban Studies, Bristol.

Galster, G. (1988) "Residential segregation in American cities: a contrary review," *Population Research and Policy Review*, 7: 93–112.

Gaubatz, P. (1995) "Changing Beijing," *Geographical Review*, 85(1): 79–96.

Griffin, E. and Ford, A. (1993) "Cities of Latin America," in Brunn, S. and Williams, J. (eds), *Cities of the World*, 2nd edn, New York, NY: HarperCollins, 225–265.

Hu, X. H. and Kaplan, D. H. (2001) "The emergence of affluence in Beijing: residential, social stratification in China's capital city," *Urban Geography*, 22(1): 54–77.

Huang, Y. Q. and Clark, W. A. V. (2002), "Housing tenure choice in transitional urban China: a multilevel analysis," *Urban Studies*, 39: 7–32.

—— (2003a) "A room of one's own: housing consumption and residential crowding in transitional urban China," *Environment and Planning A*, 35(4): 591–614.

—— (2003b) "'Renters' housing behavior in transitional urban China," *Housing Studies*, 18(1): 103–126.

Knox, P. (1993) "Cities of Europe: Western Europe," in Brunn, S. and Williams, J. (eds), *Cities of the World*, 2nd edn, New York, NY: HarperCollins, 88–131.

Li, S. M. (2000a) "The housing market and tenure decision in Chinese cities: a multivariate analysis of the case of Guangzhou," *Housing Studies*, 15(2): 213–236.

—— (2000b) "Housing consumption in urban China: a comparative study of Beijing and Guangzhou," *Environment and Planning A*, 32: 1115–1134.

Lieberson, S. and Carter, D. (1982) "A model for inferring the voluntary and involuntary causes of residential segregation," *Demography*, 19: 511–526.

Logan, J. R., Bian, Y. J. and Bian, F. Q. (1999) "Housing inequality in urban China in the 1990s," *International Journal of Urban and Regional Research*, 23(1): 7–25.

Ma, L. J. C. (1981) "Urban housing supply in the People's Republic of China," in Ma, L. J. C. and Hanten, E. W. (eds), *Urban Development in Modern China*, Boulder, CO: Westview Press, 222–259.

—— and Xiang, B. (1998) "Native place, migration and the emergence of peasant enclaves in Beijing," *China Quarterly*, 155: 546–581.

Massey, D. and Denton, N. (1993) *American Apartheid: Segregation and the Making of the Underclass*, Cambridge, MA: Harvard University Press.

Sit, V. F. S. (1995) *Beijing: The Nature and Planning of a Chinese Capital City*, Chichester: John Wiley & Sons.

—— (2000) "A window on Beijing: the social geography of urban housing in a period of transition, 1985–1990," *Third World Planning Review*, 23(3): 237–258.

Skinner, G. W. (1977) "Introduction: urban social structure in Ch'ing China," in Skinner, G. W. (ed.), *The City in Late Imperial China*, Stanford: Stanford University Press.

Solinger, D. J. (1999) *Contesting Citizenship in Urban China: Peasant Migrants, the State and the Logic of the Market*, Berkeley, CA: University of California Press.

State Council (1994) "Guowuyuan guanyu shenhua chengzhen zhufang zhidu gaige de jueding (A decision from the State Council on deepening the urban housing reform)," State Council Documentation No. 43. Beijing: State Council.

—— (1998) "Guowuyuan guanyu jinyibu shenhua chengzhen zhufang zhidu gaige jiakuai zhufang jianshe de tongzhi (A circular from the State Council on further deepening the reform of urban housing system and accelerating housing construction)," State Council Documentation No. 23. Beijing: State Council.

State Statistics Bureau (SSB) (1999) *Comprehensive Statistical Data and Materials on 50 Years of New China (Xin zhongguo wushi nian tongji ziliao huibian)*, Beijing: China Statistics Press.

—— (2001) *Zhongguo tongji nianjian 2001 (China Statistical Yearbook, 2001)*, Beijing: China Statistics Press.

—— (2002) *Zhongguo 2000 nian renkou pucha ziliao (Tabulation on the 2000 Population Census of the People's Republic of China)*, Beijing: China Statistics Press.

Tolley, G. S. (1991) *Urban Housing Reform in China: an Economic Analysis*, Washington, DC: International Bank for Reconstruction and Development/World Bank.

Wakeman, F., Jr. (1985) *The Great Enterprise: Manchu Reconstruction of Imperial Order in Seventeenth-Century China*, Berkeley, CA: University of California Press.

White, P. (1984) *The West European City: A Social Geography*, London: Longman.

World Bank (1997) Sharing *Rising Income: Disparities in China*, Washington, DC: World Bank.

Wu, F. L. (1996) "Changes in the structure of public housing provision in urban China," *Urban Studies*, 33(9): 1601–1627.

Wu, W. P. (2002) "Migrant housing in urban China: choices and constraints," *Urban Affairs Review*, 38(1): 90–119.

Zhang, L. (2001) "Contesting crime, order and migrant spaces in Beijing," in Chen, N., Clark, C., Gottschang, S. and Jeffery, L. (eds), *China Urban*, London: Duke University Press.

Zhang, X. Q. (1998) *Privatisation: A Study of Housing Policy in Urban China*, New York: Nova Science Publishers.

# 12 Migrant residential distribution and metropolitan spatial development in Shanghai

*Weiping Wu*

## Introduction

Patterns of migrant settlement have a major influence on urban spatial development, as exemplified in such developing cities as Rio de Janeiro, Mexico City and Jakarta. Congregation of migrants in urban villages aggravates existing spatial segregation, as many migrants are never fully integrated and become a permanent urban underclass. On the other hand, concentration may help migrants maintain previously established social relations and assert their group identity. Given the persistence of China's migration trends over the last two decades, migrants have begun to assert their influence on cities' spatial structure.

This chapter is a preliminary attempt to understand the potential impact of migrant residential distribution on Shanghai's spatial development.[1] Such understanding is likely to fill a gap in the existing literature on the spatial development of Chinese cities. Three research questions motivate the chapter: (a) How is migrant residence distributed geographically across Shanghai? (b) How does this spatial pattern compare to distribution patterns of the local population and employment activities? (c) What are the key environmental factors underlying the spatial distribution of migrants? Results of this chapter are based on empirical analyses at the subdistrict level. Data are drawn primarily from the 2000 Population Census and 1996 Basic Establishment Census, supplemented by the 1995 1 Percent Population Survey.[2] Metropolitan-wide analysis is done to determine aggregate spatial patterns of migrant residence and demonstrate how migrant spatial distribution compares with that of the local population. Information collected during field visits in Shanghai's periphery where a large number of migrants live also helps understand how neighborhood dynamics affect migrant residential distribution.

## Theories of migrant settlement in the Chinese context

Turner (1968) suggests a two-stage process for rural–urban migrants in urbanizing countries: initial settlement in central city slum rental units and subsequent intraurban relocation to peripheral self-built or squatter settlements. Inner-city slums are the major receiving areas for new migrants who view proximity to employment as the highest priority. As migrants improve their income level, they move to build peripheral informal shanties for residential stability or ownership and then upgrade shanty dwellings over time into more substantial houses. Turner's notion of upward housing mobility of migrants, from slum renters to squatter owners, concurs with the popular belief contrasting "central-city slum of despair" and "peripheral shantytown of hope" (Conway 1985; Eckstein 1990; Ulack 1978).

But significant disagreement has risen regarding the residential location of new migrants (e.g. Conway 1985; Gilbert and Varley 1990; UNCHS 1982). In a number of countries with continuing urbanization, inner-city slums are no longer found to be the major receiving areas for new migrants due to the expansion and redevelopment of the commercial core and in turn the rapid rise of land costs. As a result, peripheral squatter settlements become the primary destinations. Large-scale squatting has occurred in cities where there are large areas of state-owned land around the city.

Migrants are attracted to different parts of a city for different reasons. Proximity to existing or potential employment is a major determinant of their locational behavior. The formation and development of any informal settlement are often linked to changes in the economic activity of the surrounding area (Conway 1985; Gilbert and Varley 1990; Klak and Holtzclaw 1993). Others point to the importance of kinship and friendship ties, acting as social institutions (Abu-Lughod 1961; Banerjee 1983; Collier 1976; UNCHS 1982). Migrants' first place of residence in the city is largely predetermined by the location of kin or friends. A typical migrant gravitates to a small area of the city where people from his home place are already living and this results in the formation of "small enclaves of ex-villagers" (Abu-Lughod 1961: 25).

Making settlement decisions, however, is only the first steps for migrants when they begin a new life course in the city. Where and how they live and work is likely to affect their general levels of satisfaction with urban living and the ease or difficulty to adapt to the new environment. Housing provides a context in which migrants make their adjustments to urban life, and residential patterns and outcomes are going to reflect their socioeconomic standing (W. Wu 2002a).

According to the Chicago School, spatial distribution of migrant/immigrant groups is a reflection of their human capital, and dispersal is considered an indicator of successful settlement and assimilation. On the other hand, cultural pluralism emphasizes the fact that migrants maintain their identities with both places of origin and destination, and no longer accepts the view of inevitable assimilation. Migrant concentration is one key manifestation of this perspective and is seen from a positive angle as group mobilization often leads to an assertion of group identity (Alba and Nee 1997; Dunn 1998; Powers and Seltzer 1998; Zhou 1997).

My previous research has shown that informal settlements are not a viable option for China's migrants, unlike in many other developing countries, largely due to municipal authorities' intolerance of migrant congregation and squatting (W. Wu 2002b). In Shanghai, for instance, there are some small clusters of temporary housing that do not seem to belong to any enterprise or institution and resemble squatter settlements. But the size of these clusters is no more than a handful of sheds on open farmland or areas undergoing development.

However, a number of large migrant settlements or communities have been in existence in Beijing for over a decade (Ma and Xiang 1998; Wang 1998). Located mostly in suburban areas where rental housing is readily available, these migrant enclaves or villages are formed by migrants from the same province or region. Unlike most migrant communities in other developing countries, Beijing's migrant villages are existing built-up areas where migrant population outnumbers the locals. Migrants rent from local residents or live in market areas constructed by local governments or private businesses. Instances of squatting are rare. Whereas the problem of residential crowding is still widespread, these also are not ghettoes of despair as migrants are optimistic and hard workers seeking economic opportunities in the city (Ma 2003).

Parallel to the process of migrant settlement in urban China are the inevitable changes brought by the transition from plan to market in China's cities at large. Any study of migrant settlement patterns requires some understanding of how existing city residential areas are distributed geographically by socioeconomic status (Vaughan and Feindt 1973). According to F. Wu (2002a), there are four major types of urban neighborhood in China: (a) traditional ones in the old-city area developed before 1949; (b) work-unit compounds, largely associated with industrial development, developed between 1949 and 1978; (c) mixed-use suburban communities or satellite towns, developed from the late 1970s; and (d) rural–urban fringe or peri-urban villages formed after the late 1970s. Recent housing reforms have allowed housing to be commodified and developed with both domestic and global capital, giving rise to a variety of residential spaces that are replacing the cellular-type of housing structure built around work-units. Some evidence in Shanghai has already suggested that housing commodification and socioeconomic differentiation brought by the plan-to-market transition is leading to the revitalization and continuation of the pre-socialist spatial division (Ma 2003; F. Wu 2002b). In the central city, the residential profile of some better locations is upgrading while that of others remains dilapidated, although the pressure is ever rising for real-estate development of locations with good accessibility. In the suburban areas, there is an increasing residential juxtaposition of rural villages, resettlement housing for central-city residents, migrant communities and commodity housing projects.

Even cities without a strong pre-socialist legacy of residential segregation are showing signs of socio-spatial differentiation in the transitional period. Beijing, for instance, has seen the formation of a small number of wealthy housing areas in the suburbs although the concentration effect is not as pronounced as in Western cities. Aggravated living conditions of the central city and real-estate development propaganda have promoted the urban affluent to move to the eastern and northern inner suburbs of Beijing (Hu and Kaplan 2001). Ironically these same areas also attract the largest number of migrants, although some migrants concentrate in the old central city (Gu and Shen 2003). In Guangzhou, the capital city of Guangdong Province, separate residential areas serve different population segments such as cadres, workers and intellectuals. These spaces show varied levels of population density and housing quality (Yeh *et al.* 1995).

## Spatial patterns of Shanghai's local population and employment

Today, with a population of over 16 million (including both locals and migrants) and a land area of 6,377 square kilometers, the Shanghai metropolitan area is governed by the Shanghai Municipal Government, equivalent to a provincial government because of Shanghai's special administrative status. There are nineteen districts/counties (before 2000 there were twenty), sixteen with urban designation (*qu*) and three rural (*xian*), in three geographic zones (Figure 12.1).[3] Continuing efforts of fiscal decentralization

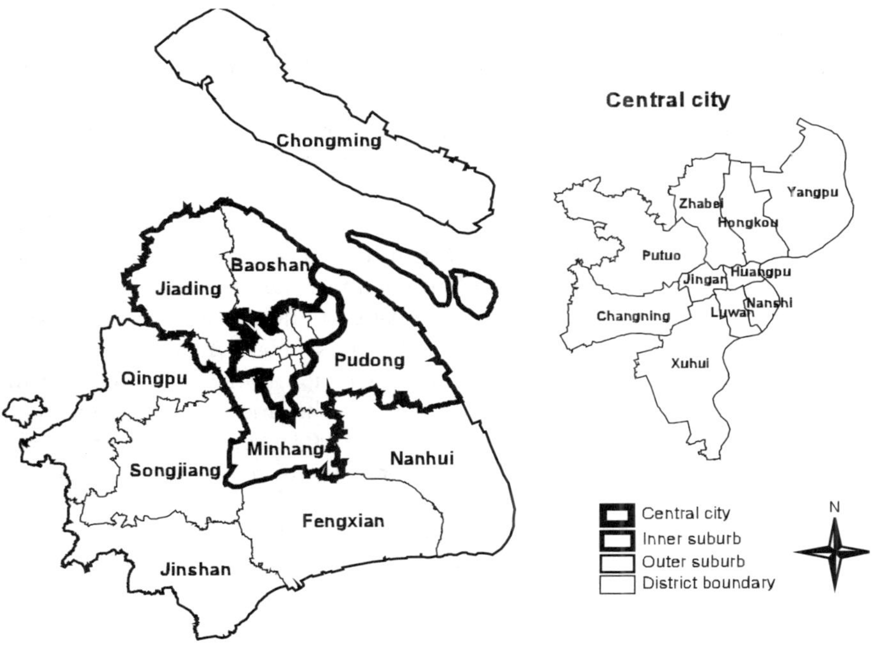


*Figure 12.1* Districts in the Shanghai metropolitan area and central city.

have offered district/county governments substantial autonomy in tax collection, budget allocation, infrastructure provision and planning.

### *Residential patterns*

More or less following a concentric pattern, Shanghai has a history of residential differentiation and migrant enclaves, dating back to the pre-1949 period. Urban space was differentiated into upper and lower ends in the central city (cf. Pan, Chapter 7, this volume). Shantytowns in this period were located along the boundaries of foreign concession areas and in areas designated for Chinese residents (Lu 1995; F. Wu 2002a). After more than thirty years of socialist development, there were still signs of such shantytowns in several districts located immediately outside of the central downtown. But residential differentiation had been reduced markedly with many years of building public housing and accelerated efforts to redevelop shanty areas. Still, in many parts of the central city, the extreme conditions of high-density, dilapidated housing, inadequate infrastructure, and mixed land use patterns made the redevelopment effort a challenging one.

The central city, with perhaps the highest population density in the world (in the range of 50,000–60,000 people per square kilometer), has been decentralized in recent decades albeit with uneven results. New housing construction outside the central city and redevelopment within are two important mechanisms of population decentralization. For instance, the development of the Pudong New Area and construction of several large housing projects (e.g. Liuli, Weifang Xincun and Jinyang Xincun) serve to accommodate close to a million residents relocated from the central city (W. Wu 1999). Between 1991 and 1998, about 400,000 households as well as 12,000 work-units were moved from downtown to the city's outskirts. The satellite town program, though constructed more for the purpose of industrial development, also has begun to attract more population since the 1980s (accommodating about two-thirds of a million residents in 1990). Five of the seven satellite towns are in the current inner suburb, including one in Baoshan, two in Jiading and two in Minhang, while the sixth in Songjiang and seventh in Jinshan are in the outer suburb (Ning and Yan 1995).

There is evidence that the importance of location, which was irrelevant in socialist cities without land markets, has led to the emergence of a land rent gradient similar to that of Western cities (Ma 2003). Some central areas previously residential are increasingly under pressure for redevelopment, largely for commercial and office uses. The prime examples of large-scale redevelopment are the area south of the commercial core of Huaihai Road in Luwan District and areas surrounding the passenger rail station in Zhabei District. But redevelopment through real estate is selective. In the central core, some residential areas with extreme dilapidation and high density have been left out of commercial redevelopment because of high costs associated with resettlement (F. Wu 2002a). As a result, there are many instances of awkward juxtaposition of flashy, high-rise commercial buildings and dilapidated, pre-1949 dwellings in the central city.

With the combined effect of satellite town program, residential resettlement, new housing construction and central-city redevelopment, Shanghai's residential patterns have changed steadily since the 1980s. The core of the central city, including Huangpu, Nanshi, Luwan, Jingan and Hongkou districts, has lost a significant number of local residents (in the range of 15–20 percent), but the periphery of the central city has seen some gains, particularly in Xuhui District. Population growth in the four districts of the inner suburb has been substantial, in the range of 20–25 percent, while the outer suburb's population level has remained stable (Gao and Jiang 2002). Specifically, inner suburban subdistricts immediately outside of the central city are accommodating a large number of local residents at a fairly high level of density, even though the central city is still the residential core (Figure 12.2).

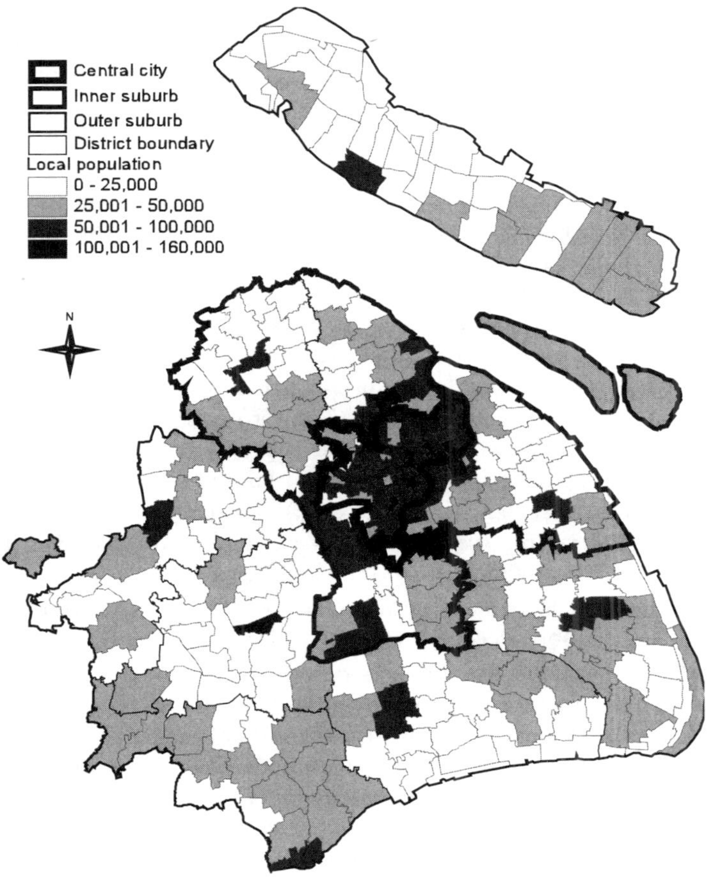


*Figure 12.2* Distribution of local population in Shanghai, 1999.

### *Spatial distribution of industrial and service establishments*

The central city was characterized by a mixed pattern of land use and industrial fragmentation in the pre-1949 period (W. Wu 1999). The lack of a proper planning framework, when foreign concessions and Chinese districts were separate jurisdictions administratively, led to a situation whereby factories and houses were located rather randomly and often encircled each other. This problem was further aggravated by Mao's policy on industrial self-sufficiency and the system of administrative allocation of land before 1979 (Hodder 1996). Even in 1990, the central city housed close to four times as many industrial establishments as the inner suburbs (Ning and Yan 1995).

To solve the problems associated with fragmented industrial land use, Shanghai has recently relied on relocating factories in the central city to new urban districts available for industrial expansion. A number of industrial parks, often called Economic and Technology Development Zone (ETDZ), have been created in the inner suburb, including Jinqiao Export Processing Zone and Zhangjiang High-Tech Zone in Pudong, Minhang ETDZ, Hongqiao ETDZ and Caohejing High-Tech Park. This process of industrial relocation, albeit slow and with mixed results, has freed up a significant amount of space in the central city and led to an industrial concentration in the inner suburb (Table 12.1). In addition, the seven industrial satellites, constructed before 1979 and now primarily in the inner suburb, continue to accommodate a large number of industrial establishments. By 1996, only about 23 percent of the city's total industrial establishments were still located in the central city.

But industrial fragmentation remains at the metropolitan level, aggravated by the chaotic location of rural and township enterprises across suburban areas with the development of rural industrialization. About 27 percent of Shanghai's land is currently for industrial use, a level much higher than the average for other large Chinese cities (about 15 percent) and comparable cities elsewhere in the world. This may be largely attributable to the fact that nearly 44 percent of Shanghai's industrial land use is scattered and not in concentrated forms (such as industrial park). Similarly, productivity level of industrial land (often measured by output value per unit of land) is lower than that in some comparable cities (Xiong and Luo 2000).

Another important force shaping Shanghai's urban space was commercial development and, more recently, services. The traditional commercial centers of the city were formed by two avenues, Nanjing Road and Huaihai Road. Though the Maoist era saw the decline of these areas, municipal authorities are now reviving the downtown business district, with the help of overseas investment. At the heart of Shanghai's new development is a set of commercial cores scattered throughout the city's old and newly developing districts. Each of these cores, including Nanjing Road, Huaihai Road, Hongqiao, Xujiahui, the Passenger Rail Station (Tianmu Road), and Baoshan Road, has become increasingly commercial in character, with shops and offices displacing residential and industrial space (Gaubatz 1999). The building of a new central business district in Pudong's Lujiazui, an area of 1.7 square kilometers on the east bank of the Huangpu River, adds to the city's service alignments as well.

The expansion of the service sector has been steady during recent decades, now counting for just over half of Shanghai's gross domestic product. In contrast to spatial patterns of industrial establishments, services tend to be more concentrated in the central city and, to some degree, outer suburb (Table 12.1). Two districts in the outer suburb, Songjiang and Qingpu, appear to house a considerably larger number of service establishments than any other suburban districts, except Pudong.

## Migrant residential distribution

Shanghai appears to resemble a number of developing countries with continuing urbanization in intra-urban migrant residential patterns. Central-city housing is becoming less attractive to migrants because of commercial redevelopment and in turn the rapid rise of housing costs. During the mid-1980s, comparable amounts of migrants lived in the central city and inner suburb (W. Wu 2002a). But with continuing urban expansion and downtown redevelopment, Shanghai's inner suburb has become the primary receiving area for migrants since the early 1990s. With the exception of a few central-city subdistricts, the largest amount of migrants now seems to concentrate exclusively in the inner suburb (see Figure 12.3). In particular, subdistricts in Minhang are attracting a very high level of migrant population. Because data on migrant and local populations at the subdistrict level are accurate only to an interval of 5,000 persons in the 2000 Population Census publications, it may not be advisable to sum district and zone totals and results shown here should be considered as merely indicative.

It appears that a number of subdistricts immediately flanking the central-city outline are now residential centers for both migrants and locals (see Figures 12.2 and 12.3). They are primarily in the districts of Minhang, Baoshan and Pudong (inner suburb) as well as Putuo, Xuhui and Yangpu (periphery of the central city). This ring area can be characterized as Shanghai's urban periphery or rural–urban transitional area (*chengxiang jiehebu*), at the edge of the city property or built-up area and as the intermediary between agricultural use in suburban areas and urban land use in the center. It is similar to the concept of peri-urban areas. The urban periphery in Shanghai's 1997 survey, for instance, consisted of subdistricts in the far edge of the central city and the four districts of the inner suburb (W. Wu 2002a).

Another important indicator measuring migrant concentration is the proportion of migrants in total population (see Figure 12.4). Twenty-seven subdistricts, primarily in the urban periphery and slightly beyond (in Songjiang and Qingpu districts), are areas where migrants outnumber local residents (see Table 12.2). The only exceptions are the two collective farms on the southern coast in Fengxian County, Wusi and Liaoyuan, which were created

*Table 12.1* Population, employment and housing distribution by city district, Shanghai

| *District* | *% migrants in total population 2000 (mean)* | *Mean density of local population 1999 (per sq. km)* | *Mean ratio of employment to local population 1996* | *Industrial establishments 1996 (total)* | *Service establishments 1996 (total)* | *Per capita housing space 2000 (sq. m.)* | *Number of subdistricts* |
|---|---|---|---|---|---|---|---|
| *Central city* | | | | | | | |
| Huangpu | 16.7 | 59,672 | 2.29 | 301 | 7,222 | 10.8 | 4 |
| Nanshi | 18.0 | 67,685 | 0.68 | 1,005 | 5,872 | 10.8 | 5 |
| Luwan | 14.3 | 54,635 | 0.90 | 617 | 3,334 | 13.4 | 4 |
| Xuhui | 25.1 | 19,788 | 1.18 | 1,826 | 4,989 | 18.4 | 12 |
| Changning | 16.3 | 24,948 | 1.04 | 1,221 | 4,902 | 17.5 | 10 |
| Jingan | 16.9 | 48,854 | 0.91 | 552 | 3,543 | 13.8 | 5 |
| Putuo | 26.1 | 28,024 | 0.71 | 1,758 | 4,204 | 16.9 | 14 |
| Zhabei | 17.4 | 36,068 | 1.10 | 1,699 | 5,132 | 15.5 | 9 |
| Hongkou | 16.5 | 39,743 | 0.68 | 1,113 | 5,891 | 15.8 | 10 |
| Yangpu | 16.8 | 27,267 | 1.10 | 1,896 | 5,684 | 15.8 | 11 |
| *Inner suburb* | | | | | | | |
| Minhang | 45.8 | 2,243 | 1.31 | 4,191 | 2,808 | 25.3 | 17 |
| Baoshan | 32.6 | 3,295 | 1.38 | 3,559 | 2,968 | 23.4 | 18 |
| Jiading | 38.7 | 1,258 | 0.69 | 3,463 | 2,541 | 31.5 | 20 |
| Pudong | 34.9 | 8,190 | 0.81 | 7,301 | 8,436 | 23.6 | 39 |

| | | | | | | | |
|---|---|---|---|---|---|---|---|
| *Outer suburb* | | | | | | | |
| Nanhui | 16.4 | 1,036 | 0.47 | 4,561 | 2,252 | 39.0 | 28 |
| Fenxian | 26.2 | 767 | 0.51 | 3,951 | 1,630 | 34.1 | 24 |
| Songjiang | 30.7 | 825 | 0.68 | 3,324 | 5,706 | 33.7 | 20 |
| Jinshan | 11.2 | 1,006 | 0.54 | 2,455 | 2,682 | 38.2 | 16 |
| Qingpu | 30.4 | 1,195 | 0.66 | 4,100 | 5,223 | 32.2 | 20 |
| Chongming | 11.3 | 805 | 0.33 | 2,590 | 1,159 | 42.6 | 33 |
| Central city | 19.4 | 35,014 | 1.00 | 11,988 | 50,773 | 14.9 | 84 |
| Inner suburb | 37.2 | 4,702 | 0.98 | 18,514 | 16,753 | 26.0 | 94 |
| Outer suburb | 20.3 | 934 | 0.52 | 20,981 | 18,652 | 36.6 | 141 |
| | ANOVA $p = 0.000$ | ANOVA $p = 0.000$ | ANOVA $p = 0.000$ | 51,483 | 86,178 | 23.6 | 319 |

Sources: SFPCO (2002) and SPECO (1997).

Notes

1 The 2000 Population Census data are available at the subdistrict level only for a small number of indicators, including migrant population and total population, and with accuracy to an interval of 5,000 persons.

2 In 2000, the districts of Huangpu and Nanshi were merged into one district named as Huangpu.

3 The one-way ANOVA is done at the subdistrict level, with a total of 319 subdistricts.

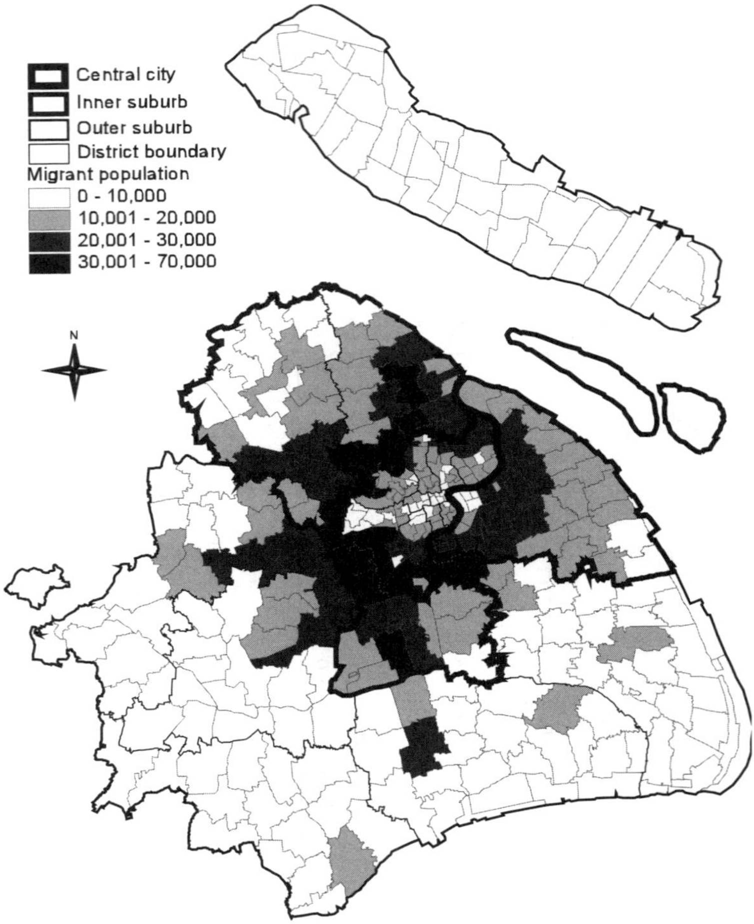


*Figure 12.3* Size of migrant population in Shanghai, 2000 (source: SFPCO 2002).

by the municipality before 1979 to attract central-city youth. The high proportion of migrants there may be attributable to the fact that more and more local farming is taken over by migrants and youth previously working on these two farms have returned to their original residence.

In general, these twenty-seven areas are suburban towns (as indicated by the name *zhen*) where the ratio of employment to local population is much higher and industrial establishments more numerous (Table 12.2). This suggests that, first of all, employment opportunities are an important attraction for migrants. Since most migrants come to Shanghai in search for work to augment agricultural income, they tend to base their locational decisions on

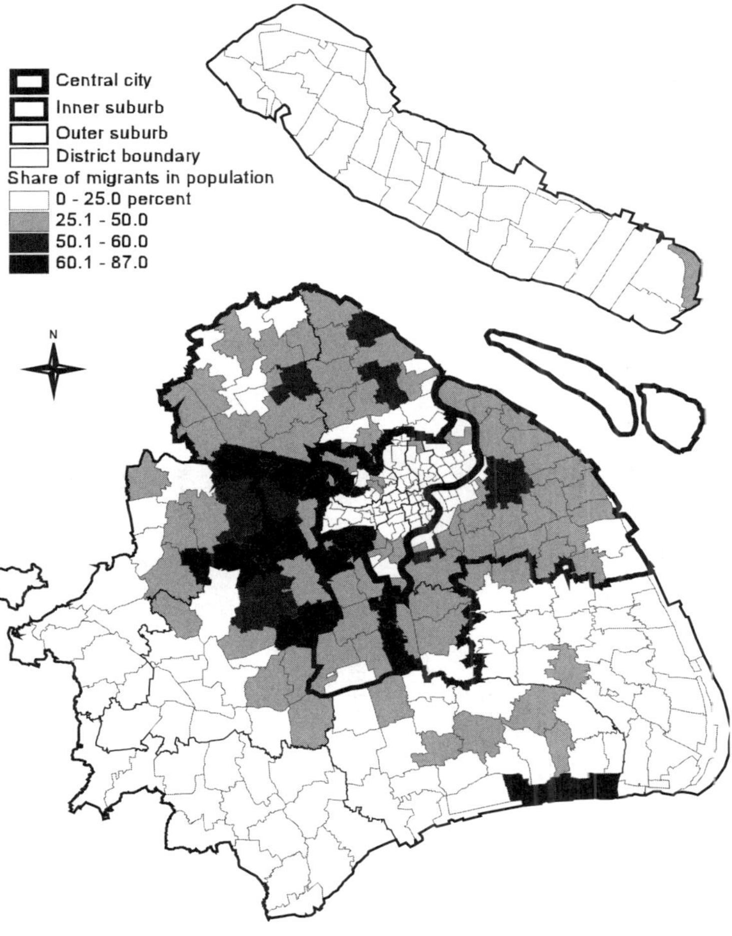


*Figure 12.4* Proportion of migrants in population in Shanghai, 2000 (source: SFPCO 2002).

where the jobs are. Although some migrants may not live in the same sub-district as where they work, my own survey indicates that three-quarters of them live within ten minutes' walking distance from their work place.[4] Accelerated urban growth during the reform period has led to increasing concentration of economic functions on the outskirts of the built-up areas, in the form of high-tech development zones, and office and industrial parks. Moreover, urban periphery often houses a large number of wholesale markets for agricultural products because of its proximity to suburban farms.

These suburban towns have diversified their economies with rural industrialization since 1979. Each houses a large number of manufacturing

*Table 12.2* Subdistricts with migrants outnumbering local population, Shanghai

| *District* | *Subdistrict* | *% migrants in total population 2000* | *Density of local population 1999 (per sq. km)* | *Ratio of employment to local population 1996* | *Industrial establishments 1996* | *Services establishments 1996* |
|---|---|---|---|---|---|---|
| *Central city* | | | | | | |
| Xiuhui | Hongmeilu | 66.7 | 5,234 | 4.50 | 320 | 177 |
| Putuo | Caoanlu | 60.9 | 26,265 | 0.35 | 65 | 277 |
| | Changzhengzhen | 60.9 | 584 | 2.01 | 308 | 218 |
| *Inner suburb* | | | | | | |
| Minhang | Qibaozhen | 60.9 | 3.401 | 1.36 | 353 | 195 |
| | Jiwangzhen | 60.0 | 983 | 1.58 | 130 | 57 |
| | Zhuzhaizhen | 60.0 | 993 | 0.83 | 118 | 31 |
| | Hongqiaozhen | 61.9 | 2,157 | 1.58 | 321 | 168 |
| | Caohangzhen | 85.7 | 980 | 1.12 | 156 | 61 |
| | Tangwanzhen | 66.7 | 660 | 0.64 | 191 | 58 |
| Baoshan | Yanghangzhen | 58.8 | 1,011 | 0.32 | 330 | 174 |
| | Shengqiaozhen | 60.0 | 1,175 | 1.23 | 122 | 67 |
| Jiading | Fengbangzhen | 66.7 | 846 | 0.93 | 264 | 103 |
| | Jibangzhen | 60.0 | 832 | 0.61 | 172 | 48 |
| | Huangduzhen | 66.7 | 909 | 0.87 | 274 | 218 |
| | Jiangqiaozhen | 66.7 | 1,316 | 0.59 | 191 | 78 |

| | | | | | | |
|---|---|---|---|---|---|---|
| Pudong | Yangsizhen | 58.3 | 5,226 | 0.73 | 145 | 142 |
| | Zhangqiaozhen | 58.8 | 1,172 | 0.84 | 214 | 151 |
| | Jinqiaozhen | 58.8 | 1,233 | 3.28 | 333 | 246 |
| *Outer suburb* | | | | | | |
| Songjiang | Sijingzhen | 60.0 | 897 | 0.78 | 234 | 330 |
| | Xingqiaozhen | 66.7 | 620 | 0.91 | 290 | 245 |
| | Dongjingzhen | 60.0 | 503 | 1.02 | 120 | 481 |
| Qingpu | Zhaoxiangzhen | 66.7 | 533 | 1.08 | 516 | 325 |
| | Xujingzhen | 66.7 | 628 | 0.92 | 472 | 273 |
| | Huaxinzhen | 60.0 | 777 | 0.70 | 245 | 199 |
| | Fengxizhen | 60.0 | 651 | 0.93 | 164 | 63 |
| Fengxian | Wusi farm | 70.0 | – | – | 46 | 24 |
| | Liaoyuan farm | 70.0 | – | – | 32 | 11 |
| *Average* | | 63.1 | 2,383 | 1.19 | 242 | 175 |
| *City average* | | 25.1 | 11,481 | 0.80 | 161 | 270 |

Sources: SFPCO (2002) and SPECO (1997).

Note
– Data not available.

establishments, particularly township enterprises. Many industrial enterprises prefer to hire migrant workers because of their willingness to work harder but for less pay. About 75 percent of migrants in the manufacturing sector work for state-owned, collective or foreign-invested enterprises (based on own survey results). On the other hand, these towns still do not have the same level of services as central-city areas (see Table 12.2). This may actually be an attraction to many migrants as opportunities to open small businesses are more abundant and competition with established local establishments are less fierce. The 2000 Population Census results show that 27.4 percent of all migrants work in Shanghai's service sector, outnumbering those in manufacturing (25.9 percent). My own survey also indicates that close to 88 percent of migrants working in the service sector are self-employed, compared to about 45 percent in all sectors and a mere 5.4 percent in manufacturing.

Housing availability is likely another important factor attracting migrants to the urban periphery, particularly the availability of rental housing. My previous research on Shanghai and Beijing shows that the majority of migrants rent (W. Wu 2002b). Private rental housing accommodates the largest number of migrants, especially in suburban areas that used to be or still are agricultural. Institutionally provided dormitories are another key housing choice for migrants. As a result of restricted access to the mainstream urban housing distribution system, ownership is minimal among migrants (under 1 percent).

In the urban periphery, local population density is significantly lower than in the central city and per capita living space is much larger (see Table 12.1). Since citywide housing data are not available at the subdistrict level, local population density may be used as a rough reverse proxy for housing space. In the twenty-seven subdistricts where migrants outnumber local population, this density is much lower than the city average (Table 12.2). In the urban periphery, private housing is more common because of the previous rural status of many neighborhoods. Local rural residents have been allocated ample land to build private living quarters. As a result, rental housing is more readily available and costs less in the urban periphery. Under less strict government supervision, local residents in these districts also have reverted to building low-quality rental housing on vacant or farmland to make a quick profit.

However, it is clear that housing availability is a secondary consideration to migrants as much less of them are residing in the outer suburb where per capita housing space is larger than that of the inner suburb (see Table 12.1). Employment opportunities are likely to be of prime importance. The urban periphery offers both, as well as good access to facilities in the central city. As a result, migrant residential distribution displays an inverted U shape across the metropolitan area. Peak concentration occurs in the periphery of the built-up area (about 15–18 kilometers from the city center), where the proportion of migrants in total population is significantly higher than in the other two zones (see Table 12.1).

## Attractions of the urban periphery – some observations

The urban periphery, often viewed by authorities as growth areas for urban expansion, is where urban and rural land uses intersect. The unique attractions it offers to migrants are clearly shown in the situation of Gaojing Village in Baoshan District (W. Wu 2002a). Inexpensive rental housing is readily available in the village. There are a number of reasons behind this. First, housing for local residents is much more spacious. An average household of four people tends to have at least eight to ten rooms and the average number of rooms rented out by a household was four rooms. Second, many of the newly "converted" urban residents have been assigned jobs in state enterprises that today are facing a great deal of pressure to restructure.[5] Because of their severed economic ties with the village, these residents tend to fair worse than the remaining rural residents in the village. As a result, they have very strong incentives to rent out rooms for extra income. Third, four of the six neighborhoods are occupied by converted urban residents without residential committees and in turn have minimal levels of administrative oversight. Therefore private rentals are under few restrictions.

The village has a good location and offers migrants employment opportunities as well. It sits at the edge of the central city area and is connected to the downtown with a new elevated expressway. On the other direction, the expressway leads to the Wusong ferry that connects with Shanghai's rural hinterland – Chongming Island and Jiangsu Province. In addition, many of the remaining rural residents in the village no longer are willing to farm themselves, as elsewhere in Shanghai's suburban areas, and instead lease land to migrants. The village enterprises also employ migrants and have built makeshift housing to accommodate them. Given all of these attractions, the village houses far more migrants than local residents.

The case of this village is only a microcosm of Shanghai's urban periphery. The largest such area under tremendous growth is the Pudong District. Between 1988 and 1998, the volume of floating population in Pudong increased about ten-fold, from around 40,000 to 395,000 (Pudong New Area Floating Population Office 1998). In addition to employment opportunities, Pudong offers migrants abundant private as well as public housing for rental. Over 40 percent of migrants rent private housing for accommodation in Pudong, well above the rate for the city as a whole (about 27 percent). In two of Pudong's townships – Qinyang and Yanqiao, migrants outnumbered local residents in 1998 and an even higher percentage of migrants rented private housing there (55 and 71 percent respectively).

Pudong's situation again speaks to the attraction of the urban periphery for migrants, based on field visits of the two townships and eighteen villages in which migrants outnumbered local residents in 1998. The foremost factors are employment opportunities and the availability of inexpensive, private rental housing. In addition to industrial facilities, there are often many recently built commercial and residential buildings, with some farming and rural housing in a number of pockets. Two adjacent villages (Maluqiao and Zhujiabang), for instance, are dotted with heavy industrial facilities, including shipyards, several chemical factories, a power plant, and a gas company. Many villagers have found jobs in nearby factories and leased out their remaining land to migrants for farming. These areas, therefore, offer plentiful opportunities for migrants to find work, in all sectors including industrial, construction and agricultural. In addition, the multitude of small, roadside businesses also could accommodate many migrants.

In reality, there is already a large rental market operating in these areas, as rents tend to stabilize by location. The incentive on the supply side is particularly strong where farmland has been acquired for urban development but old village neighborhoods still remain. In some of these neighborhoods local villagers have been relocated. Their rural housing, vacated temporarily before being torn town to make room for development, has become concentrated living quarters for migrants. Administrative reorganization during the urbanization process and subsequent neglect or incapacity also have allowed the rental market to operate unregulated much more easily than the city's established urban areas. This administrative fluidity is less prevalent in rural areas where traditional village communities are more intact.

Other attractions to migrants include good transport services and easy access to Shanghai's central areas. One Pudong township official suggested several reasons for the high concentration of migrant population in his town (personal interview). First, the area was well served by a road network constructed when Pudong was first developed. If the Pudong government had had more funds, the entire area would have already been entirely urbanized. Second, this area sat right at the edge of Lujiazui central business district and offered good access to a large number of employment opportunities. Its location also had attracted enterprises to the township. Third, the township, although the smallest in size in Pudong, had a number of township enterprises that would employ migrant workers.

## Conclusion

The significant rise in migratory flows during the last two decades is a prominent feature of China's economic transition, reflecting a rapid process of industrialization and urbanization. Shanghai alone has received in excess of three million migrants, most of whom come to the city to seek employment opportunities. Their settlement patterns are superimposing upon an already changing spatial structure in the metropolitan area. Many local residents have been moving away from the central city even though density there remains much higher than the suburbs. Accompanying such moves is industrial relocation. As a result, there is an increasing juxtaposition of rural towns, resettlement housing for central-city residents, new commodity housing projects, high-tech development zones and industrial parks in the urban periphery.

Migrant residential distribution appears to coincide with this trend of decentralization. Replacing a somewhat even distribution between the central city and inner suburb during the 1980s, migrants are now increasingly settling in the latter. A number of subdistricts in the outskirts of the built-up area are now residential centers for both migrants and locals, primarily in the districts of Minhang, Baoshan and Pudong as well as Putuo, Xuhui and Yangpu. The overall distribution resembles an inverted U shape, with migrant population peaking in the urban periphery and outnumbering local population in some subdistricts.

This distribution pattern is not accidental as migrants are attracted to different parts of a city for good reasons. One key factor likely to account for migrant concentration in the urban periphery is opportunities to work in existing industrial enterprises and to open up small service establishments. The central city also is within an easily accessible distance, between 15 and 20 kilometers. Housing availability adds another attraction, albeit secondary. Given that ownership is yet to become an attainable goal, housing comfort and tenure are less relevant as a motivation for migrants in making settlement decisions, unlike migrants in many other developing countries.

If restrictions on migrant settlement in Chinese cities are lifted in light of the new state policy, housing considerations may become more paramount and migrant residential distribution may display new patterns. In early January 2003 the State Council issued a circular to promise migrants legal rights to work in cities, propose an end to making some urban jobs off limits to migrants, and call for local authorities to abolish discriminatory policies. Although this non-binding circular may be some distance away from actual implementation (since no specifics regarding the policy have been made public), it begins to address the costs and hardship migrants bear. A new Rural Land Contracting Law also has gone into effect since March 2003, making it easier for farmers to transfer or lease their land and giving them an extra incentive to move to cities.

With more tolerant migration policies, over time urban ties will surpass rural ties and many migrants will settle permanently at urban destinations. Their residential patterns are going to be a major influence on urban development. Although some migrants may adjust to urban life well, others will have no choice but to become the first of an emerging group of poor in cities with increasing spatial segregation. If no concerted municipal actions are taken, Shanghai may see the return of some pre-1949 divisions where migrants dominate poorer neighborhoods in less desirable locations.

## Acknowledgments

Support from the National Science Foundation (BCS-9974540) and the United States Department of Education (P019A80016) is deeply appreciated.

I would like to thank members of the research team at Shanghai Academy of Social Sciences' Institute of Population and Development Studies for survey support and Professor Jianping Wu at the East China Normal University for informational support. This chapter would not be complete without the able research assistance by Kimberly Klinker, Mary Reynolds and Irayda Ruiz.

## Notes

1 Migration takes place in two forms: through permanent migration (*qianyi*) with formal changes of household registration (*hukou*) and through temporary movement (officially called "floating population" or *liudong renkou*) without official changes of *hukou* from the origin to the destination. The latter group, which makes up the bulk of China's internal migration, is the focus of this chapter.

2 The main purpose of the 1996 Basic Establishment Census was to survey the urban economic structure, basic economic activities, labor force and distribution of production factors.

3 Two districts in the central city, Huangpu and Nanshi, have been consolidated into one (named as Huangpu).

4 My own migrant housing survey was conducted between December 1998 and March 1999 in twenty-two neighborhoods of Shanghai's seventeen districts/counties (out of a total of twenty) and eleven enterprises/institutions. It generated complete questionnaires for 1,789 temporary migrants, eighty permanent migrants, and 137 local residents. Multi-stage stratified cluster sampling procedures were used for selecting respondents. First, a number of districts were selected in three stratified geographical zones: central city, inner suburb and outer suburb. Within each selected district in the central city and outer suburb, one neighborhood was randomly picked. Because of the large size of and migrant concentration in the inner suburb, two neighborhoods were picked in each selected district there. Second, the total number of migrants in each of these neighborhoods was obtained from official sources. Population-proportionate-to-size procedures were then used to determine the exact number of migrants to be surveyed in each neighborhood. In addition, a small number of migrants were selected from representative work-units in different sectors.

5 Local rural residents are converted to urban status (*nong zhuan fei*) once their land has been acquired for urban development. They are often assigned jobs in state enterprises and their economic ties with the village severed after conversion. But many converted urban residents continue to live in old, privately owned rural housing as development authorities have tended to avoid acquiring rural neighborhoods due to high costs of relocation. See W. Wu (2002a).

## References

Abu-Lughod, J. (1961) "Migrant adjustment to city life: the Egyptian case," *The American Journal of Sociology*, 67(1): 22–32.

Alba, R. and Nee, V. (1997) "Rethinking assimilation theory for a new era of immigration," *International Migration Review*, 31(4): 826–974.

Banerjee, B. (1983) "Social networks in the migration process: empirical evidence on chain migration in India," *Journal of Developing Areas*, 17(2): 185–196.

Collier, D. (1976) *Squatter Settlements and the Incorporation of Migrants into Urban Life: The Case of Lima*, Cambridge, MA: Center for International Studies, MIT.

Conway, D. (1985) "Changing perspectives on squatter settlements, intraurban mobility, and constraints on housing choice of the Third World urban poor," *Urban Geography*, 6(2): 170–192.

Dunn, K. M. (1998) "Rethinking ethnic concentration: the case of Cabramatta, Sydney," *Urban Studies*, 35(3): 503–527.

Eckstein, S. (1990) "Urbanization revisited: inner-city slum of hope and squatter settlement of despair," *World Development*, 18(2): 165–181.

Gao, X. D. and Jiang, Q. Z. (2002) "*Dui Shanghai chengshi renkou fenbu biandong yu jiaoquhua de tantao* (Changes in Shanghai's urban population distribution and suburbanization)," *Chengshi renkou (Urban Population)*, 26(1): 66–69.

Gaubatz, P. (1999) "China's urban transformation: patterns and processes of morphological change in Beijing, Shanghai and Guangzhou," *Urban Studies*, 36(9): 1495–1521.

Gilbert, A. and Varley, A. (1990) "Renting a home in a Third World city: choice or constraint?," *International Journal of Urban and Regional Research*, 14(1): 89–108.

Gu, C. L. and Shen, J. F. (2003) "Transformation of urban social–spatial structure in socialist market economies: the case of Beijing," *Habitat International*, 27: 107–122.

Hodder, R. (1996) "Industrial location," in Yeung, Y. M. and Sung, Y. W. (eds), *Shanghai: Transformation and Modernization under China's Open Policy*, Hong Kong: Chinese University Press, 225–248.

Hu, X. H. and Kaplan, D. H. (2001) "The emergence of affluence in Beijing: residential social stratification in China's capital city," *Urban Geography*, 22(1): 54–77.

Klak, T. and Holtzclaw, M. (1993) "The housing, geography, and mobility of Latin American urban poor: the prevailing model and the case of Quito, Ecuador," *Growth and Change*, 24(2): 247–276.

Lu, H. C. (1995) "Creating urban outcasts: shantytowns in Shanghai, 1920–1950," *Journal of Urban History*, 21(5): 563–596.

Ma, L. J. C. (2003) "Some reflections on China's urbanization and urban spatial restructuring," paper presented at the Urban China Research Network Workshop on Urban Studies and Demography in China, April 30, Minneapolis, MN.

Ma, L. J. C. and Xiang, B. (1998) "Native place, migration and the emergence of peasant enclaves in Beijing," *China Quarterly*, 155: 546–581.

Ning, Y. M. and Yan, Z. M. (1995) "The changing industrial and spatial structure in Shanghai," *Urban Geography*, 16(7): 577–594.

Powers, M. G. and Seltzer, W. (1998) "Occupational status and mobility among undocumented immigrants by gender," *International Migration Review*, 32(1): 21–55.

Pudong Floating Population Office (1998) "Shanghai Pudong Xinqu liudong renkou tongji zilao (Floating Population in Shanghai Pudong New Area – Statistical Data)," unpublished.

Shanghai Basic Establishment Census Office (SBECO) (1997) *Shanghaishi jiben danwei pucha zilao huibian 1996 (Census Results on the Basic Establishments of Shanghai, 1996)*, Beijing, China: China Statistical Publishing Housing.

Shanghai Fifth Population Census Office (SFPCO) (2002) *Shanghaishi diwuci renkou pucha dituji (Atlas of Shanghai's Fifth Population Census)*, Shanghai, China: Shanghai People's Press.

Turner, J. F. C. (1968) "Housing patterns, settlement patterns, and urban development in modernizing countries," *Journal of the American Planning Association*, 34(6): 354–363.

Ulack, R. (1978) "The role of urban squatter settlements," *Annals of the Association of American Geographers*, 68(4): 535–550.

United Nations Center for Human Settlements (UNCHS) (1982) *Survey of Slum and Squatter Settlements*, Dublin: Tycooly International Publishing Limited.

Vaughan, D. R. and Feindt, W. (1973) "Initial settlement and intracity movement of migrants in Monterrey, Mexico," *AIP Journal*, 39(6): 388–401.

Wang, H. S. (1998) "'Zhejiang Village': a unique way for Chinese peasants to move into the cities," *Social Sciences in China*, 19(1): 22–37.

Wu, F. L. (2002a) "Real estate development and the transformation of urban space in Chinese transitional economy: With special reference to Shanghai," in Logan, J. R. (ed.), *The New Chinese City: Globalization and Market Reform*, New York: Blackwell, 151–166.

—— (2002b) "Sociospatial differentiation in urban China: evidence from Shanghai's real estate markets," *Environment and Planning A*, 34: 1591–1615.

Wu, W. P. (1999) "City profile: Shanghai," *Cities*, 16(3): 207–216.

—— (2002a) "Temporary migrants in Shanghai: housing and settlement patterns," in Logan, John R. (ed.), *The New Chinese City: Globalization and Market Reform*, New York: Blackwell, 212–226.

—— 2002b. "Migrant housing in urban China: choices and constraints," *Urban Affairs Review*, 38(1): 90–119.

Xiong, L. X. and Luo, Z. (2000) "Shanghaishi gongye yongdi de xiaolu yu buju (The efficiency and pattern of Shanghai's industrial land use)," *Chengshi Guihua Huikan* (*Urban Planning Forum*), 126: 22–29.

Yeh, A. G. O., Xu, X. Q. and Hu, H. Y. (1995) "The social space of Guangzhou city, China," *Urban Geography*, 16(7): 595–621.

Zhou, M. (1997) "Segmented assimilation: Issues, controversies, and recent research on the new second generation," *International Migration Review*, 31(4): 975–1008.

# 13 Migrant enclaves and impacts of redevelopment policy in Chinese cities

*Li Zhang*

## Introduction

Dynamic urbanization during the reform period has led to the emergence of a new spatial and social space in Chinese cities, so-called urban villages (*chengzhongcun* in Chinese).[1] During the process of urbanization, many villages adjacent to the city have been spatially encompassed or annexed by the expanding city, forming a distinctive rural landscape within the city's boundaries. Their rural status remains unchanged, however. This kind of geo-social landscape has commonly appeared in Chinese cities, especially in those that have experienced a rapid pace of urbanization.

For the large cities at least, such peri-urban villages are accommodating millions of rural migrants who have relocated from rural areas to cities in search of jobs since the economic reforms. As institutional and economic barriers have limited migrant options for housing in other parts of the city, urban villages are attractive to rural migrants because they provide accommodations that are socially accessible and economically affordable. Nonetheless, Chinese policy-makers and journalistic reports have viewed urban villages in a rather negative way, attributing to them much of the urban problems such as poor living conditions, social disorder and the deterioration of urban environment.

Based largely on a study of policy documents pertaining to the redevelopment of urban villages in various cities[2] and on interviews with those whose work is closely related to urban villages, this chapter seeks to understand the role of urban villages in China's urbanization and to assess government strategies for the renewal of such migrant enclaves. While not denying their social problems, this chapter argues that urban villages in fact play a positive role to promote urbanization in today's China by housing massive rural migrants and assimilating them into cities without using government resources. However, the supportive role of urban villages in China's urbanization has thus far received no acknowledgment. The current government strategy for urban village redevelopment, characterized by "demolition and redevelopment," is good for nobody and may indeed not be workable. Politically, it has generated a wide range of conflicts of interest and confrontations between the government and the stakeholders in the urban village. Financially, it may not be able to meet the high demand for investment required by urban village redevelopment. Socially, it makes city governments run into problems in their attempt to accomplish their aim of eliminating the problematic urban villages because little attention has been paid to the basic housing needs of the low-income migrants. With these concerns, the chapter concludes that politically acceptable and economically feasible policy options are needed for urban village redevelopment in the Chinese road to urbanization.

## Urban villages as migrant enclaves

Much has been said regarding the basic nature of migrant enclaves in Chinese cities. Most scholarly works have viewed urban villages as a social place, focusing on the processes of radical social change occurring in such settlements (such as Xiang 1993; Wang 1995; Piante and Zhu 1995; Beja and Bonnin 1995; Liu and Liang 1997; Ma and Xiang 1998; Chen and Zhao 1998; Xiang 1999; Smith 2000; Zhang 2001; Wu 2002). Ma and Xiang (1998) and Zhang (2001), respectively, provide ethnographic investigation on the creation of social networks based on migrants' places of origin and the formation of non-state spaces produced in the interaction between migrants and government. Fan and Taubmann (2002), as well as Gu and Liu (2002), offer case studies of migrant enclaves in Shanghai and Beijing, giving greater emphasis to emerging social inequality and spatial segregation in the Chinese city. By examining the housing and settlement patterns of migrant households, Wu (2002) points out that migrants are excluded from the formal housing distribution system and, as a result, an incipient migrant housing market on the fringe of the city is in the making.

Despite their different focuses of specific research questions, there is one common point among the studies of urban villages. All studies explicitly recognize that urban villages are a hot spot of living for migrants who seek a new life in cities. Comprehensive surveys on urban villages at the national level are unavailable. Even the latest census data (year 2000) do not offer a clear picture because of the lack of detailed itemization. But many surveys have revealed that urban villages where the low-income migrants are concentrated have provided practical housing space for them because they are denied access to urban public housing. Based on the data from different sources, Table 13.1 demonstrates that urban villages are a kind of notable migrant enclaves. Though there is limitation in data consistency and comparability across studies as each study has its own definition of migrants, Table 13.1 points to the fact that the number of rural migrants staying in urban villages has been quite significant. In several cases, migrants outnumber native residents by a ratio of 2/3:1. In the case of Ruibao Village, the ratio rises to 30:1. About one mile from Guangzhou's downtown, this village of 2,000 indigenous peasants has leased houses to 60,000 migrants from Sichuan, Hunan, Guangxi and other provinces.

*Table 13.1* Ratio of migrants to indigenous residents in selected urban villages

| *Urban village* | *Surveyed year* | *No. of migrants* | *No. of indigenous residents* | *Ratio (indigenous residents as 100)* |
|---|---|---|---|---|
| *Beijing city* | | | | |
| Wabian cun | 1997 | 2,162 | 2,086 | 104:100 |
| 917 district | 1997 | 287 | 318 | 90.3:100 |
| *Guangzhou city* | | | | |
| Shipai cun | 2000 | 12,000 | 9,234 | 130:100 |
| Sanyuan li | 2000 | 11,000 | 4,200 | 262:100 |
| Tangxi cun | 2000 | 9,534 | 4,656 | 205:100 |
| Ruibao cun | 2000 | 60,000 | 2,000 | 3,000:100 |
| *Dongguan city* | | | | |
| Shangyuan cun | 1998 | 8,000 | 2,842 | 281:100 |
| Yantian cun | 1998 | 70,000 | 2,877 | 2,433:100 |

Sources: Data on Beijing are from Chen (1998). Data on Guangzhou and Dongguan are from the author's interviews with local cadres of the surveyed sites.

Categorization of urban villages based on migrants' native places and occupations can be complicated. Though many migrant enclaves in Chinese cities are often formed by people from the same native places and are well known for their residents' occupational specialization, as pointed out by many studies such as Ma and Xiang (1998), not all urban villages are necessarily native-place based with specialized employment. Migrants as a group in a specific urban village are not always homogeneous in terms of social origins and occupations or income levels (Lan 2001; Chan *et al.* 2003; Wu 2003).

Urban villages attract migrants to stay because of several advantageous conditions. The proximity to workplaces and shopping areas is one of them (Wu 2003). Being a part of the city, urban villages are well connected with different places in a city with good urban transportation. There is a housing rental market in urban villages that can provide accommodations with comparatively low rental rates which, very importantly, meet the financial capacity of the migrants. Most houses in urban villages are made of durable building materials. A substantial number of dwellings have essential facilities like tap water, electricity, private kitchen, flush toilet and sewage drainage that offer better living conditions than many distant and isolated rural areas. One survey indicated a considerably high level of satisfaction by those living in urban villages (Zhang *et al.* 2003).

Of course, urban villages are by no means "garden settlements." To many migrants the locational aspect and affordability may be more important than housing conditions, as they must face many uncertainties in cities with respect to job security and the right of abode. As part of the strategy to tackle the uncertainties, they maximize their savings by minimizing their living costs, of which the cost for housing is a significant part. They seek cheap but usually substandard housing as a second and temporary home to save money. Housing sharing is also a common practice among migrants, very often leading to bed-sized living space per dweller. Many urban villages are plagued by aged facilities with poor maintenance, very narrow pathways between rows of terraced buildings, intensive use of space without appropriate planning, and high residential density that are beyond the capacity of infrastructure services. There are also such social problems as violence, pornographic activities, burglary and robbery in addition to building dilapidation. To some extent, urban villages share some of the worst features of shantytowns in many cities of the developing world. Such conditions were hardly tolerated by socialist ideology.

## Migrants' barriers to urban public housing

As mentioned above, urban villages have become migrant enclaves because most rural migrants are not eligible for urban housing. The state sector monopolized the urban housing market in the pre-reform period after the means of production were nationalized. Urban housing was dominated by the public sector. Once the Chinese society had been organized on the basis of public ownership, urban housing was no longer a simple commodity but a welfare benefit to be provided to urban dwellers. Whatever the property rights in practice, all urban housing was unexceptionally constructed with government investment and then bureaucratically allocated to employment units. Allocation of public housing was both locale based and enterprise controlled. Among the crucial criteria for housing entitlement were official urban household registration (*hukou*) and permanent employment in a state enterprise. Persons without the permanent rights of abode and employment in cities were excluded from public housing.

An urban non-public housing market has emerged since the mid-1980s as a result of housing reform. Housing reform is orientated to commodification and marketization of urban housing, which is intended to change the overwhelming role of the state in urban housing provision and to transform urban housing from welfare goods to a commodity in several respects. First, the state has decided not to provide free housing to urbanites any more (Lee 2000; Wang 2000), and individual dwellers are required to amortize at least one-third of the dwelling's replacement cost (Li 1999; Li 2000). Second, the state has gradually privatized its housing stock by selling it off to the current tenants at heavily discounted prices (Ministry of Construction 1999). Third, the development of commercialized housing, which is built essentially for profit, is encouraged. Private or joint venture developers are allowed to build and sell or lease residential properties for profit on the open market (Wang and Murie 1999a). Along with such developments, a new form of low-cost public housing, known as the *anju* (literally, "peaceful living") project, has been sponsored by the state for urban low-income groups who cannot purchase housing at full market prices.

Commercialization of urban housing has basically provided no homes for the majority of rural migrants who are employed in urban low-paid positions. One study found that rural migrant laborers, with an annual income of about RMB 6,000–7,000, generally earned less than that of urban workers (Knight *et al.* 1999). Commercialized housing is also very expensive by Chinese income standards. The income situation of urban residents and the prices of commercialized housing shown in Table 13.2 suggest that the cost of commercialized housing is simply too high for rural migrants, as most of them properly belong to the urban poverty group (Smith 2000).

Apart from the development of commercialized housing, the other restructuring of the urban housing market is also irrelevant to most rural migrants who are continuously denied legitimate urban status, a major prerequisite for obtaining state-subsidized urban housing. Despite a decrease in restrictions on migrant employment, the complexity and rigidity of the existing *hukou* registration and the various types of permits required for living and working in the city still make it difficult for many rural laborers to obtain permanent working and living status in the city (Solinger 1999). As they lack the permanent right of abode in the city, rural migrants are not eligible to buy public housing privatized by the state. For the same reason, they are also not entitled to urban low-cost public housing. Urban housing is largely beyond the reach of rural migrants because of their limited purchasing power and the lack of urban citizenship.

## The rural status of urban villages and the formation of migrant enclaves

The emergence of urban villages as popular residential areas for rural migrants cannot be properly understood without reference to the rural status of these peri-urban villages. Administratively urban villages are rural places,

*Table 13.2* Personal incomes and commercialized housing prices in selected cities, 2002

| *City* | *Annual income per capita (yuan)* | *Commercialized housing price* | | *Ratio between housing price and annual personal income* |
|---|---|---|---|---|
| | | *Yuan/m²* | *Total amount of yuan per house unit of 70 m²* | |
| Beijing | 10,349 | 4,557 | 318,990 | 31:1 |
| Shanghai | 11,718 | 3,297 | 230,790 | 20:1 |
| Guangzhou | 13,967 | 4,269 | 298,830 | 21:1 |
| Shenzhen | 21,626 | 5,498 | 384,860 | 18:1 |
| Chengdu | 7,649 | 1,281 | 89,670 | 12:1 |
| National average | 6,280 | 1,952 | 136,640 | 22:1 |

Source: *Economy Daily*, January 2, 2002.

a fact that makes it possible for them to provide affordable housing for rural migrants. First, indigenous villagers, on account of their rural *hukou* status, have the legal right to cost-free land for their own housing, which, unavailable to urban residents and migrants, is the most valuable resource to finance housing construction. Second, because the governance of urban villages is outside the urban administration system, there is fluidity in enforcing urban planning and development control regulations, making housing additions and modifications for renting purpose largely unimpeded. This helps suppress the rent levels. Third, as housing in urban villages is overwhelmingly privately owned, market forces have undermined government restrictions on the housing rental, thus making the housing in urban villages fully open to rural migrants.

Housing construction in urban villages should not be seen as independent of the property rights of rural land. The institution of using collectively owned land for housing is conducive to the construction of low-cost housing in urban villages. Indigenous residents in urban villages are organized into village committees and classified as rural population by the system of *hukou* registration, regardless of their actual occupations. This population, who live within the city boundary but are officially regarded as rural, receive no financial support from the state for their housing needs. However, under the collective land ownership, the villagers, rich or poor, do have the same legal right to a certain size of land on which they can build their housing at their own expense. Land for housing is allocated free (with a nominal fee very recently) by the village committee which represents the government to control collective land ownership. An individual's rights to use the land for housing on a continuing basis are safe as long as his/her membership (qualified by having local *hukou*) in the village is kept.

Land is not only a prerequisite to build houses but also a resource to finance housing construction. Having access to a piece of free land for housing constitutes an important element to build houses at lower costs in urban villages than in other parts of the city. With the rapid expansion of the city, much farming land of urban villages is requisitioned by the state for urban development. With income from the distribution of the compensation for land requisition, indigenous villagers often have the necessary capital for their housing construction.

Housing has become a financial asset that could be a subsidiary source of income for many indigenous residents of an urban village. Indigenous villagers are not allowed to alienate their allocated land to others under the current regime of collective land ownership. However, they can take advantage of the opportunity of increasing demand for hosing resulting from accelerated rural–urban migration and can benefit from renting out parts of their dwellings to city newcomers for income. The large size of cityward migrants has produced a high demand for urban housing. Because renting is lucrative, indigenous villagers are keen to build as many living units for rent as they can on their housing land. Renting out housing to migrants is a common business among indigenous villagers. As housing land can be obtained free, the cost of housing construction is therefore low. Indigenous villagers are able to provide housing at rents lower than could be offered by other housing providers in the city.

Although urban villages are located within city boundaries, their official rural status largely protects them from the intervention of urban planning control which is mainly concerned with the urban built-up areas. Under the city administration system, it would take months or even years to get approval for construction plans from the dozens of city authorities involved. Traditionally, Chinese rural settlements have developed without formal planning and regulations. Land and housing development in urban villages are in practice beyond the enforcement power of city planning authorities who are primarily responsible for providing planning services to and enforcing development control of the city proper, not the rural areas within the city's administrative boundaries. In many cases there is no formal government organization in charge of rural housing development (Wang and Murie 1999b). Housing construction in urban villages can easily evade planning regulations and indigenous villagers often built housing at a very high density which is not allowed by the building codes.

Under the regime of collective land ownership, rural communities have direct power over land use rights and development control. The urban village authority often takes action in the interests of the indigenous residents for land use and housing construction that may not conform with the statutory city plan and building codes. It can be easily observed in urban villages that many houses are built outside the planning regulations. Encouraged by potential rental income, more housing is built on speculation through increasing housing density and the number of floors.[3] Unit cost of housing construction can be easily lowered by circumventing planning restrictions. In fact, the investment in housing construction is in large part recovered by a substantial reduction in the official minimal requirements concerning plot ratio, density limits, permissible building materials, road widths and compulsory infrastructure. Though their physical appearance looks disarrayed, urban villages can provide low-cost housing for those depending on inexpensive accommodations.

For the interests of public security and social order, urban housing rental is codified in a series of municipal regulations. However, government intervention in housing rental in urban villages is not always effective, given the dominance of private housing there and the economic incentives of the owners and tenants. Under existing regulations, both landlords and tenants are required to register with the local police and to complete necessary paperwork for housing rental. At the level of implementation, these requirements are perceived as costly and troublesome, regardless of their specific aims. Government supervision of rental housing in urban villages is ineffective because of loopholes in the regulations. Landlords can easily get around the regulations. Most houses rented out do not follow municipal regulations but are based on a bilateral agreement (written or verbal) between landlord and tenant. The number of tenants without registration with the local police is substantial in urban villages.

Although urban villages are not the only location receiving rural migrants, migrant options for rental housing in other parts of the city are limited by the acute housing shortage in many Chinese cities, in addition to the speculative prices in the commercialized housing already mentioned. As of the year 2000, the average per capita living space in urban China was 10.3 square meters (*Economy Daily*, January 2, 2002). Many urban families simply have no space to rent out even though they may be willing to do so. The provision of rental housing is further constrained by ambiguous property rights. Under the old welfare housing system, urban housing was primarily allocated by work-units and organized around work-unit compounds (Zhou and Logan 1996). Although currently many people have bought their own housing (Wu 2002), there are many *de facto* regulations restricting the buyers to put their homes on the urban housing market (at least for a certain period after buying) for sale or for rent. As a result, migrant enclaves are hard to form in the city except for urban villages.

## Government strategy towards urban villages

Urban villages as migrant enclaves have increasingly become a matter of policy concern. Notwithstanding their positive role of making affordable housing available for migrants, urban villages are officially regarded as slums or shantytowns with chaotic land use, dilapidated housing, severe infrastructure deficiency, intensified social disorder, and unsightly urban eyesore. Journalistic reports and Chinese scholarly works have also perceived urban villages as undesirable, attributing to them the root of social and environmental problems that are seen as jeopardizing urban development. City governments believe that such slums will not go away without some sort of government intervention and massive renewal projects.

The earlier government action commonly taken in many cities was simply to cleanse and deport migrants from urban villages under the slogan of "maintaining social order and cracking down on crimes" while keeping housing there basically unchanged. The "cleansing-deportation" campaigns, though harsh and even brutal, were carried out whenever city governments perceived them as necessary. It was reported that, in November 1995, there was an armed confrontation between rural migrants and officials of public security in Dahongmen District of Beijing, a large migrant enclave consisting of twenty-six urban villages with about 100,000 migrant residents (well known as Zhejiang Village) (*Beijing Daily*, November 21, December 12, 1995; Beja and Bonnin 1995; Liu and Liang 1997; Gu and Shen 2003). The bloodshed had led to a citywide forced deportation of the migrants who had no proper living documents. Like other places, however, such cleansing and deportation could not stop the flow of migrants into cities. Soon after the deportation, migrants returned with even a larger number (Cai *et al.* 2001: 316; Gu and Shen 2003: 119). Some even brought their family members, friends, or fellow villagers. In most cases, the larger the scale of deportation, the greater the number of migrants who came back. Migrant enclaves became bigger and bigger.

In 2002 and 2003, there was a strong signal from city governments to rebuild urban villages under urban redevelopment plans. Albeit varied in operational details, government strategies for urban village redevelopment among cities were basically identical, which can be summarized as a "demolition–redevelopment" model (*China Construction Daily*, March 31, 2003).[4] The basic ideas of the "demolition–redevelopment" model involve urbanizing urban villages, dismantling the existing structures and rebuilding. Urban villages located near the inner city will be designated as urban renewal districts to be gradually rehabilitated through a series of initiatives, including administrative reorganization, land use restructuring and housing reconstruction (*Nanfang Daily*, June 9, 2002; *People's Daily*, June 16, 2002; Cheng 2003). This will be done in two phases. The first phase is characterized by institutional transformation of the rural status of urban villages. All collectively owned land in urban villages will be compulsorily turned over to state ownership. The rural *hukou* of indigenous villagers will accordingly become urban, and village committees will be reorganized into urban residents' committees. In the second phase, in compliance with city regulations and city planning, urban villages will be reconstructed through real-estate development. Illegal dwellings will be eradicated without any compensation. Management of land development will be unified in terms of development application and land transactions. Housing construction requires an official permission and will be strictly controlled with respect to building size and density. Existing precarious housing will be replaced by decent apartments. Some of these apartments will be allocated to the indigenous villagers free as a form of compensation and the others will be sold at market prices. Private and joint venture developers are encouraged to invest in the renewal projects. Redevelopment programs will be financed in accordance with the principle of "beneficiaries pay." City governments will, if necessary, subsidize the provision of urban infrastructure and amenities.

Under such a policy framework, the redevelopment of urban villages will be done by both administrative and market forces, without offering redevelopment choices to indigenous villagers. The key to the "demolition–redevelopment" model is to change the rural status of urban villages. Such forcible urbanization of indigenous villagers by transforming their rural *hukou* identity into the urban one will deprive them of their rights and capacities to develop housing for rent in the future. The land use rights of urban villages will be first taken back by the city government and then be granted to government-designated developers if needed. While reconstruction can be facilitated through projects of real-estate development, the redevelopment of urban villages should be under the full control of city planning. By not giving alternatives to indigenous villagers, city governments hope that urban villages will be completely redeveloped and the future of Chinese cities will be free of shantytowns (Zhang *et al.* 2003).

## Impacts of government strategy for urban village redevelopment

For the time being, the full effect of urban village redevelopment is hard to evaluate comprehensively because the redevelopment is still in an initial stage in many cities. However, some impacts can be anticipated based on the interviews with indigenous villagers, developers and migrants.

As migrant enclaves, urban villages have created conflicts of interest between city governments and stakeholders of urban villages. In the reform period, "waves of rural labor" are escaping state controls and flooding into the cities. They need to find urban housing that they can access and afford. From the perspective of local village administration, the presence of migrants, who can be easily imposed various kinds of management fees and therefore financially exploited, has been regarded as a bonanza to expand local revenue. In order to accrue financial benefits by attracting more migrants living in urban villages, local administration may tacitly allow some illegal activities by not strictly enforcing laws and municipal regulations. Nonetheless, economically, all stakeholders of urban villages have benefitted. So at the very local level, village administration, indigenous villagers and migrants all wish to make urban villages work as migrant enclaves. However, due to its informal and illegitimate characteristics, urban villages have been seen in the public media in the negative light projecting the image of social "disorders" which makes the formal urban citizens (those with formal urban *hukou*) and urban authorities uncomfortable. The dissatisfaction of the formal urban citizens has generated strong political pressure on the municipal government who is very sensitive to the social consequence of urban disorder and the "bad image" of its city. Urban policies are therefore often adopted for the sake of political expediency rather than economic rationality. Rural migrants, who are politically and economically powerless, are most likely to be assaulted whenever confrontation between municipal government and the various parties in urban villages becomes unavoidable.

As providers of inexpensive housing, indigenous villagers will be negatively affected by urban village redevelopment. Albeit the indigenous villagers will be compensated for their relinquished property and will be resettled to suitable new sites, they will no longer be able to benefit from housing rental as before. With reorganization referred to above, indigenous villagers will lose their land use right, one of the few entitlements granted to rural citizens. Their income from housing rental will be lost or reduced as their housing is redeveloped and their major tenants – migrants – are forced out.

Redevelopment is likely to have significant impact on the supply of affordable housing for migrants. Such redevelopment inevitably means an amelioration of housing qualities, a decrease in inexpensive housing stock, and an increase in rent at the renewal sites. As old houses are required to be rehabilitated and new housing construction must meet higher technocratic standards, renewal projects are no doubt a costly exercise. Migrants will not be able to afford new houses produced without generous government subsidies. Shenzhen's experience of urban village redevelopment shows that private property developers need to compensate homeowners at least RMB 1,500/$m^2$ (Table 13.3). In addition to resettlement compensations and construction costs, developers have to pay for various charges and administrative fees, such as land transfer fee, environmental sanitation fee and construction taxes. Therefore, one urban village renewal project may require millions or even billions of investment, depending on the actual dwelling density of the urban village in question and the renewal scale. From the perspective of developers, a consequence of huge investment is that the threshold price of per square meter housing that can make the urban village project profitable cannot be lower than RMB 6,800. Large investment, whoever will provide it, will finally be passed onto housing prices which will impose heavier financial burden on the buyers or renters. Predictably, a low-cost housing stock in the cities would further lag behind the housing needs of the urban poor. Rural migrants, who are denied any state housing allowance, will be forced to increase their housing expenses. So far, Chinese cities have already had trouble meeting the migrants' demand for affordable accommodations. Urban village renewal might make the problem even worse.

Affordability problems could be further exacerbated by employment policies that discriminate against migrant laborers. In the pre-reform period, state control over rural-to-urban residency conversion was strict and sought to minimize personal movement into cities. Economic reforms have dramatically altered the very basis of state control over migration, leading to a large number of rural migrants in cities. At the same time, China's cities, where there was already a large pool of laid-off workers waiting for employment, have little hope for fully accommodating the flood of rural migrants. Such a situation compelled city governments to maintain their effort to control rural-to-urban migration. Restrictions on the employment rights of migrant

*Table 13.3* The compensation scheme for urban village redevelopment in Shenzhen city (unit: yuan/$m^2$)

| | *Resettlement compensation* | *Housing decoration compensation* | *Moving allowance* | *Total* |
|---|---|---|---|---|
| Private apartment | 1,080 | 420 | 12 | 1,512 |
| Public apartment | 288 | – | – | 288 |

Source: Interviews with scholars from the Association of Property Valuation of Shenzhen City.

laborers are a sign reflecting city governments' continued attempts to exert migration control. Rural migrants are not allowed to work in the cities on equal terms with the locals. Many cities have issued regulations restricting the types of jobs that migrant workers can undertake and setting a maximum percentage of migrant workers that a given employment sector can hire. The jobs that rural migrants are allowed to take are usually manual ones and poorly paid that require a great deal of physical strength or have long-term adverse health effects. As their opportunities are restricted by government-imposed constraints on employment, rural migrants constitute an economically exploited underclass in cities.

Thus, rural migrants as a socially underclass group are big losers in urban village redevelopment. Although the employment right of migrant laborers is partially legalized under the rubric of "temporary" population, their residency right in the cities is uncertain at best. Many "temporary" residents have lived in cities for years, but they have not been granted the formal urban *hukou* which is essential for such urban entitlements as employment without restrictions, housing benefits and children's education. For now, there is no sign that the government intends to remove the differentiation in housing provision between rural and urban populations. Without legitimate residence in cities, rural migrants must take housing matters into their own hands. The current state of urban villages as migrants' temporary home base is the outcome of social inequality between urban and rural population. In the absence of any policy consideration for migrants' housing needs, the renewal of urban villages into more orderly urban spaces will no doubt be at the expense of rural migrants.

Without significant reference to the intensification of housing shortage for rural migrants, the renewal of urban villages may not necessarily result in the eradication of unsightly migrant enclaves that city governments seek to do. Despite the large numbers of peasant workers in cities, their housing rights and needs have never been properly addressed. Beyond the political imperative that the physical and social environment of urban villages must be improved, the policy framework of "demolition and redevelopment" is much less explicit as to how the millions of existing migrants should be relocated. The growing housing needs of migrant workers are left unattended. It can be expected that only a fairly small number of the rich migrants can afford to live in the redeveloped residential districts while most of the poor will have to survive on marginal sites. Unless the housing issue of rural migrants is properly addressed in the policy of urban village renewal, migrants will face serious problems of finding suitable housing. They may be forced to form squatter settlements elsewhere in or around the cities.

From the analysis above, it seems that the proposed policy framework of urban village renewal aiming at the "demolition of cheap housing and the development of new estate" would benefit neither the indigenous villagers nor the migrants. Although it will offer indigenous villagers formal urban *hukou* status and provide compensations (in whatever forms) for their relinquished properties, the renewal policy would eliminate their opportunity to receive rental income. Their contribution to housing migrants remains under-appreciated. It generates only costs and suffering but no benefits to most migrant inhabitants in urban villages as they will lose their affordable accommodations and be forced to locate elsewhere, though this could not be called forced deportation. City government itself would also run into problems as the policy does not address the basic issues of migrants' full citizenship rights in general and their housing needs in particular.

## Conclusions

This chapter is concerned with issues of habitation of rural migrants in Chinese cities with special reference to urban villages. An attempt is made to understand the underlying rationale of urban villages as migrant enclaves and the problems city authorities face in dealing with such migrant settlements. The findings demonstrate that urban villages provide necessary housing for the low-income rural migrants under the institutional constraints of China's urban housing market. While spatially urban villages are a result of city expansion, the existence of urban villages as migrant enclaves is an economic response to the demand for inexpensive housing as a result of growing rural-to-urban migration and a consequence of the existing urban housing system that excludes those without legal urban status. While such places are officially viewed as dirty and problematic corners in Chinese cities, they have produced multiple benefits for China's urbanization. Far from jeopardizing urban development, they have offered low-cost albeit marginal livable space with a degree of security for those who have to be self-sufficient in the cities in the face of many systemic constraints and institutional discriminations. The reality of urban villages thus should be viewed in a more positive light, although they may not be the best solution for the problem of migrant housing demand. Urban villages provide inexpensive shelters for low-income migrants, allowing them to get around the institutionally biased housing impediments and to settle in the cities. Also, very importantly, urban villages have reduced the pressure of city governments in instituting costly programs of housing migrant laborers during rural–urban transformation. In effect, urban villages have allowed city governments to take advantage of cheap and flexible rural labor without bearing the extra costs associated with labor relocation and without risking social stability. As a result, such enclaves represent an inexpensive and practical solution to the problem of migrant housing demand. They should be viewed as a positive element in China's urbanization during economic transition.

Morphologically as well as socially, the development of urban villages as migrant enclaves adds a new dimension of spatial segregation, one of the most visible aspects of the changing Chinese urban space in the reform era. During the socialist period the spatial organization of Chinese cities were characterized by uniform and self-sufficient urban cells. Urban land use was mainly work-unit based and urban districts were generally undifferentiated by functions. Urban space was structured around large, independent and walled work-unit compounds (Gaubatz 1999). The living space of urban residents was seldom beyond the vicinity of their work-units. This work-unit based urban space has undergone restructuring in the reform era. Changes in the spatial structure of Chinese cities have been characterized by specialization of land use and spatial segregation. Urban space has been reorganized around multiple business and service centers. Many manufacturing industries have been relocated to development zones on the urban fringe. A set of rebuilt or new business and shopping centers, scattered throughout the city's old and newly developing districts, have become the foci of the city. At the same time, housing becomes increasingly separated from workplaces and newly developing residential districts have sprawled outwards toward the urban periphery. While a fairly small number of old urban neighborhoods have been reconstructed into luxury residential districts, many dilapidated and overcrowded neighborhoods within the city proper (including the inner suburbs) have not yet been scheduled for demolition and rebuilding. One general result of these restructuring processes is the creation of segregated residential districts for different classes, spatially separating luxury housing for the rich from the basic shelters for the ordinary urban residents. The emergence of urban villages as migrant enclaves represents a new type of urban low-class clusters within the city's built-up areas or at the edge of the city. Urban villages are not only out of sync with those luxurious districts in terms of living environment, but also worse than those decayed old neighborhoods with respect to urban governance. The government strategy of "demolition–redevelopment" towards urban villages discussed above is the latest policy response to the "problems" (defined and interpreted by government) of migrant settlements in cities. Nonetheless, as demonstrated in this chapter, the strategy remains problematic and may not be feasible because it does not take migrants' need for housing into account.

Given the positive role of urban villages on China's road to urbanization, city governments need to be careful about the social consequences of the current strategy for urban village redevelopment. As discussed earlier, such redevelopment would be disastrous for the migrant inhabitants in the urban villages. For rural migrants, urban village redevelopment may involve destroying large amounts of inexpensive housing stock near the places of their employment. They have to resettle elsewhere in the course of its implementation. In the absence of appropriate resettlement plans for the migrants living in urban villages, one may expect confrontation rather than collaboration during urban village renewal. With increasing numbers of rural-to-urban migration, the demand for urban housing is likely to be very high for years to come. Without a policy that would lead to the provision of affordable housing for migrants, urban village redevelopment is likely to encounter considerable resistance from migrants. Another possible adverse consequence is the reproduction of shanty settlements in other parts of the city, when low-income migrants are forced to move out from urban villages to search for alternative places to live.

Instead of simple demolition, what should be done for urban villages is to implement the kinds of projects or programs that respond directly to the needs of migrants. International experience has shown that enabling rather than simply eliminating slum settlements would be a better policy option for developing economies (United Nations Center for Human Settlements 1991 and 1996). The notion of "enablement" in the context of housing the poor entails the provision of better administrative support and essential services (Pugh 1995). It takes advantage of self-help by the local community that gives rise to cost reductions, rapid response to changing demands and a diverse range of housing available for sale or rent (United Nations Center for Human Settlements 1996: 337). While it requires well-developed partnerships and interdependence among government agencies, markets, non-government agents and individuals to build up an affordable and reasonable human settlement for the urban poor, the "enabling" approach also strongly emphasizes the role of government in poverty alleviation and service provision that can enhance both human dignity and social justice. Much could be achieved to improve the living environment in urban villages if government would more actively support people's own efforts to alleviate the problem of housing shortage for themselves. A shift in government emphasis from controlling the development of urban villages to facilitating their future growth would be a right direction toward finding a more satisfactory solution to the housing problems created by the increasingly large number of rural-to-urban migrants in Chinese cities.

## Acknowledgment

This research is supported by the Direct Grant of Chinese University of Hong Kong (project code 2020714).

## Notes

1 The term *chengzhongcun* literally means "village within city." For brevity and clarity, I use the term "urban villages" in this chapter to refer to such settlements that are administratively classified as villages but are located in the peri-urban or even inner-city areas. These "villages," formerly in suburban areas, have been swallowed up by urban sprawl. With the extension of urban areas, many of these villages have been spatially absorbed into dense built-up areas. In the Chinese case, urban villages in the expanded city represent an anomaly of living space that differs in many respects from their surrounding areas.

2 These cities include Guangzhou, Dongguan, Zhuhai and Shenzhen in Guangdong Province, Shijiazhuang in Hebei Province, Qingdao in Shandong Province and Xi'an in Shaanxi Province.

3 Most structures in urban villages are four stories or more in height, built with concrete blocks in simple rectangular shapes. They cannot meet the government requirements of building safety and fire control standards.

4 There are many reports on the policy of urban village redevelopment in different cities. For the case in Shenzhen, see Cheng 2003; for Zhuhai's case, see *People's Daily*, March 3, 2002; for the case in Xi'an, see *China Construction Daily*, April 9, 2003; for Guangzhou's case, see *Southern Weekend*, October 31, 2002; for Beijing's case, see *Nanfang Daily*, September 6, 2002; for Qingdao's case, see *Qingdao Daily*, June 19, 2002.

## References

Beja, J. P. and Bonnin, M. (1995) "The destruction of the Zhejiang village," *China Perspectives*, 2: 21–25.

Cai, F., Guo, H. Y. and Gao, J. L. (eds) (2001) *Zhongguo renkou liudong fangshi yu tujing (1990–1999) (The Ways and Channels of Chinese Migration, 1990–1999)*, Beijing: Social Science Literature Press.

Chan, R. C. K., Yao, Y. M. and Zhao, X. B. (2003) "Self-help housing strategy for temporary population in Guangzhou, China," *Habitat International*, 27: 19–35.

Chen, T. (1998) "Wailai renkou jujiqu guifanhua guanli de changqi duice (Long-term policy for a well-ordered administration of non-local settlements)," unpublished manuscript, Institute of Geography, Chinese Academy of Sciences.

Chen, T. and Zhao, X. B. (1998) "Liudong renkou jujiqu xingcheng yu yanhua jizhi de weiguan toushi (Micro-analysis of the mechanisms for the formation and evolution of the floating population's settlements)," unpublished manuscript, Institute of Geography, Chinese Academy of Sciences.

Cheng, J. L. (2003) "Shenzhen tequ Chengzhongcun gaizao kaifa moshi yanjiu (Research of the reconstruction models of villages inside the city in the special economic zone of Shenzhen)," *Chengshi guihua huikan (Urban Planning Forum)*, 3: 57–60.

Fan, J. and Taubmann, W. (2002) "Migrant enclaves in large Chinese cities," in Logan, J. R. (ed.), *The New Chinese City: Globalization and Market Reform*, Oxford: Blackwell, 183–197.

Gaubatz, P. (1999) "China's urban transformation: patterns and processes of morphological change in Beijing, Shanghai and Guangzhou," *Urban Studies*, 36(9): 1495–1521.

Gu, C. L. and Liu, H. Y. (2002) "Social polarization and segregation in Beijing," in Logan, J. R. (ed.), *The New Chinese City: Globalization and Market Reform*, Oxford: Blackwell, 198–211.

—— and Shen, J. F. (2003) "Transformation of urban socio-spatial structure in socialist market economies: the case of Beijing," *Habitat International*, 27: 107–122.

Knight, J., Song, L. and Jia, H. B. (1999) "Chinese rural migrants in urban enterprises: three perspectives," *Journal of Development Economics*, 35(3): 73–104.

Lan, Y. W. (2001) "Chengzhongcun: cunluo zhongjie de zuihou yihuan (Chengzhongcun: the last lap of the village ending circle)," *Zhongguo shehuikexueyuan yanjiushenyuan xuebao (Journal of the Graduate School, the Chinese Academy of Social Science)*, 6: 100–105.

Lee, J. (2000) "From welfare housing to home ownership: the dilemma of China's housing reform," *Housing Studies*, 15(1): 61–76.

Li, L. H. (1999) *Urban Land Reform in China*, UK: Macmillan Press Ltd.

Li, S. M. (2000) "The housing market and tenure decision in Chinese cities: a multivariate analysis of the case of Guangzhou," *Housing Studies*, 15(2): 213–236.

Liu, X. L. and Liang, W. (1997) "Zhejiangcun: social and spatial implications of informal urbanization on the periphery of Beijing," *Cities*, 14(2): 95–108.

Ma, L. J. C. and Xiang, B. (1998) "Native place, migration and the emergence of peasant enclaves in Beijing," *China Quarterly*, 155: 546–581.

Ministry of Construction (1999) "*Guanyu shenhua gongfang gaige de jueding* (*A directive concerning further reforms of public-owned housing*)," issued on August 13, 1999.

Piante, C. and Zhu, H. B. (1995) "A law unto itself – Beijing's Zhejiang village," *China Perspectives*, 2: 12–15.

Pugh, C. (1995) "The role of the World Bank in housing," in Aldrich, B. C. and Sandhu, R. S. (eds), *Housing the Urban Poor: Policy and Practice in Developing Countries*, New Delhi: Vistaar Publications, 34–92.

Smith, C. J. (2000) "The floating population in China's cities a new ethnic underclass?," in Cannon, T. (ed.), *China's Economic Growth: The Impact on Regions, Migration and the Environment*, New York: St Martin's Press, 91–114.

Solinger, D. (1999) *Contesting Citizenship in Urban China*, Berkeley, CA: University of California Press.

United Nations Center for Human Settlements (Habitat) (1991) *Evaluation of Experience with Initiating Enabling Housing Strategies*, Nairobi.

—— (1996) *An Urbanizing World: Global Report on Human Settlements, 1996*, New York: Oxford University Press.

Wang, C. G. (1995) *Shehui liudong yu shehui zhonggou – jingcheng zhejiangcun yanjiu (Social Mobility and Restructuring – A Case Study of Beijing's Zhejiang Village)*, Hangzhou: Zhejiang People's Press.

Wang, Y. P. (2000) "Housing reform and its impacts on the urban poor in China," *Housing Studies*, 15(6): 845–864.

—— and Murie, A. (1999a) "Commercial housing development in urban China," *Urban Studies*, 36(9): 1475–1494.

—— (1999b) *Housing Policy and Practice in China*. New York: St Martin's Press.

Wu, W. P. (2002) "Temporary migrants in Shanghai: housing and settlement patterns," in Logan, J. R. (ed.), *The New Chinese City: Globalization and Market Reform*, Oxford: Blackwell, 212–226.

Wu, X. (2003) "Bianyuan shequ tansuo (An investigation into marginal communities)," *Chengshi guihua (City Planning Review)*, 27(7): 40–45.

Xiang, B. (1993) "Beijing youge zhejingcun: shehui zhuanxingzhong yige zifa chengshihua quanti de chubu yanjiu (Beijing has a Zhejing village – a preliminary study on a spontaneous urbanization group during the social transition)," *Shehuixue yu shehui diaocha (Sociology and Social Investigation)*, 3: 68–74; 4: 48–54 and 5: 51–54.

—— (1999) "Zhejiang village in Beijing: creating a visible non-state space through migration and marketized traditional networks," in Pieke, F. N. and Mallee, H. (eds), *Internal and International Migration: Chinese Perspectives*, Richmond, Surrey: Curzon Press, 215–250.

Zhang, L. (2001) *Strangers in the City: Reconfigurations of Space, Power, and Social Networks within China's Floating Population*, Stanford: Stanford University Press

Zhang, Li, Zhao, X. B. and Tian, J. P. (2003) "Self-help in housing and *chengzhongcun* in China's urbanization," *International Journal of Urban and Regional Research*, 27(4): 912–936.

Zhou, M. and Logan, J. R. (1996) "Market transition and the commodification of housing in urban China," *International Journal of Urban and Regional Research*, 20(3): 400–421.

# 14 The Chinese city in transition

## Towards theorizing China's urban restructuring

*Fulong Wu and Laurence J. C. Ma*

## Introduction

The theme of this book is China's urban restructuring, focusing on the changes in society, economy and space. In this final chapter, the social, economic and spatial changes are framed through the notion of "transitional cities." Here "transition" is defined as the process of change from one condition or a set of circumstances to another without implying "convergence" or "switching" towards a preconceived and well-defined model. This redefinition differs from the use of "transition" in "transitional economies" which often suggests the convergence from the socialist economy to a capitalist market one.

Our redefined concept of "transitional cities" is meaningful exactly because the processes of change never stop, because it is not a complete transformation and because the pre-existing historical, institutional and local contexts will continue to interact with the growing ideological thinking of "neoliberalism" brought about by economic globalization to generate hybrid urban processes and multiple spatial forms. In this sense, while the concrete manifestations of transition are distinctively embedded in local institutions, urban restructuring in "transitional cities" constitutes the changes in contemporary human settlements in general.

This is *not* to suggest that Chinese urban transformation is merely a manifestation of some global/external process imposed on the Chinese city. Rather, many transition processes are indigenously driven. As will be shown later, they are very much rooted in the regime of accumulation, but are manifested as part of contemporary urban transformations. Without an understanding of the urban processes in the cities outside the Western market economies, knowledge is always partial. We argue that the experience of Western capitalist cities is *not* adequate to understand transitional cities and that the understanding of the Chinese cities as well as other cities outside the extensively researched Western economies will inform the debate of the nature of contemporary urban restructuring in general.

"Urban restructuring" is framed in different ways in the literature to suggest that in the later decades of the twentieth century Western societies have entered a stage of profound economic, political, social and cultural transformations (Beauregard and Haila 1997). There are several dimensions of urban restructuring. First, the economic structure has seen a process of de-industrialization and restructuring towards the post-industrial economy (Lever 2001). In terms of the regime of accumulation, the regulation school emphasizes urban restructuring as the transition from the Fordist to post-Fordist economy and accordingly from the Keynesian welfare state to a post-Keynesian workfare regime (Jessop 1998). From the cultural perspective, urban restructuring has been characterized by some scholars as the turn towards "postmodern urbanism" (Dear and Flusty 1998).

No matter how the transition that occurs in Western societies is framed, there are multiple forces at work. Social, economic and spatial aspects of transition are interrelated. For example, increasing capital mobility has transcended the fixed scale of the national state that is produced in the Fordist regime. This economic transition has led to the transition of governance from the managerial and interventionist to entrepreneurial governance (Harvey 1989). More recently, the literature of neoliberalism (Brenner and Theodore 2002) suggests the global spread of the phenomenon for the market and minimalist state. But the experience has been mainly drawn from North America and Western Europe. To what extent can China's urban restructuring contribute to the understanding of worldwide urban transformation? It is useful to examine both the processes and forms of urban restructuring so as to go beyond the similarities of spatial manifestations (such as "gated communities," see Webster *et al.* 2002, for further discussion of their global spread; and Dick and Rimmer 1998 for the replicas in Southeast Asian cities) in different local contexts.

## Changing urban processes

Urban restructuring in the Western economies has been understood as a process of the changing regime of accumulation (Amin 1994). Very few attempts have been made to examine changing urban processes in "transitional cities" from a similar perspective. Most studies have paid attention to concrete policy changes (such as the policies of "economic reform" and "open door") and to changing conditions of urban development (for example, foreign investment and trade). This level of analysis is useful but not adequate for understanding the "evolutionary" and "internal" logic of production which defines the process of urban development.

The political economic approach has been extensively applied to the analysis of capitalist cities. But this could be equally useful for understanding the (post-)socialist city so as to understand the change at the policy level as a response to the internal contradiction of the urban process. Revisiting the logic of production under socialism helps to understand post-reform urban conditions. In short, the transition is understood through a fix on the impasse/crisis of accumulation. In Western economies, urban restructuring has been understood through "flexible accumulation" driven by the force of production and the attendant political/institutional fix on this new regime of accumulation. Here, we argue that the "urban" scale has been foregrounded and "substantialized" in post-reform China and that, consequently, "urbanization" has become a state strategy and solution to the sustenance of the new regime of accumulation (see Figure 14.1).

### *Socialist state-led industrialization*

The key thread is to understand the economic conditions of extensive industrialization in the socialist countries (Andrusz *et al.* 1996; Pickles and Smith 1998). Socialist "crisis management" was necessitated by the Cold War, and the consequent economic isolation, military competition, national defense and prioritization of heavy industries. There are some apparent similarities between the state socialist regime and Fordist regime; however, the mode of regulation reveals different solutions to different crises. For the Keynesian welfare state, the market principle is the fundamental mechanism of the economic system, while state intervention, as a last-ditch cure for market failures, is to combat the crisis of over-accumulation and thus inherent economic instability. By comparison, the socialist state regime was created to remove the barriers of accumulation and to facilitate rapid state-led industrialization. But both regimes uphold a belief in the comparative advantage of scale economies and advocate the role of state intervention, albeit to varying degrees and in different ways.

Industrialization has been fostered in state socialism with the aim of establishing an efficient factory system and eventually building up the national economic system. As such, the city under socialism was no more than a physical site for containing industrial production. While urban

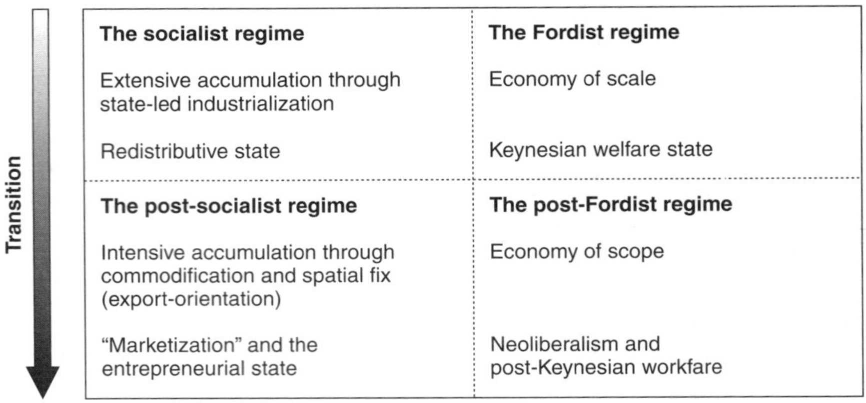


*Figure 14.1* Changing regimes of accumulation affecting urban restructuring (source: Wu 2003a).

squares were often used for political rituals, the city itself was *not* the basic unit for organizing consumption and urban space was organized around industrial production. The scale of the national state was predominant. Hence, the socialist economy was called the "centrally planned economy," although the actual implementation of the plan was carried out in *ad hoc* ways and regulated through scattered production work-units.

The effective extraction of social surplus gave a powerful role of "redistribution" to the state. In fact, the ability of the state to channel the surplus into new arenas of extensive accumulation is the precondition for avoiding over-accumulation, as mass consumption was not the driving force for economic growth. Instead, factories relied on the state as the ultimate purchaser of their goods in order to sustain and even to expand production.

The rationale of "forced growth" and shortage economies are elaborated by Kornai (1992) and the social consequences are reviewed by Pickvance (2002). The regulatory process that sustained such a regime resulted in the dominant role of the state in organizing production and consumption (Hausner *et al.* 1995). One of the implications for urbanization has been summarized in the so-called distinctive socialist "under-urbanization" model in which the level of urbanization lags behind the level of industrialization (Szelenyi 1996).

This difference between socialist and Fordist capitalist regimes in treating consumption as the force of economic growth has led to different manifestation of urban landscapes. The conceptualization of the city as the site of production rather than as a conglomeration of a heterogeneous population has on the one hand led to constraining "urbanism" – high population density, diversity and anonymity as the way of life perceived by Louis Wirth – and on the other hand formulated urban-biased policies such as the systematic favors given to the state factory system, which is inherently urban based, and the privileges such as comprehensive welfare coverage based on life-time employment given to urban residents. Despite some debate, there is now a consensus that there is a need to read the various socialist practices within the process of urban development (Ma 2002; Solinger and Chan 2002; Lin 2004).

There are two major problems associated with the socialist accumulation regime. First, the mode of regulation is incompatible with the regime along with increasing complexity in industrial production and difficulty in restraining resource mobility. Overwhelming state dominance did not at the end of the day meet the demand for regulation. Castells (2000) argues that such a difficulty comes from the incoming network/information society. The failure of state socialism was due to the impossibility of rigorous planning and the lack of effective incentives to mobilize subordinate production units, to such an extent that it lost competitive advantage. As complexity grows along with the increase in the scale of the economy, planning became astronomically difficult. The Chinese experience suggests there was a variety of *ad hoc*, cellular, self-contained modes of regulation carried out by the units under "soft budgetary constraint." This effectively defeats the ability of planning.

Second, related to the lack of incentive at the local level and the difficulty in constraining the thrust of self-expansion driven by the soft budgetary constraint, there has been a tendency towards an over-accumulation crisis, although it is different from the one in the capitalist economy. The socialist state had to avoid this tendency by constantly channelling capital out of the existing production circuit into a new production circuit. Because of the suppression of consumption, there was no alternative outlet, for the built environment was not the arena for diverting the concentrated social surplus. But the state's ability to redirect the surplus into military-type and low-efficiency industrial production became increasingly ineffective.

Moreover, the removal of the "city" as a means for organizing collective consumptions meant the state ultimately had to undertake the function of labor reproduction. State-organized labor reproduction did lead to the supply of low-cost labor forces but the policy of full employment meant that enterprises had little incentive to advance technology to replace labor. The compulsory and rigid labor system led to hidden redundancy of employment in state enterprises, which suffocated technological innovations.

### *Post-reform accumulation regime*

The major transition in the post-reform era is the reconceptualization of the city, which has been increasingly used as a means to overcome the constraint of accumulation. This change involves three major fixes: first, the foregrounding of the city as the preferred scale for the development of intensive accumulation through urbanization, which is the spatial fix (cf. Ma 2005); second, the commodification of urban development through urban land use reform allowing land use rights to be acquired through negotiation with land occupiers and through commodified housing, which is the institutional fix through marketization; and third, the adoption of export-oriented growth and integration into economic globalization as a production fix. The relaxation of control over labor production has introduced consumer demand for light industrial goods. In contrast to the import of advanced technology immediately after the death of Mao which failed as it did not solve the problem of the accumulation regime, township and village enterprises (TVEs) successfully flourished under the condition of ambiguous property rights. By exploiting the regulation gap between the urban and rural economic systems and targeting light industries, TVEs grew and occupied some temporary market niches during the 1980s and the early 1990s, but only to decline since then with further marketization and increasing competition from firms established by foreign investors.

Further, the commodification of urban development began to reveal its powerful effect on economic growth. Since the mid-1980s, the Chinese economy has struggled to find an engine of growth. Real estate was one such engine. The decline in the state industrial system is in sharp contrast with profitability in real-estate projects. Selling factory sites to real-estate and commercial projects is a common practice. Preliminary evidence suggests that there was a shift of capital from manufacturing industries to the real-estate business during the building boom of the mid-1990s. Since the 1990s, real-estate development has boosted urban-based consumption and opened up the built environment for further accumulation. In other consumption areas, niche markets have been expanded. With the emergence of a middle class, urban China is now truly experiencing a "consumer revolution" (Davis 2000).

Economic restructuring started in the late 1970s. But the process had been slow until the mid-1990s when the state enterprise system itself was remodelled. Large-scale command-based industries, which produce non-consumer products, are gradually giving way to hybrid economic entities, which produce "commodities." Thus the whole process of marketization aims to adjust the organization of production around the demand for commodities. In order to achieve this aim, property rights are rebundled, including but not limited to the land-leasing system and share-holding companies. Such a quasi-privatization process, which separates the state as the ultimate owner from the leaseholder or share-holder who can draw legitimate benefits, often leads to a transfer of state resources to private hands. The so-called gradual reform in China is thus politically slow but *economically radical*.

The state-centered accumulation regime came under severe challenge first by the mushrooming of township and village enterprises in the 1980s and then from foreign-funded ventures in the 1990s. The impact is most severely felt in mining, military industries, and light industries such as textiles. Large-scale restructuring has generated a nationwide total of 6.6 million redundant workers in 2000. Rapid economic restructuring has also led to increasing social stratification. With the emergence of the new rich, the demand for luxury goods is robust at the high end of the market. For example, buoyant housing prices tagged with over half a million yuan are seemingly incomprehensible, given the present low development stage of the overall Chinese economy. All these changes indicate a fundamental shift in how the city is conceptualized and how it is functioning.

This evidence indicates that the more closely an activity is associated with the system of commodity production, the more profitable it may be. In other words, with the shift in the logic of production, profitability is measured against whether it adopts more market-oriented strategies, targeting those who can afford the product and thus exploiting the niche market where effective demand lies. The restructuring of tertiary industry can be understood in the same way. The cultural industries, presented in innovative forms such as spectacular urban space and glorified consumption, at present function similarly.

Recent development strategy in urban China has put forward a slogan of developing the city "on the basis of human need." This essentially treats the city as a conglomeration of population, and as a spatial form of higher urbanism which is the engine for economic growth. In the post-reform era, the city is recognized as the node for economic organization. The city's status has been strengthened by enlarging its jurisdiction, especially through so-called "city-led counties" in the 1980s. This process has recently been intensified, with a new wave of counties and smaller cities being annexed into central city districts (Ma 2005). At the same time, the changing fiscal regime accompanied by decentralization has mobilized local states into entrepreneurial agents. Such localism or "local state corporatism" in rural China is extensively documented, but entrepreneurial governance in the cities has been studied to a lesser extent.

The change in the accumulation regime has also led to the search for new institutional changes to suit the new regime (Hausner *et al.* 1995). At this point, the contextual difference comes into effect. In the Chinese city, the state plays an active role in facilitating such a transition, partially due to the historical legacy of strong intervention and the availability of instruments of intervention, and partially because it is influenced by the mentality of the "development state" in East Asia. Such a transition is impossible without the state as the builder of market institutions. In this primitive stage of capital accumulation, effective regulation is still the key for market certainty as well as the way of opening up new arenas for accumulation.

The change from the production site to the entrepreneurial city is materialized through a sequence of institutional re-engineering: besides the change in the fiscal regime which has given greater autonomy to the locality, urban land and housing are "commodified," allowing rent/profit to be legitimately drawn. The components of the productive infrastructure – airport, deep-water port, metro system, elevated roads and highway, fast-rail, info-ports – are becoming the indispensable elements for building the entrepreneurial city. As the city is staged at the center of accumulation, the outcome is severe economic competition between cities and within the city. This phenomenon of place promotion is not unfamiliar to the rest of the world.

There is no doubt that globalization and marketization are powerful characterizations of the post-reform changes in China (Logan 2002). In this chapter, the attempt has been made to further examine the urban process according to the regime of accumulation. Furthermore, the changes in development strategies are linked to this transition of regimes. Through this analysis, it has been shown that transitional cities could be treated both theoretically and methodologically in ways similar to their Western counterparts. The process of globalization in transitional cities is as much an indigenously generated process as an externally imposed one. In this sense, globalization in transitional cities should not be seen as Westernization or convergence towards the "Western market model." The new sphere of accumulation is expanded through adopting an export-oriented strategy as a spatial fix and consequently through the urbanization of the whole coastal region of China as world factory sites. As a result, the Chinese economy has thus become integrated into the global commodity production. Associated with reinventing the urban sphere as the means for expanding accumulation, the built environment has become the medium of achieving such an aim. The mode of regulation is adjusted in order to meet such a requirement (for example, commodification of urban space through reforming land and housing systems, see Yeh, Chapter 4, this volume). The local "entrepreneurial agent" is generated alongside this process of the shifting mode of regulation who in turn facilitates such a shift.

The history and dynamics of the establishment of the land-leasing system partially reveal the change in the regime of accumulation. The reform of the land use system is not a trivial change of management but rather a critical step towards post-reform urban restructuring (Wu 1997, 1999; Yeh, Chapter 4, this volume). The marketization of housing and land has begun to introduce the built environment as a means through which capital has been diverted from state-owned industrial enterprises, which formed the core of the socialist economy, into the real-estate sector and city building. With an over-accumulated industrial sector (ironically due to the effective extraction of social surplus through the SOEs) and the inability of the SOEs to shift gears in response to the market conditions, the demand for industrial goods produced by SOEs has reached saturation point (not so much because there is a lack of "need" as because of consumer confidence). The introduction of commodity housing and the accompanying imaginative concept of "home" are capturing the demand of the upwardly mobile social stratum. It is no wonder that real estate has been utilized from time to time (around the 1992–1994 building boom, also in 1998 to cope with the Asian financial crisis to "stimulate domestic demand" (*kuoda neixu*), and more recently in connection with globalization and the WTO fever) to stimulate economic growth.

Recently, Chinese cities have been the subject of new theoretical analysis. Studying the regime of accumulation in Hong Kong, a global Chinese city, Smart and Lee (2003) suggest that the imperative of a property-based mode of regulation appears together with a finance-led regime of accumulation, and emphasize the role of real estate as a key driving force in its economy. This represents a fruitful application of an important theoretical perspective while fully recognizing local distinctiveness. Yeung and Lin (2003) have called for "theorizing economic geographies of Asia" in a similar vein, which can be argued as a way of studying transitional cities. We support their call and hope the views we have expounded in the introductory chapter of the volume and in this concluding chapter can make a contribution towards theorizing China's urban restructuring.

## Changing urban spatial forms

New urban spatial forms have been produced through urban restructuring. These new urban spaces have been discussed in detail in previous chapters and they include: glittering office buildings and skyscrapers in the financial district and CBDs, foreign retail outlets and shopping complexes, clustered suburban villas and downtown condominiums, development zones, high-tech, university and science parks, dilapidated workers' villages and migrant enclaves. These spatial elements can be read individually but they together piece out a large picture of post-reform spatial production.

### *The active role of spatial transformation*

Rather than treating spaces as the object that evolves by its own course, the previous chapters provide rich theoretical inspiration and empirical evidence to situate spatial transformation within the changing Chinese economy and society. No matter whether urban space is addressed implicitly (as in the reorganization of administrative hierarchy, changing scale, illegal building, place-making, uneven urban development and housing inequality) or explicitly (as in the internal spatial structure, new city center and CBDs, land use restructuring, residential segregation and relocation, and distribution of migrants and migrant settlements), these analyses point to the broad political and economic processes underlying spatial restructuring. These include economic decentralization, globalization and global orientation, marketization and reform of economic institution, social stratification and residential differentiation, and migration. These multiple forces have generated different spatial manifestations and produced new spatial elements. Beyond the narrow focus on space itself, it has been shown that the city is conceptualized as a meaningful space within which these multiple processes interact and are entangled.

Space production is more than a passive outcome of social and economic restructuring. It is more than the imprint left by some "forces" on the built environment. Spatial restructuring is an integral process of urban restructuring. Space is more than a container: manipulating it paves the way to sustain and expand the post-reform regime of accumulation. Spatial formation is thus more than a passive receiver of social changes. It is instrumental in the accumulation process. In other words, transition is a whole set of evolutions that cannot be separated into the transformation of political economic systems and changes in the built environment.

### *New urban spaces*

First, new *more urban and globally oriented spaces* in addition to rural areas annexed through administrative changes (Shen, Chapter 3, this volume and Ma 2004) and global-oriented financial zones and CBDs (Gaubatz 1999 and in this volume) are created to facilitate the expansion of urban-based accumulation. The expansion includes new economic zones and rescaling of the state towards the city level (Cartier, Chapter 2, this volume). Through understanding the spatial "reorganization" of the Chinese urban hierarchy

(Cartier, Chapter 2, this volume; Shen, Chapter 3, this volume) as well as within the city (Wu 2002a), it is shown how new space production is linked with the conversion of suburban counties to urban districts, and the merging of cities and towns to form a gigantic metropolis. Making the city bigger reflects the need to forge stronger agglomeration to capture "mobile resources" now unleashed by market reforms. The term "great international city" shows how "globalization" has captured the policy makers' imagination (Gaubatz, Chapter 6, this volume). Thus large-scale urban redevelopment through, for example, rebuilding CBDs, is critical towards maintaining the thrust of economic growth and further legitimizing the state's involvement in economic intervention. This intervention, however, is more aggressively market-oriented and neoliberal in its intention to support the market, rather than to remedy "market failure" through redistributive social objectives. Entrepreneurial governance treats place-making as a state project, but in essence is re-oriented towards "competitiveness" rather than redistribution (Wu 2003b).

Second, *informal spaces* such as migrant enclaves have been produced through illegal building and spontaneous self-construction. While the remaking of the global city has introduced modernist landscapes into the Chinese city, some unregulated and illegal building activities persistently show another side of the "Third World" (Drakakis-Smith 2000; Potter and Lloyd-Evans 1998) not found in industrialized Western economies. Such "gray" activities cannot be made "white" (Smart and Tang, Chapter 5, this volume), and it is difficult to clearly redefine ambiguous property rights (see also Zhu 2002). Informal and illegal constructions have been strengthened by economic reform and marketization. This ambiguity is the source of "illegal building" and leads to the struggle between urban villagers and the government (Li Zhang, Chapter 13, this volume). Tension arises when the local government strives to take control of encroaching urban villages while the residents attempt to make a profit by renting houses to rural migrants. While the socialist city has seen residents' illegal construction of housing extensions to increase living space in the old city neighborhoods, the scale and nature of migrant enclaves are beyond comparison. The migrant population is concentrated in the periphery of the city. Such a skewed distribution is related to the attraction of industrial employment and cheaper housing in the peri-urban area (Weiping Wu, Chapter 12, this volume).

Third, *consumption spaces* have been created through converting land uses inside the urban area. Land use conversion from industrial to commercial use plays an important role in facilitating the transformation of the industrial city (Yin *et al.*, Chapter 9, this volume). Together with suburbanization and the relocation of urban residents from the inner areas, land use changes have led to a new urban internal structure (Gaubatz 1999 and Chapter 6, this volume; Wu and Yeh 1999; Zhou and Ma 2000). Indeed, for many urban districts with concentrated industrial areas, the request for more autonomy to convert industrial land into commercial uses is so strong that it can hardly be denied by city planners. Thus land use conversion and urban sprawl, though often criticized for their negative impact on historical preservation and sustainable urban development, are indispensable for China's urban restructuring, which is witnessing simultaneous transition from a rural to urban society and from socialist industrial to post-industrial economy.

Fourth, especially in recent years, the production of *recreational and amenity space* becomes the tactic of place-making which extends beyond the economic arena (e.g. favorable tax and land deals). The cultural production of space into "place," for example, the property developers' resorting to "Shanghai nostalgia" and building the "imagined community" of colonial upper quarters (Pan, Chapter 7, this volume) not only reveals the division in actually existing socialism but also capitalizes such spatiality into the new identity of "gated community" dwellers. The social–spatial differentiation, as shown in the variations in housing prices (Wu 2002b), is critical in exploiting niche markets in the midst of housing overproduction and high housing vacancy. Indeed, the "gated community" should be seen as a new form of space where elitist consumption is manifested rather than as a purely residential form (for gated community, see also Huang, Chapter 11, this volume). There is no need to have a fixed tactic in the manipulation of the built environment. Besides exploiting the historical memory in the temporal sense, it is possible to look for geographical solutions, i.e. transplanting the imagined "global" (read, Western) cityscapes such as "townhouses" and gated villa compounds (Wu 2004a), which serve the purpose of overcoming the constraints of post-reform accumulation and of opening up a new venue for capital investment.

Fifth, new *residential spaces* are created, differentiated from the old, and fragmented into the fabric of the urban structure. The concentration of low-status migrants coincides with luxury commodity housing estates there (Huang, Chapter 11, this volume), which has created greater fragmentation of social space in the suburbs. The process of housing inequality continues to shape spatial inequalities (Logan *et al.* 1999). Residential segregation as well as other spatial unevenness is therefore a natural and expected spatial form at the core of the new accumulation regime. Although urban villages provide a cheap way of accommodating the migrants and thus a "low-cost" method of urbanization, the local state is very anxious to grasp development control over these villages (Li Zhang, Chapter 13, this volume). As this spontaneous form of city building is regarded as chaotic and unsuitable for the city's image, the redevelopment of these areas into real-estate projects has thus become inevitable.

## *Politics of spatial development*

Spatial restructuring has become instrumental – the "magic" use of land leasing not only brings revenue to the local government but also modernizes urban landscapes. The incentive of land development has driven local government to continue to intervene in urban development. In this case, ambiguous property rights lay down the foundation for the growth-oriented regime. Such pragmatism in the post-reform political landscape reflects the wisdom of "decentralization." Intra-urban politics between urban districts, as shown in three urban districts in Shanghai, creates differentiated capacities due to historical divisions between these districts and subsequently different political leadership (Tingwei Zhang, Chapter 8, this volume). This imbalance in development capacities has resulted in uneven urban development, and the relatively homogeneous intra-urban space has evolved into a more heterogeneous spatial order.

Perhaps nowhere shows the imperative of urban redevelopment better than the derelict inner urban land. But redevelopment does not happen necessarily in the most dilapidated area because there is market demand to be considered. The nostalgia behind the redevelopment of Xintiandi, a stone-portal-gate housing area at the edge of "upper quarter" (see also, Tianshu Pan, Chapter 7, this volume) fits very well into a futurist view of Shanghai's re-globalization strategy, because it is there the expected rise in rent exceeds the current rent. From the residential relocation (Si-ming Li, Chapter 10, this volume; Li and Wu 2004), it becomes clear that a new sorting mechanism has now been established to relocate residents according to their varied socioeconomic status. "Museumized" heritage development such as Xintiandi, recalling the motif of "stone-portal-gate" housing and alleyways in the colonial past, can only host ostentatious consumption rather than accommodating the needs of the "petty urbanities (*xiaoshimin*)" because of the need for recovering intensive investment.

Through understanding how these new spaces are natural constituent elements in the regime of accumulation, it is evident that the production of new spaces is neither totally "new" nor the result of the logic of economic globalization. In fact, it is linked to very "old" institutional roots. For example, the migrant enclave is a response to the state's incapacity to provide shelter for these newly mobilized rural workers. The formal urban land market excludes "informal" users from the space of production (such as in clothing production in Beijing's Zhejiang Village, Ma and Xiang 1998).

### *Contested urban space: migrant enclaves*

The formation of the migrant enclave is thus a low-cost, albeit less than elegant, approach to urbanization, through which new sources of labor are made available for intensified accumulation in the urban arena. In fact, the whole coastal region and now increasingly inland regions as well have become the production zone for global commodities. The weak "citizenship" of peasants, imposed by institutional constraints, was tactically designed to enforce state-led industrialization in the pre-reform era and is still being exploited to suppress labor costs so as to maintain attractiveness for labor-intensive commodity production. As a significant number of these newly freed migrant workers are in non-state sectors, especially overseas-funded ventures, their citizenship is defined in a different framework from that of state employees in the socialist regime.

While the state recognizes the contribution of migrants to overall economic growth, in practice it is impossible and unwise for the state to shoulder the responsibility of labor reproduction. Labor reproduction has now been "commodified": the migrants are supposed to be paid enough to make a living. In theory, migrants could buy "commodity housing," because nowadays an urban household registration status is not required for commodity housing purchase. While a few wealthy migrants do become the patrons of new real-estate projects, most of them are simply too poor or unwilling to do so (Weiping Wu, Chapter 12, this volume).

It is therefore interesting to ask what the discriminatory household registration system in fact protects. While the view of universal citizenship could be used to attack the discriminatory policy of the state towards rural migrants, the institutional reason for the formation of migrant enclaves is not due to household registration *per se*. Historically, the registration system was established to prevent the rural population from accessing subsidized urban facilities. In fact, except for public transit systems, sewage and trash collection, there were few "universally existing" *urban* facilities because such facilities were provided by individual workplaces. The ability of the state to provide such comprehensive yet occupationally based welfare benefits which also include housing was due to the effective extraction of social surplus under the work-unit system. Because the newly freed rural workers are outside this loop of extraction, to add them to the over-stretched state welfare system would devastate the situation of the urban "proletariats" who are now laid-off and have lost the state's commitment to lifelong security (Wu 2004b). In fact, being a rural household is not necessarily a disadvantage, as former farmers in Dongguan and Baoan near Shenzhen Special Economic Zone won a fortune in windfall land profit, and urban villagers are not willing to give up their rural status so as to continue to possess rural land (Li Zhang, Chapter 13, this volume).

The tension between local government and rural villagers should be understood against a set of institutional factors. Rural village land abutting the city is collectively owned farm land and cannot, technically speaking, be converted into urban uses without compulsory acquisition by the state. To achieve this conversion, the villagers have to give up their claim to the land, although under-the-table deals are frequently made to compensate them with a higher standard of compensation than is stipulated by the regulation. Moreover, once this land is placed under urban control, development rights become subject to planning control. While such a monopolistic state supply of land can be questioned, the current situation of migrant settlements exists because of the regulatory loophole: while the land remains rural, it is beyond urban planning controls. Chaotic development is thus the outcome. Behind this is the competition for ownership and development profits between different land systems. While the legitimacy of the official demolition policy can be challenged, the solution has to be based on the law. Indeed, the existence of migrant space poses two profound challenges to the transitional city: governance and social exclusion.

The discriminating feature is that the hegemonic accumulation process continues to create uneven development across developed and developing nations and across urban and rural areas within developing nations. To the migrants, their spatiality is a result of a dual "marginalization" process – becoming the subject of global capitalism and being denied full citizenship because of the malign neglect of the state. The fact that the household registration system is used to exclude migrants from accessing subsidized/welfare-type urban benefits means it will ultimately reach its demise with the diminution of such benefits. When the gap between subsidized labor reproduction and commodified production comes to an end, household registration will become meaningless. The residential preference of migrants (Weiping Wu, Chapter 12, this volume) shows that clustering at migrant enclaves is a tactical choice made by the migrants themselves, although such a decision has been made in the institutional environment of unequal and deprived citizenships.

The literature rightly points to power contestation between the state's attempts to control the growth of migrant enclaves and migrants constantly attempting to carve out a better transactional space and living environment (Solinger 1999; Li Zhang, Chapter 13, this volume). Migrants without household registration, migrants arriving through formal channels and local registered urban households are urban residents with differentiated status (Fan 2002). Although on the surface the state resorts from time to time to very aggressive measures of demolition, the position of the state is passive and weak in dealing with migrant enclaves. While the instrument used by the state to demolish migrant settlements may be old, the motivation is new: to eliminate "urban eyesores" to better sustain capital accumulation in the place as a whole. When the spontaneous constructions of migrant settlements reach a large scale or create a negative "externality," the city government begins to intervene, either by simply demolishing the "illegal" buildings or more positively by providing planned housing compounds to migrants.

What has been imposed on the migrants is unrestrained capitalism, global and domestic, in urban China. From this perspective, it is perhaps unfair to blame the state for being solely responsible for all of the problems of the migrants. Capitalism not only creates the logic for intervention, such as maintaining "market order" and curtailing market failure but also stimulates the development of "structural competitiveness" through various entrepreneurial strategies and the remaking of urban images. In fact, the justification of migrant housing demolition is often a mix of both, associated with the motivation of controlling land use rights in order to benefit from future urban development.

The above analysis has been applied to the spatiality of migrant enclaves; however, it is possible to analyze each spatial element in a similar way. What has been consistently shown in the dissection of each aspect of restructuring in this book is the *radical* nature of urban structuring and its new spatiality, despite the mixture of the old and new. If it is appropriate to characterize economic reform in China as gradualist, different from the "shock therapy" in Central and Eastern Europe, such a gradualism has lost its validity since the late 1990s. As the Chinese economy becomes more and more liberalized, the city has also developed some sociospatial characteristics similar to those in the cities of the West that have been affected by neoliberalization, including increasing social exclusion, intensified inequality and gated communities, although this trajectory should not be seen as converging to a universal model as these characteristics are merely parts of a larger complex whole. Nor should China's transitional city be seen as a prototype of something that is qualitatively different from the neoliberal city. The concrete form of any urban spatiality is based on historical contingencies and geographical contexts. A question that merits further analysis is the extent to which China's transitional city has been affected by neoliberalism whose impact seems global in scale. Peck and Tickell (2002: 47–48) define neoliberalism as a set of ideological concepts which comprises several ways of re-orienting the city: promoting and normalizing a "growth-first" approach, using pervasive naturalization of market logic justified on the grounds of efficiency, advocating lean government, privatization and deregulation, licensing a reflexive and aggressive posture on the part of local elites and states, associating with an extremely narrow urban policy repertoire based on capital subsidies and place promotion, building punitive institutions, social surveillance and authoritarian governance. It can be seen from this book that the Chinese city does not follow all these generic elements but it does show some sociospatial developments that resemble those in the cities of the West undergoing neoliberalization. Work is needed to sort out the salient sociospatial features of neoliberal cities and to compare the Chinese city with them.

### *Social–spatial inequalities*

It would perhaps be less appropriate to develop a "spatial model" because the Chicago school model is very much a product of "organized capitalism." The new spatial pattern does not form a clearly defined "urban structure" or land use model. The Chinese city shares the feature of greater fragmentation found in the "post-Fordist city," partially because structural change is piecemeal and change in the built environment is much slower than economic restructuring (Beauregard and Haila 1997), and partially because the new growth regime often exploits various market niches, thus forming enclave-like spatial fragments.

The uniformity of social space in the "socialist city" and its internal division has been contested in Central and Eastern European cities (Pickvance 2002). In the Chinese city under state socialism the division was produced between the inner core which was constrained by the pre-revolutionary legacy and the suburbs which were developed by state-led industrialization. The former had been deliberately under-invested, while the latter contained more integrated work-unit compounds and comprehensive workers' residences (workers' villages). The biggest spatial division, however, was between the urban (designated under urban household registration) and the rural by the household registration system (Solinger and Chan 2002; Li Zhang, Chapter 13, this volume). Such division is a product of China's dualistically organized regime of accumulation: placing the urban regime in the orbit of the state system while leaving the rural regime outside the state system (*tizhi nei* versus *tizhi wai*), and by extension the separation of the industrial from the agricultural sector. The state urban sector was more privileged at the core of state-led industrialization, while the non-state urban sector was supplementary to the state sector, although its existence was based on being a pragmatic solution for the livelihood of the residual urban population. The rural sector was exploited and surplus social value was extracted to support state-led industrialization.

Such a defined pattern, similar to the model of the industrial city depicted by the Chicago School, has been transformed by the complex, spontaneous and opportune changes in social structure. "Allowing some to become rich first," relaxation of the control over rural population, and selective real-estate development are creating enclave-like spaces and greater spatial fragmentation. Recent studies on residential relocation have begun to show the sorting of population (Li and Wu 2004; Li Zhang, Chapter 13, this volume; Wu 2004c). But such a sorting process is unlikely to establish a defined metropolitan-wide land use model in the short term. Moreover, urban redevelopment coexists with suburbanization. Most affordable housing estates and the poorest migrant enclaves are both located in suburbs, often next to the most luxury gated communities such as Purple Jade Villas (Huang, Chapter 11, this volume) or Orange County in Beijing (Wu 2004a). Therefore, the suburban areas of the Chinese city have seen the highest heterogeneity. The spatial juxtaposition not only opens up the possibility for comparing "quartered city" or "layered city" in different contexts (Marcuse 1997) but also has laid down the materiality of "gating," "surveillance" (Davis 1990) and the discourse of fear (Low 2003) and service provision (Webster *et al.* 2002) in urban China.

## Perpetual urban restructuring

Urban restructuring in the Chinese city as well as in other places is a perpetual process. There is a constant temptation to claim that spatial novelty is born out of urban restructuring. Such temptation is clearly reflected in the (mis)use of "transition" in sense of "transitional economies." The extensive and in-depth research on restructuring the Chinese city suggests that "transition" should not be regarded as a once-for-all convergence towards some universal model – often implied to be the "market economies." This prototype of "market economies" ignores that the capitalist world itself is changing and that the classic *laissez-faire* state has been surpassed by the Keynesian welfare state and further by the entrepreneurial and workfare state (Jessop 1998). Transition is a long-lasting process in which old and new, internal and external forces interact with each other and produce spatial forms of urban restructuring.

Moreover, many spatial forms may be similar to those seen in the Western world but the underlying process does not have a purely global logic. The restructuring of the Chinese city partially reflects the same trend of (neo)liberalism at work. This is because the transition, as argued in earlier chapters, is not a simple shift of "regulation form" (state socialism versus free-market capitalism) but rather a change deeply rooted in the "regime of accumulation" (extensive and state-enforced industrialization versus the post-Fordist accumulation). By tracing the transition back to this root, it is possible to transcend ideological labels (socialism versus capitalism), thus coming closer to a firmer grasp of the complexity of China's post-reform urban restructuring.

The restructuring of the Chinese city is a local process that exploits and constitutes "global" processes. It holds an exciting intellectual challenge for understanding global processes in different geographical contexts. The presence of the world's largest McDonald's in Beijing is thus more than a simple confirmation of a known phenomenon of economic globalization. It reveals how the possibility is created by a series of changes in the economy, in society and in space. Aggressive municipal entrepreneurialism is an indispensable component in facilitating the production of space in post-reform urban China. Just as there is no "global spread" of such a phenomenon as the "gated community," China's luxury gated residences have appeared within the suitable milieu of "private governance" (for example, treating citizens as consumers in housing commodification).

While there have been greater diversities and fragmentations forged by changing institution and economic transition in the Chinese city, these new fragments are not "occurring on a quasi-random field of opportunities" like postmodernism (Dear and Flusty 1998: 66). Seen from the perspective of these integrated economic, social and spatial changes, the "consumer revolution" (Davis 2000) is not necessarily antagonistic to state hegemony, and in fact both helps to legitimize the regime of accumulation and opens all sorts of imaginative venues for capital accumulation, while at the same time forming spaces of resistance. Neoliberalism is a double-edged sword with strong proponents as well as opponents at the global, national and local scales in different parts of the world. Briefly, it can raise the level of economic growth, according to its proponents, but it also intensifies inequality and polarization, as emphasized by its opponents. The demand for regulation in the process of freeing market functions has been underestimated. The Chinese city might well suggest the stage of "rolling out" neoliberalism, which emphasizes the state's support for the market, in contrast to "rolling-back" neoliberalism, which advocates deregulation (Tickell and Peck 2003). Further research is needed to test the validity of this hypothesis.

## Acknowledgment

We wish to thank John Logan, Henry Yeung, Michael Leaf, Alan Smart, Anthony Yeh, Cindy Fan and George Lin for enlightening us on many of the issues discussed in this chapter.

## References

Amin, A. (ed.) (1994) *Post-Fordism: A Reader*, Oxford: Blackwell.

Andrusz, G. M., Harloe, M. and Szelenyi, I. (eds) (1996) *Cities after Socialism: Urban and Regional Change and Conflict in Post-Socialist Societies*, Oxford: Blackwell.

Beauregard, R. and Haila, A. (1997) "The unavoidable incompleteness of the city," *American Behavioral Scientist*, 41: 327–341.

Brenner, N. and Theodore, N. (eds) (2002) *Spaces of Neoliberalism: Urban Restructuring in North America and West Europe*, Oxford: Blackwell.

Castells, M. (2000) *End of Millennium*, 2nd edn, Oxford: Blackwell.

Davis, D. S. (ed.) (2000) *The Consumer Revolution in Urban China*, Berkeley, CA: University of California Press.

Davis, M. (1990) *City of Quartz: Excavating the Future in Los Angeles*, London: Verso.

Dear, M. and Flusty, S. (1998) "Postmodern urbanism," *Annals of the Association of American Geographers*, 88: 50–72.

Dick, H. W. and Rimmer, P. J. (1998) "Beyond the Third World City: the new urban geography of South-east Asia," *Urban Studies*, 35: 2303–2321.

Drakakis-Smith, D. (2000) *Third World Cities*, 2nd edn, London: Routledge.

Fan, C. C. (2002) "The elite, the natives, and the outsiders: migration and labor market segregation in urban China," *Annals of the Association of American Geographers*, 92: 103–124.

Gaubatz, P. (1999) "China's urban transformation: patterns and processes of morphological change in Beijing, Shanghai and Guangzhou," *Urban Studies*, 36: 1495–1521.

Harvey, D. (1989) "From managerialism to entrepreneurialism: the transformation in urban governance in late capitalism," *Geografiska Annaler*, 71B: 3–18.

Hausner, J., Jessop, B. and Nielsen, K. (eds) (1995) *Strategic Choice and Path-Dependency in Post-Socialism*, Aldershot: Edward Elgar.

Jessop, B. (1998) "The rise of governance and the risks of failure: the case of economic development," *International Social Science Journal*, 155: 29–45.

Kornai, J. (1992) *The Socialist System: the Political Economy of Communism*, Princeton, NJ: Princeton University Press.

Lever, W. F. (2001) "The post-fordist city," in Paddison, R. (ed.), *Handbook of Urban Studies*, London: Sage, 273–283.

Li, S.-M. and Wu, F. (2004) "Contextualizing residential mobility and housing choice: evidence from urban China," *Environment and Planning A*, 36: 1–6.

Lin, G. C. S. (2004) "Toward a post-socialist city? Economic tertiarization and urban reformation in the Guangzhou metropolis, China," *Eurasian Geography and Economics*, 45(1): 18–44.

Logan, J. R. (ed.) (2002) *The New Chinese City: Globalization and Market Reform*, Oxford: Blackwell.

Logan, J. R., Bian, Y. J. and Bian, F. Q. (1999) "Housing inequality in urban China in the 1990s," *International Journal of Urban and Regional Research*, 23: 7–25.

Low, S. (2003) *Behind the Gates: Life, Security, and the Pursuit of Happiness in Fortress America*, New York: Routledge.

Ma, L. J. C. (2002) "Urban transformation in China, 1949–2000: a review and research agenda," *Environment and Planning A*, 33: 1545–1569.

—— (2004) "Economic reforms, urban spatial restructuring and planning in China," *Progress in Planning*, 61(3): 237–260.

—— (2005) "China changing urban administrative system: spatial restructuring and local economic development," forthcoming in *Political Geography*.

Ma, L. J. C. and Xiang, B. (1998) "Native place, migration and the emergence of peasant enclaves in Beijing," *China Quarterly*, 155: 546–581.

Marcuse, P. (1997) "The enclave, the citadel, and the ghetto: what has changed in the post-fordist US city," *Urban Affairs Review*, 33: 228–264.

Peck, J. and Tickell, A. (2002) "Neoliberalizing space," in Brenner, N. and Theodore, N. (eds), *Spaces of Neoliberalism: Urban Restructuring in North America and West Europe*, Oxford: Blackwell, 33–57.

Pickles, J. and Smith, A. (eds) (1998) *Theorising Transition: The Political Economy of Post-Communist Transformations*, London: Routledge.

Pickvance, C. (2002) "State socialism, post-socialism and their urban patterns: theorizing the Central and Eastern European Experience," in Eade, J. and Mele, C. (eds), *Understanding the City: Contemporary and Future Perspectives*, Oxford: Blackwell, 183–203.

Potter, R. B. and Lloyd-Evans, S. (1998) *The City in the Developing World*, Harlow: Pearson Education.

Smart, A. and Lee, J. (2003) "Financialization and the role of real estate in Hong Kong's regime of accumulation," *Economic Geography*, 79: 153–171.

Solinger, D. J. (1999) *Contesting Citizenship in Urban China: Peasant Migrants, the State, and the Logic of the Market*, Berkeley, CA: University of California Press.

Solinger, Dorothy J. and Kam Wing Chan (2002) "The China difference: city studies under socialism and beyond," in Eade, J. and Mele, C. (eds), *Understanding the City: Contemporary and Future Perspectives*, Oxford: Blackwell, 204–221.

Szelenyi, I. (1996) "Cities under socialism – and after," in Andrusz, G. M., Harloe, M. and Szelenyi, I. (eds), *Cities after Socialism: Urban and Regional Change and Conflict in Post-Socialist Societies*, Oxford: Blackwell, 286–317.

Tickell, A. and Peck, J. (2003). "Making global rules: globalisation or neoliberalisation?," *Globalization and World Cities Study Group and Network (GAWC) Research Bulletin*, 102: 1–20.

Webster, C., Glasze, G. and Frantz, K. (2002) "The global spread of gated communities," *Environment and Planning B*, 29: 315–320.

Wu, F. (1997) "Urban restructuring in China's emerging market economy: towards a framework for analysis," *International Journal of Urban and Regional Research*, 21: 640–663.

—— (1999) "The 'game' of landed-property production and capital circulation in

China's transitional economy, with reference to Shanghai," *Environment and Planning A*, 31: 1757–1771.
—— (2002a) "China's changing urban governance in the transition towards a more market-oriented economy," *Urban Studies*, 39: 1071–1093.
—— (2002b) "Sociospatial differentiation in urban China: evidence from Shanghai's real estate markets," *Environment and Planning A*, 34: 1591–1615.
—— (2003a) "Transitional cities," *Environment and Planning A*, 35: 1331–1338.
—— (2003b) "The (post-) socialist entrepreneurial city as a state project: Shanghai's reglobalisation in question," *Urban Studies*, 40: 1673–1698.
—— (2004a) "Transplanting cityscapes: the use of imagined globalization in housing commodification in Beijing," *Area*, 36(3): 227–234.
—— (2004b) "Urban poverty and marginalization under market transition: the case of Chinese cities," *International Journal of Urban and Regional Research*, 28(2): 401–423.
—— (2004c) "Intra-urban residential relocation in Shanghai: modes and stratification," *Environment and Planning A*, 36: 7–25.
—— and Yeh, A. G. O. (1999) "Urban spatial structure in a transitional economy: the case of Guangzhou, China," *Journal of the American Planning Association*, 65: 377–394.
Yeung, H. W. and Lin, G. C. S. (2003) "Theorizing economic geographies of Asia," *Economic Geography*, 79: 107–128.
Zhou, Y. and Ma, L. J. C. (2000) "Economic restructuring and suburbanization in China," *Urban Geography*, 21: 205–236.
Zhu, J. M. (2002) "Urban development under ambiguous property rights: a case of China's transition economy," *International Journal of Urban and Regional Research*, 26: 41–57.

# Index